Sybase ASE 15.X IN ACTION
Sybase ASE 15.X 数据库全程实战

何雄/著

中国水利水电出版社
www.waterpub.com.cn

内 容 提 要

Sybase 在中国相当普及，近年来少有相关图书出版。随着 SAP 并购并整合 Sybase，Sybase ASE 数据库迎来了巨大的发展机会，SAP 对数据库研发的投入已经大大加强，其目标是数据库和 ERP 成为同等的利润增长点。这对于 Sybase ASE 数据库的从业者来说，也带来了更多的机会。

本书的作者多年直接从事 Sybase 数据库相关领域的开发工作，同时长期担任 Sybase 数据库相关论坛版主，熟悉初学者和有一定经验技能的学员迫切需要掌握的相关技能知识。

本书是作者长期工作经验的系统性总结，系统地介绍了三部分内容：①管理篇：Sybase ASE 数据库的安装（静默安装、图形界面安装）、License 的使用、物理设备的创建、ASE 的网络连接、交互命令行工具、Transact SQL 的使用、字符集、数据库空间管理、用户安全管理、事务处理、事务日志、备份与恢复；②开发篇：Open Client 接口编程、ESQL（嵌入式 SQL 编程）、ODBC 接口开发、Java 接口开发、Python 访问 ASE、ADO.NET 访问 ASE、PHP 访问 ASE；③发布应用篇：介绍如何将 ASE 的功能组件重新打包，如何打包 Open Client、Sybase Central、JUtils 等。

通过系统学习和不断实践，学员既能够胜任 ASE 的应用开发工作，也能胜任 Sybase DBA 的工作。成为 Sybase DBA，意味着你将有更多的机会成为 SAP 实施顾问。

图书在版编目（CIP）数据

Sybase ASE 15.X数据库全程实战 / 何雄著. -- 北京：中国水利水电出版社，2013.2
 ISBN 978-7-5084-9150-9

Ⅰ. ①S… Ⅱ. ①何… Ⅲ. ①关系数据库系统－程序设计 Ⅳ. ①TP311.138

中国版本图书馆CIP数据核字(2013)第027178号

策划编辑：周春元　责任编辑：张玉玲　加工编辑：孙 丹　封面设计：李 佳

书　　名	Sybase ASE 15.X IN ACTION——Sybase ASE 15.X 数据库全程实战
作　　者	何　雄 著
出版发行	中国水利水电出版社 （北京市海淀区玉渊潭南路 1 号 D 座　100038） 网址：www.waterpub.com.cn E-mail: mchannel@263.net（万水） 　　　　sales@waterpub.com.cn 电话：（010）68367658（发行部）、82562819（万水）
经　　售	北京科水图书销售中心（零售） 电话：（010）88383994、63202643、68545874 全国各地新华书店和相关出版物销售网点
排　　版	北京万水电子信息有限公司
印　　刷	北京市蓝空印刷厂
规　　格	184mm×240mm　16 开本　33 印张　767 千字
版　　次	2013 年 2 月第 1 版　2013 年 2 月第 1 次印刷
印　　数	0001—3000 册
定　　价	68.00 元

凡购买我社图书，如有缺页、倒页、脱页的，本社发行部负责调换

版权所有·侵权必究

前 言

Sybase 是一家老牌独立数据库厂商，它在中国的市场份额一直位居第二，仅次于 Oracle。但近年来少有相关图书出版。随着 SAP 并购并整合 Sybase，Sybase ASE 数据库迎来了巨大的发展机会，SAP 对数据库研发的投入已经大大加强，其目标是企业数据库服务和它原有的 ERP 成为同等重要的利润增长来源。这为 Sybase ASE 数据库的从业者也带来了更多的机会。

Sybase ASE 作为一个老牌 DBMS，无论是在国际还是国内，依旧发挥着很大的魔力作用，在金融证券、电信、铁路交通、政府等多个行业和领域占据了重要的市场，尤其是在中国。在融入 SAP 以后，ASE 更是作为后端最重要的企业级事务型数据库，与 Sybase IQ 分析型数据库一起，为 SAP 的 ERP 等企业信息系统提供强大的支持。

本人有幸长期从事 Sybase 数据库相关领域的开发，同时在 CSDN Sybase ASE 及 SQLAnywhere 两个数据库论坛长期担任版主，熟悉初学者和有一定数据库经验的数据库技术人员迫切需要掌握的技能知识。在 Sybase 数据库论坛相关版面上，初学者经常出现以下问题：

- 不知道从哪里获取 Sybase ASE 相关的技术资料，手头没有相关的纸质技术图书，而从网上获取的联机文档也不足以直接入门。
- 不知道如何使用 Sybase 自带的工具集进行常规的管理和开发。
- 没有各个层级的数据库接口应用开发经验，从最底层的 C-API、ODBC API、OLEDB、ADO.NET 及脚本开发语言 PHP、Python 开发接口，这也是当前数据库开发图书中相应介绍比较匮乏的。作为数据库应用开发人员而言，掌握的数据库调用接口越全面，其开发能力也越全面，便很容易从一种数据库切换到另一种数据库。

本书为何取名为 Sybase ASE 15.x In Action？15.x 意指 ASE 数据库从 15.0.1→15.0.3→15.5→15.7 系列，取名 In Action（实战）是因为书中的内容大多是技能实战的总结，通过大量的实例来介绍 ASE 数据库相关技术，内容从基本的数据库管理到各种类型的数据库接口访问技术，都有相关介绍。

全书的内容涵盖了如下几方面的内容：

- 管理篇：Sybase ASE 数据库的安装（静默安装、图形界面安装）、License 的使用、物理设备的创建、ASE 的网络连接、交互命令行工具、Transact SQL 的使用、字符集、数据库空间管理、用户安全管理、事务处理、事务日志、备份与恢复；
- 开发篇：Open Client C-API 接口编程、ESQL（嵌入式 SQL 编程）、ODBC 接口开发、Java 接口开发、Python 访问 ASE、ADO.NET 访问 ASE、PHP 访问 ASE。通过详细的实例介绍这些开发接口的调用技术。
- 组件发布篇：介绍如何将 ASE 的功能组件重新打包，如何打包 Open Client、Sybase Central、JUtils、ODBC、OLEDB、ADO.NET、JConnect 驱动等。

通过系统的实战学习，读者可以从一个非数据库应用开发人员变成一个专业的数据库开发人员，也可以转变为一名 Sybase ASE 数据库的 DBA，这取决于个人的兴趣。成为（SAP）Sybase DBA，意味着将有更多的机会成为 SAP 实施顾问。

本书在写作和出版过程当中，得到 SAP 亚太区数据库解决方案技术总监卢东明先生（原 Sybase 中国 CTO）和 SAP 亚太区数据库解决方案中国市场部其他诸位同事的大力帮助，同时得到中国水利水电出版社万水分社策划编辑周春元的大力支持，在此表示衷心的感谢。

目 录

前言

第 1 章 搭建 Sybase ASE 环境

1.1 什么是 Sybase ASE ······ 1
   1.1.1 ASE 名称的来历 ······ 1
   1.1.2 Sybase ASE 的体系结构 ······ 1
1.2 安装 Sybase ASE ······ 3
   1.2.1 获取安装文件 ······ 3
   1.2.2 准备工作 ······ 5
   1.2.3 使用图形界面安装 ······ 7
   1.2.4 ASE 的静默安装 ······ 11
   1.2.5 安装完成时 ASE 的目录结构 ······ 11
   1.2.6 手动创建服务器 ······ 14
   1.2.7 验证服务器是否在运行 ······ 25
   1.2.8 修改 sa 用户口令 ······ 26
   1.2.9 Runserver 文件 ······ 27
1.3 如何卸载已经安装的 Sybase ASE ······ 29
1.4 忘记了 sa 用户密码 ······ 30
1.5 预装本书用到的 iihero 数据库 ······ 32

第 2 章 License 的使用

2.1 评估版 License ······ 33
2.2 License 的正式获取及使用 ······ 34

第 3 章 定义物理设备

3.1 物理设备管理 ······ 36
   3.1.1 创建设备 ······ 37
   3.1.2 删除设备 ······ 40
   3.1.3 裸设备与常规文件 ······ 41
   3.1.4 Dsync 选项 ······ 41
3.2 设备（文件）的限制条件 ······ 42
3.3 创建 master 设备 ······ 42
3.4 设备镜像 ······ 43
3.5 与设备信息相关的存储过程 ······ 48
3.6 与设备相关的系统表 ······ 49

第 4 章 连接 ASE

4.1 ASE 客户端概述 ······ 53
4.2 网络连接 ······ 53
   4.2.1 interfaces 文件的内容 ······ 54
   4.2.2 interfaces 文件的工作原理 ······ 55
   4.2.3 配置网络连接 ······ 56
4.3 使用 ASE 客户端 ······ 59
   4.3.1 连接 ASE ······ 59
   4.3.2 创建数据库设备 ······ 61

4.3.3 创建数据库 …… 64	6.4.2 TOP 限定记录及 distinct 消重 …… 104
4.3.4 创建登录账户和数据库用户 …… 67	6.4.3 Like 通配符模糊查询 …… 106
4.3.5 使用 Interactive SQL 客户端 …… 71	6.4.4 NULL 值及其含义 …… 109
4.4 启动和关闭服务器 …… 73	6.4.5 SQL 查询的标准格式 …… 112
4.4.1 启动 Adaptive Server …… 73	6.5 创建表的索引 …… 115
4.4.2 关闭服务器 …… 74	6.5.1 索引简介 …… 115

第 5 章 ASE 的交互命令行工具

5.1 SQL 交互命令 isql …… 76
  5.1.1 启动和停止 isql …… 76
  5.1.2 isql 的命令选项 …… 77
  5.1.3 指定 interface 文件、语言、
      字符集、数据库名 …… 79
  5.1.4 改正输入 …… 79
  5.1.5 性能统计信息收集与更改命令
      终结符 …… 80
  5.1.6 设置 isql 的网络包大小 …… 81
  5.1.7 设置输入和输出文件 …… 81
5.2 导入/导出数据 bcp …… 82
  5.2.1 使用 bcp 导出数据 …… 83
  5.2.2 使用 bcp 导入数据 …… 83

第 6 章 使用 Transact-SQL

6.1 数据库对象 …… 86
  6.1.1 T-SQL 中的数据类型 …… 86
  6.1.2 系统数据类型 …… 87
6.2 数据库对象的创建 …… 88
  6.2.1 使用和创建数据库 …… 88
  6.2.2 使用和创建表 …… 91
6.3 操纵数据库对象（DML）…… 98
  6.3.1 插入记录 …… 98
  6.3.2 更新操作 …… 100
  6.3.3 删除操作 …… 102
6.4 SQL 查询操作（DQL）…… 104
  6.4.1 使用 "*" 查询所有记录 …… 104

  6.5.2 创建索引 …… 116
  6.5.3 聚簇索引和非聚簇索引 …… 117
  6.5.4 创建索引的几个选项 …… 117
  6.5.5 索引删除与索引统计信息的更新 …… 120
6.6 ASE Transact-SQL 中的内置函数 …… 121
  6.6.1 获取数据库系统信息的
      系统函数 …… 121
  6.6.2 字符串相关函数 …… 125
  6.6.3 操作 TEXT/IMAGE 的文本函数 …… 128
  6.6.4 集合函数 …… 130
  6.6.5 数学函数 …… 131
  6.6.6 时间日期函数 …… 132
  6.6.7 数据类型转换函数 …… 136
  6.6.8 随机数据的生成 …… 139
6.7 ASE 中的存储过程 …… 142
  6.7.1 创建并执行存储过程 …… 143
  6.7.2 存储过程的参数 …… 144
  6.7.3 存储过程选项 …… 146
  6.7.4 执行存储过程的方式 …… 147
  6.7.5 以参数形式作为返回值 …… 149
  6.7.6 存储过程的限制 …… 150
  6.7.7 删除、重命名存储过程 …… 150
  6.7.8 游标的使用 …… 151
6.8 ASE 中的触发器 …… 155
  6.8.1 触发器的工作原理 …… 155
  6.8.2 创建触发器 …… 156
  6.8.3 ASE 中触发器的限制 …… 157
  6.8.4 触发器的禁用及删除 …… 158
  6.8.5 获取触发器的相关元信息 …… 159

第 7 章　Sybase ASE 的字符集

7.1　字符集的基本知识……………………162
7.2　中文字符集……………………………164
7.3　Sybase ASE 中的字符集文件…………167
7.4　Sybase ASE 的字符集设置……………169
　　7.4.1　直接设置字符集………………173
　　7.4.2　有重要用户数据的情况下
　　　　　如何调整…………………………174
7.5　乱码的产生……………………………175

第 8 章　ASE 中的空间管理

8.1　安装完 ASE 后的物理空间调整………178
8.2　用户数据库的容量管理………………183
8.3　使用段管理数据库空间………………192
　　8.3.1　段与其他数据库对象的关系……192
　　8.3.2　创建数据库段…………………194
　　8.3.3　改变数据库段的指定…………195
　　8.3.4　在段中存放数据库对象………198
　　8.3.5　使用 Sybase Central 客户端工具
　　　　　管理段……………………………202

第 9 章　ASE 的用户及安全管理

9.1　操作系统级别的安全…………………209
9.2　ASE 服务器级别的安全………………209
　　9.2.1　调整修改登录用户……………211
　　9.2.2　密码的强化管理………………214
　　9.2.3　ASE 中的特殊登录用户………215
　　9.2.4　ASE 中的标准角色（role）……218
　　9.2.5　查看已连接用户………………221
9.3　数据库级别的安全……………………223
　　9.3.1　新建数据库用户………………223
　　9.3.2　guest 用户………………………224
　　9.3.3　别名……………………………225
　　9.3.4　访问检查顺序…………………226
　　9.3.5　数据库访问的设置途径………227
　　9.3.6　组 group………………………228
　　9.3.7　用户名对传输数据库的影响…229
9.4　数据库对象级别的安全………………230
　　9.4.1　系统中的默认角色……………232
　　9.4.2　角色方式授权…………………234
9.5　对 SSL 协议的支持、配置管理
　　及使用……………………………………237
　　9.5.1　服务器端 SSL 的配置…………237
　　9.5.2　ASE 客户端 SSL 配置…………241

第 10 章　Sybase ASE 中的事务

10.1　设置事务模式和隔离级……………247
10.2　读未提交（level 0）…………………251
10.3　读已提交（level 1）…………………252
10.4　可重复读（level 2）…………………254
10.5　可串行化（level 3）…………………256
10.6　如何在事务中允许 DDL 操作………258

第 11 章　ASE 数据库的事务日志

11.1　事务…………………………………260
11.2　事务日志……………………………261
11.3　事务提交（commit）…………………261
11.4　检查点………………………………262
11.5　恢复（recovery）……………………263
11.6　恢复间隔……………………………264
11.7　日志填满……………………………265

第 12 章　ASE 数据库的备份、恢复及数据迁移

12.1　备份权限及周期……………………266
　　12.1.1　备份需要的权限………………266
　　12.1.2　备份周期（策略）……………267
12.2　简单备份……………………………267
12.3　远程备份……………………………270
12.4　dump/load 命令的使用………………272

12.5 用户数据库的备份与恢复 ································ 276

第13章 应用 Open Client 库编程

13.1 环境搭建 ··· 280
    13.1.1 Windows 下的环境 ································ 280
    13.1.2 UNIX/Linux 下的环境 ···························· 281
    13.1.3 验证连接 ··· 282
    13.1.4 开发环境 ··· 283
13.2 编程模型 ··· 284
13.3 连接数据库 ··· 285
    13.3.1 创建连接 ··· 285
    13.3.2 处理命令 ··· 291
    13.3.3 关闭连接 ··· 296
13.4 SQL 中的 DDL 操作 ···································· 297
13.5 获取 SQL 查询结果集 ································· 299
    13.5.1 简单结果集获取 ···································· 299
    13.5.2 类型绑定 ··· 305
    13.5.3 获取表的元信息 ···································· 307
13.6 数据的插入、更新与删除操作 ················ 312
    13.6.1 不带任何参数的 CUD 操作 ················ 312
    13.6.2 带动态参数的 CUD 操作 ···················· 315
    13.6.3 BLOB/CLOB 值的读写 ························· 318

第14章 嵌入式 SQL 编程

14.1 基本原理 ··· 329
14.2 一个简单的示例 ·· 330
14.3 NULL 值及特殊字段类型的处理 ············· 336
14.4 存储过程调用 ·· 340
14.5 插入/更新数据 ··· 343
    14.5.1 直接 Insert/Update ······························· 343
    14.5.2 通过游标来更新数据 ·························· 347
14.6 BLOB/CLOB 数据处理 ······························ 349

第15章 使用 ODBC 开发 ASE 应用

15.1 ODBC 简要介绍 ··· 355

15.1.1 ODBC 介绍 ··· 355
15.1.2 ODBC 体系结构 ···································· 356
15.2 ASE 中的 ODBC 环境 ······························· 357
15.3 连接 ASE ··· 358
    15.3.1 连接 ASE 的过程 ·································· 359
    15.3.2 配置及编译运行 ···································· 363
    15.3.3 一种增强的连接方式 ·························· 367
15.4 错误处理 ··· 369
15.5 一个 CRUD 的综合示例 ··························· 376
    15.5.1 Insert/Update 操作 ······························· 388
    15.5.2 Select 查询操作 ···································· 390

第16章 使用 Java 访问 ASE

16.1 环境和工具 ··· 392
    16.1.1 DBISQL ·· 392
    16.1.2 JUtils ·· 396
    16.1.3 DBeaver ·· 399
    16.1.4 JDBC 驱动 Jconnect 6.0.5
           简介 ·· 404
16.2 通过 JDBC 连接 ASE 数据库 ················· 405
16.3 使用 JDBC 操作 ASE 表数据 ················· 411
    16.3.1 Select 查询操作 ···································· 412
    16.3.2 Insert/Update/Delete 操作 ··············· 416
    16.3.3 事务的提交 ··· 420
16.4 BLOB/CLOB 读写 ····································· 420
    16.4.1 TEXT 字段的读写 ································ 420
    16.4.2 IMAGE 字段的读写 ···························· 423
16.5 调用存储过程 ·· 425
16.6 使用 JDBC 访问 ASE 元信息 ················· 428
16.7 JDBC 中的 ASE 数据库连接池 ············· 433
    16.7.1 数据库连接池的基本原理 ················ 433
    16.7.2 开源连接池在 ASE 数据库上
           的应用 ·· 434
16.8 使用 Java 直接支持 ASE 中的面向
        对象 SQL 访问 ··· 441

第 17 章 应用 PHP 访问 ASE

17.1 PHP 运行环境搭建 ································447
  17.1.1 Apache + PHP 运行环境 ············447
  17.1.2 Nginx + PHP 运行环境 ··············451
  17.1.3 PHP 环境对 ASE 数据库
     的支持 ································453
17.2 php_sybase_ct 模块介绍 ·······················454
17.3 一个访问 ASE 数据库的 PHP
  简单实例 ···458
  17.3.1 数据库数据准备 ·······················458
  17.3.2 系统实现 ································459

第 18 章 应用 Python 访问 ASE

18.1 安装 python-sybase 模块 ······················464
18.2 使用 Python 连接 ASE ························469
18.3 使用 Python 访问 ASE 数据库表 ·········470

第 19 章 使用 ADO.NET 访问 ASE

19.1 ASE ADO.NET 运行时环境 ·················474
19.2 连接 ASE 数据库 ································476
19.3 创建删除表 ··480
19.4 插入数据 ··481
  19.4.1 使用 DataSet 类来插入数据 ······482
  19.4.2 使用 Insert 语句来插入数据 ······488
  19.4.3 BLOB/CLOB 数据的
     插入操作 ·····························492
19.5 更新数据 ··494
19.6 调用存储过程 ······································495
19.7 获取结果集或表的元信息 ···················499
19.8 ASE ADO.NET 应用程序的发布 ·········505

第 20 章 Sybase ASE 功能包生成

20.1 JUtils 工具包生成 ·······························507
20.2 ODBC、OLEDB 及 ADO.NET 包 ······508
20.3 Open Client 库 ·····································509
20.4 Sybase Central 客户端工具生成 ···········510

第 21 章 Sybase ASE 发展历史及版本演进

1

搭建 Sybase ASE 环境

1.1 什么是 Sybase ASE

1.1.1 ASE 名称的来历

很多人以为 Sybase 只有一种数据库，这种看法是错误的。实际上，Sybase 有三种类型的数据库产品。ASE 是它的企业级数据库，其全称是 Adaptive Server Enterprise，它主要面向企业级应用。ASE 最早的时候并不叫做 ASE。1987 年 5 月，Sybase 公司推出 Sybase SQL Server 1.0，它基于 Client/Server 架构，是业界最早实现此架构的数据库产品。其间，Sybase 与微软有过合作，共同开发基于 Windows 平台的 SQL Server。1994 年，合作终止，Windows 上的最后一个版本是 Sybase SQL Server 4.21。其后，他们各自开发独立版本的 SQL Server。于是出现了两个 SQL Server，一个是 Microsoft SQL Server，一个是 Sybase SQL Server，后来，Sybase 为了与微软的 Microsoft SQL Server 以示区分，将 SQL Server 改名为 ASE。它除了是 Adaptive Enterprise Server 的首字母缩写外，同时也是 Sybase 的后三个字母。这就是 Sybase ASE 这个名称的来历。它也是通常人们所说 Sybase 数据库的真实产品名称。

Sybase 还有另外两个数据库产品，一个是 IQ，主要面向数据仓库；另一个是 ASA，主要面向移动和嵌入式数据库，它占领了移动和嵌入式数据库的绝大部分市场。

目前 ASE 的正式版本已经发展到 15.7 版本。

1.1.2 Sybase ASE 的体系结构

Sybase ASE 是一个支持分布式 Client/Server 体系的企业级数据库服务器。其基本体系结构如图

1-1 所示。

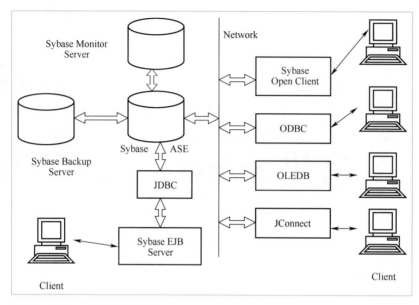

图 1-1　Sybase ASE 体系结构

ASE 包含多个部件，用于有效的管理维护和监控系统性能。例如，你可以使用 Backup Server 来进行数据的备份与恢复；使用 Monitor Server 来监控 Server 端的活动。另外，ASE 支持分布式事务处理，它是由 ASE 的分布式事务管理器部件来实现的。此外，ASE 15.0 还支持全文检索服务。

ASE 还支持 Java 应用程序的创建和使用，你可以使用 Sybase EJB Server 来创建 EJB 应用，并可以使用 Sybase EJB Server 来访问 ASE 中的数据。

目前，ASE 数据库支持包括 Windows NT/2000/XP/2003、Linux 以及各种 UNIX（HP UNIX、IBM AIX、Solaris 等）的几乎所有主流操作系统。

同时，ASE 支持各种形式的客户端访问，如 Open Client 客户端应用、Java 客户端应用，还有其他形式的应用：ODBC、OLEDB、JDBC。ASE 使用 JConnect 作它的 JDBC 驱动。这些客户端与 ASE 服务器之间通过采用特定的网络协议进行通信。ASE 目前采用的就是 TDS 5.0 协议（TDS 协议在 Microsoft SQL Server 中也被使用，不过，它们自定义的版本是 TDS 8.0）。

ASE 服务器包括几个重要的部件，如 Backup Server 和 Monitor Server。服务器当中最核心的部件就是 Sybase ASE，实际上内部仍叫做 Sybase SQL Server。这些部件的主要功能如下：

1. Sybase Backup Server：备份和恢复数据库中的数据。
2. Sybase 分布式事务管理器：管理 ASE 环境中的分布式事务。
3. Sybase XP Server：支持在 ASE 环境中运行扩展存储过程。
4. Sybase EJB Server：管理 EJB 应用中到 ASE 的数据连接。

5．Sybase 全文检索专用数据存储：用于对 ASE 中存储的数据进行全文检索。

1.2 安装 Sybase ASE

在安装 ASE 数据库之前，我们应该知道 Sybase ASE 数据库是一个支持很多不同平台的产品。对于各种不同的平台，不管是 Windows、Linux、Solaris 或是其他 UNIX，它们都有各自不同的地方，安装起来也就有少许不同的安装细节。但在各种平台下安装 ASE 仍然有很多相同点，值得您考虑。

首先，最重要的一点是安装 ASE 的计划。要想成功地安装 ASE，需要有一个比较完备的安装计划，在安装过程当中，许多问题的决定都会影响到将来 ASE 的运行。也许对于开发测试环境来说影响不大，可是对于生产环境而言，这些决定无疑将影响深远。在安装之前，您需要预先考虑清楚如下问题的答案：

- Sybase ASE 需要安装到哪个分区的哪个目录下？
- 要使用裸设备还是要使用文件系统（非 Windows 平台下）？
- Master 物理设备创建的位置和大小确定了吗？
- Sybsystemprocs 物理设备的位置和大小确定了吗？
- 如果您准备安装审计功能，那么 Sysaudits 物理设备的位置和大小确定了吗？
- 您打算在 ASE 数据库里使用哪种字符集和排序方式？
- 您打算让 ASE 数据库支持什么样的网络协议和网络地址？
- 您打算让 ASE 数据库使用多大的页大小？

上面有些问题在安装完以后还会再次发生，比如，创建数据库要使用裸设备还是使用文件系统？而另一些问题基本上只发生在安装时刻。比如，使用哪种字符集和排序方式？

下面，我们就分步介绍如何安装 ASE 数据库。

1.2.1 获取安装文件

如果已经购买了 ASE 产品，自然就有完整的安装文件，还可以得到相应的 license。对于普通的学习者而言，虽然从外界很难找到 Sybase ASE 的下载地址，但是我们可以直接从 Sybase 网站（http://www.sybase.com）上下载。其基本过程如下：

Step 1 进入 http://www.sybase.com/developer/downloads，右端有一个 login，会提示你注册一个 SDN 用户，注册时用户名使用邮箱名即可。

Step 2 使用刚注册的用户名登录，进入 http://www.sybase.com/developer/downloads，会发现有一个列表，内容如图 1-2 所示。

直接进入第一个链接：Adaptive Server Enterprise。

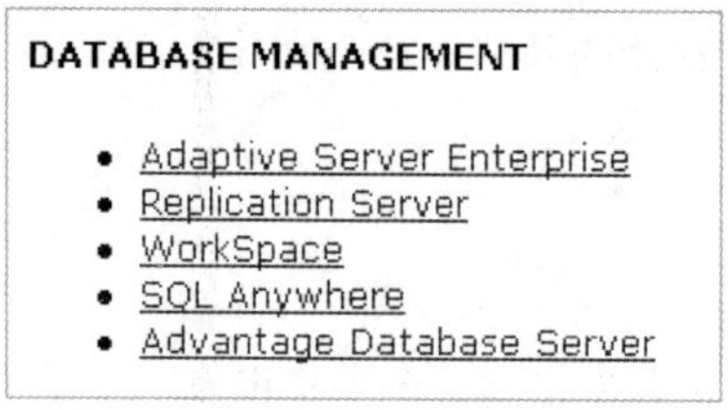

```
DATABASE MANAGEMENT

• Adaptive Server Enterprise
• Replication Server
• WorkSpace
• SQL Anywhere
• Advantage Database Server
```

图 1-2 数据库管理相关软件

Step 3 在后续的页面中，再进入链接 Adaptive Server Enterprise Evaluation Options。

```
Adaptive Server Enterprise Evaluation Options

• Download ASE Express Edition
• Experience ASE Labs with TrySybase
• Developer Edition
```

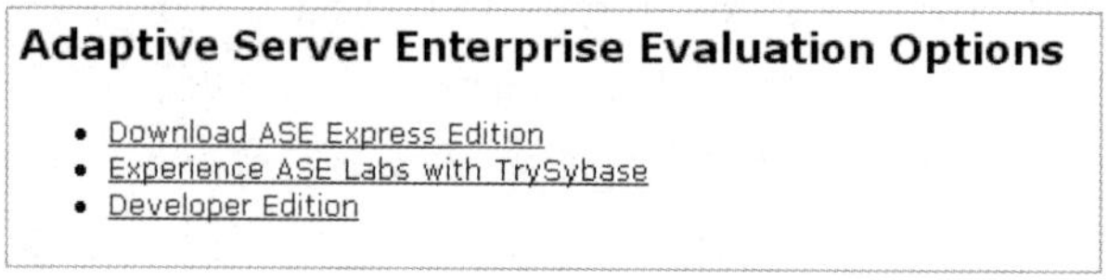

图 1-3 ASE Evaluation 选项

我们选择 Developer Edition 选项，会出现如图 1-4 所示的内容。

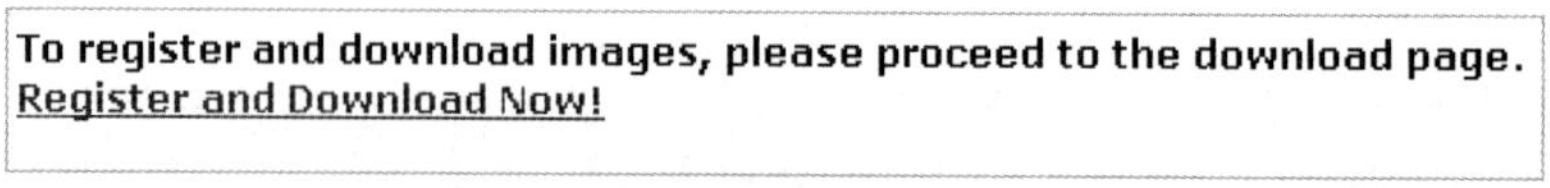

```
To register and download images, please proceed to the download page.
Register and Download Now!
```

图 1-4 ASE Developer Edition 注册下载

这时我们单击"Register and Download Now!"链接，有两个选框：是否同意使用的声明，都选上。

Step 4 接着会让你填一个表单，把必填项都填上，不是美国的用户，省份信息可以不填。提交以后，在新的结果页面末尾会出现类似如图 1-5 所示的内容。

```
Adaptive Server Enterprise 15.0.2 Developer's Edition or Sun Solaris 64-bit

• ASE 15.0.2 Solaris 64-bit Server

Adaptive Server Enterprise 15.0.2 Developer's Edition on Windows

• ASE 15.0.2 Windows Server

Adaptive Server Enterprise 15.0.2 Developer's Edition on Linux

• ASE 15.0.2 x86 Linux Server
• ASE 15.0.2 x86 64 bit Linux Server
• ASE 15.0.2 x86 64 bit Linux on POWER
```

图 1-5 各平台下的 ASE Developer Edition

Step 5 最后，我们可以从上图中找到自己感兴趣的 ASE 版本执行下载。图 1-5 中对 Linux 提供了三个版本。默认的 x86 是指 32 位的 x86 体系。Windows 版本提供的是 32 位的 ASE（ase1502_winx86_dev.zip）。

最终下载开发版 ASE，Windows 下选择 Windows 版本，32 位 Linux 选择 x86 Linux Server 即可。当然，Linux 还有一些准备工作。经过 ASE 认证的 Linux 主要有两个版本，一个是 SuSE 9.0 企业版，另一个是 RedHat 3.0 或者 4.0 的企业版。

从功能上来说，开发版与企业版几乎没什么区别，唯一的区别是它没有有效的 license，只有 60 天的有效使用期限。不过对于普通的学习者而言，60 天的使用周期已经足够。对于获取其他 Sybase 产品都可以使用相同的方法。

1.2.2 准备工作

如果有文档，仔细阅读 Sybase ASE 在目标平台下边的安装文档，各个平台下的准备工作还是有较大差异的。

接着确定下述内容：

- 逻辑页大小。
- 系统数据库的物理设备信息（类型、大小、位置）。
- 服务器名字。
- 网络相关信息（IP 地址或主机名、端口号、命名管道地址）。
- SySAM（Sybase 软件财产管理器）的激活码或者 License 文件，这对于已经购买了 Sybase 产品的用户来说，肯定不是问题，如果没有也没关系，可以使用试用版本。
- 字符集、排序方式和默认语言集。
- 备份服务器/XP 服务器/监控服务器/历史服务器相关信息（服务器名、网络信息）。
- 安全配置（是否需要使用 SSL）。

事先决定好上述内容，可以做到安装过程事半功倍。

逻辑页大小：ASE 允许在安装的时候确定逻辑页的大小，默认值为 2KB，可供选择的值还有 4KB、8KB、16KB。它要求各系统数据库的最小页大小，如表 1-1 所示。

表 1-1 系统最小页大小需求

页大小	master		sybsystemprocs		其他系统数据库（model 等）	
	设备大小	数据库	设备大小	数据库	设备大小	数据库
2K	24MB	13MB	120MB	120MB	3MB	3MB
4K	45MB	26MB			6MB	6MB
8K	89MB	52MB			12MB	12MB
16K	177MB	104MB			24MB	24MB

物理设备：在安装过程当中，系统将初始化主设备，包括为 master、sybsystemdb、tempdb 提供公共默认的设备，同时也会让你选择 sybsystemprocs 的设备信息相关参数。

从 ASE 12.0 开始，用户可以自己选择将设备创建在裸分区还是文件系统上，这取决于目标平台是否支持裸分区。所有 UNIX（包括 Linux 平台）都支持裸分区，而在 Windows 平台上，目前还不支持裸分区，通常推荐基于 NTFS 文件系统创建文件系统设备。

基于表 1-1，我们推荐 master 设备最小值为 50MB，便于系统扩展和升级。sybsystemprocs 数据库也可以存放到 master 设备上，但它最好还是与 master 数据库分开来放，可将它放到一个单独的设备上。

由于 Sybase 官网在不断整合，可能原下载地址：http://www.sybase.com/developer/downloads 无法得到您想要的下载。为方便读者下载，在本书源码的 readme 文件里列出了 CSDN 上的下载地址。另外，您还可以尝试从百度网盘里得到下载：① http://pan.baidu.com/share/link?shareid=200108&uk=2047101729（Linux 平台：Sybase/ase1503_linx86_32_2.tgz）；② http://pan.baidu.com/share/link?shareid=200109&uk=2047101729（Windows 平台：Sybase/ase1503_winx86_32_2.zip）。

> **注意**
> 以上资源只可用于测试、学习或者开发环境，切不可直接用于商业生产环境，正式的商业生产环境需要购买 SAP/Sybase 公司的正式授权。

需要提醒的是，master 设备不要创建得过大，master 设备过大将会延长恢复数据库的时间。最好将 master 设备专用于系统数据库（除 sybsystemprocs 数据库外，它需要单独的一个设备），并且不要将用户数据库放到 master 设备上。这样的设置将有利于系统管理和数据库的恢复。

服务器名：对主服务器则言，UNIX 下的默认服务器名形式是"SYB_<机器名>"，NT 下，默认的服务器名就是机器名。

对于备份服务器而言，默认服务器名是"SYB_<机器名>_BACK"（NT 下，则为<机器名>_BS）。对于 XP 服务器和监控服务器，都有类似的命名方式。服务器名最好是全部采用大写形式。

网络信息：ASE 支持两类网络：TCP/IP 和命名管道（适用于 NT 下边）。TCP/IP 是 ASE 客户端最常用的通信协议。各服务器进程都需要单独的端口号，主服务器默认端口为 5000，备份服务器为 5001，监控服务器为 5002，而 XP 服务器为 5004。

SySAM：主要为 ASE 提供 license 检查，系统启动的过程中，会自动检查所安装的 ASE 有哪些特性有对应的 license，只有具有对应 license 的那些特性才被使用。比如，我们如果什么 license 都没有，那么安装的应该是开发版，那么一些重要的高级特性便被禁止使用，如分布式事务、Java 存储过程等。对一般用户而言，我们只要理解 SySAM 的这个功能就好了。

排序方式：ASE 严格按照指定的排序方式对表数据进行排序，所有的 order by 及 group by 查询都会用到排序。ASE 默认的排序方式是二进制排序，简单地按照字节的 ASCII 码对应的数值进行大小比较，因此大写字母会排在前头，而小写字母会排在后边，因为所有的大写字母的 ASCII 码值比小写字母要小。为此，ASE 还支持多种不同的排序方式，以满足不同的应用需求。

1.2.3 使用图形界面安装

在获取 ASE 安装文件之后,将其解压,运行其中的 setup.exe,即可启动图形界面进行安装(如图 1-6 所示)。

单击"下一步"按钮,可以选择 Evaluation Edition 或者 Developer Edition,如图 1-7 所示。前者的功能要全一些,后者功能会有些限制。我们这里选择 Evaluation Edition。单击"下一步"按钮,选择地区,如图 1-8 所示,再单击"下一步"按钮。

图 1-6 安装的首页面

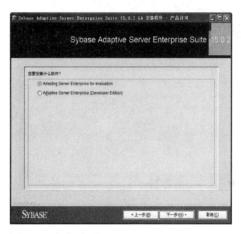

图 1-7 安装选项

进入图 1-9 以后,指定要安装的目录,这里指定安装目录为 D:\sybase。单击"下一步"按钮进入图 1-10。

图 1-8 选择地区

图 1-9 指定安装目录

如果指定"典型"安装,安装完以后,可能会发现有些内容并没有安装进去。所以,最好的方式是选择"定制",进入图1-7,你会发现,默认情况下,Sybase Server下的 ASE Web Service、Job Schedule、Connection下边的 XA Interface 库、Embedded SQL/*、ASE Data Providers 下边的 OLEDB、ADO.NET、jConnect for JDBC 下边的 Free Utilities for jConnect(jiSQL, RIBO)、Shared 目录、ASE Administrator Tools 下的 Microsoft Cluster Server Resource,这些模块都不会安装到系统当中。我们将这些重要的模块都选上,以便于后续的实践练习,如图1-11所示。单击"下一步"按钮。

图1-10 指定安装类型

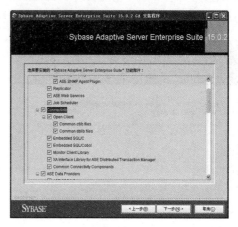

图1-11 自定义安装的组件

进入图1-12之后,可以选择不进行电子邮件警报。接着单击"下一步"按钮,进入图1-13,选择是否在安装过程中配置 ASE 中的几个 Server 部件(Adaptive Server、Backup Server、Monitor Server、Job Scheduler、Web Service 等)。如果不选择,可以在安装完软件之后手动创建。比较方便的是一次性都选上。单击"下一步"按钮。

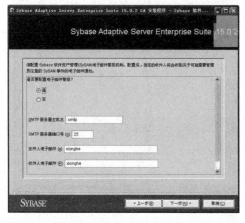

图1-12 设定邮件警报

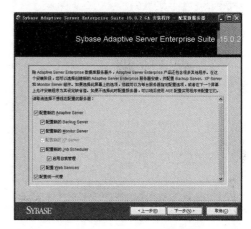

图1-13 是否在安装过程中配置服务器

进入图1-14,选择系统默认的配置。单击"下一步"按钮。

进入图 1-15，发现系统为我们设置了默认的参数值。后续过程是不断地单击"下一步"按钮，直至安装顺利结束。

图 1-14　是否自定义配置 Server　　　　图 1-15　系统为 ASE 安装设置默认的参数值

在 Windows 平台上，从图 1-15 开始，后续的创建 ASE 中几个 Server 的工作都会比较顺利。对于 Linux 用户而言，安装之前，还需要一些额外的准备工作。

对于大多数 Linux 版本而言，操作系统共享内存默认值为 32MB。对于具有 2K 页的默认服务器，Adaptive Server 要求的最小操作系统共享内存为 64MB。如果打算增大 ASE 的内在使用，则需要一个更高的值。

使用 sysctl(8) 方法可检查和调节操作系统共享内存参数。

若要检查当前共享内存大小，请输入：

/sbin/sysctl kernel.shmmax

若要调节当前共享内存大小，请输入：

/sbin/sysctl -w kernel.shmmax=nnn

其中 nnn 是以字节为单位的新的内存大小（至少 64MB，即 67108864 字节）。若要保证每次启动系统时都应用所设置的值，应将上述命令行添加到 /etc/rc.d/rc.local 文件中。或者直接修改 /etc/sysctl.conf 文件，添加内容 kernel.shmmax=67108864，重启以后即生效。

在 SuSE 系统上，该文件为/etc/init.d/boot.local。

对于 SuSE 10.0 或者 RHEL 4.0 及以上版本，有一个环境变量需要设定，即 LD_POINTER_GUARD，它的值必须为 1。

在 Linux RHEL Update 3.0 及更高版本上，在多个引擎上运行的 Adaptive Server 15.0 版本会要求禁用安全功能 Exec-Shield。

禁用 Exec-Shield 的方法是：

1. 在 /etc/sysctl.conf 中添加以下命令行：
   kernel.exec-shield=0
   kernel.exec-shield-randomize=0

2. 以超级用户身份输入：
   /sbin/sysctl -P
   有关其他信息，请参见位于 http://www.redhat.com/f/pdf/rhel/WHP0006US_Execshield.pdf 上的 Red Hat web 站点。

如果调整 SHMMAX 参数后，服务器重新启动失败，可能还需要增加另一个内核参数 SHMALL 的值，该参数为可分配的共享内存的最大量。其值位于文件 /proc/sys/kernel/shmall 中。此任务需要 root 用户权限。

同时，如果系统安装了防火墙，一定要开放系统默认的几个端口，客户端无法远程访问 ASE 的几个 Server。

还需要注意安装用户使用的语言问题，建议在安装时，将环境变量 LANG 值设为 C，待安装完以后，再设定为目标值。

以中文为例，可能在 Linux 系统里得到下述 locale 值：

```
iihero@seanlinux:~> locale
LANG=zh_CN.UTF-8
LC_CTYPE="zh_CN.UTF-8"
LC_NUMERIC="zh_CN.UTF-8"
LC_TIME="zh_CN.UTF-8"
LC_COLLATE="zh_CN.UTF-8"
LC_MONETARY="zh_CN.UTF-8"
LC_MESSAGES="zh_CN.UTF-8"
LC_PAPER="zh_CN.UTF-8"
LC_NAME="zh_CN.UTF-8"
LC_ADDRESS="zh_CN.UTF-8"
LC_TELEPHONE="zh_CN.UTF-8"
LC_MEASUREMENT="zh_CN.UTF-8"
LC_IDENTIFICATION="zh_CN.UTF-8"
LC_ALL=
```

这里建议在安装完 ASE 之后，在用户环境变量里，将 LANG 设为 zh_CN.GBK。因为 ASE 15.0 目前并不支持 zh_CN.UTF-8 这个 LANG，可以从安装完以后的$SYBASE/locales/locales.dat 里的 [linux]下与中文有关的几个 locale 项得到印证：

```
locale = CHINESE, chinese, eucgb
locale = zh_CN, chinese, eucgb
locale = zh_CN.gb18030, chinese, gb18030
locale = zh_CN.GB18030, chinese, gb18030
locale = zh_CN.gbk, chinese, eucgb
```

```
locale = zh_TW, tchinese, big5
locale = zh_TW.euctw, tchinese, euccns
```

1.2.4　ASE 的静默安装

作为企业级数据库，ASE 同样支持静默安装（Silent Install）方式。静默安装方式是使安装所有的参数都形成一个文件，作为安装命令的输入。如何得到这个文件呢？

首次安装时，使用 setupconsole.exe-options-record c:\ASE_silentinstall.txt 命令得到静默安装的参数文件。安装到末尾之前，不要选择安装数据库服务，其他所有的构件都选上，这样会得到一个完整的参数文件。

安装完成以后，就得到了样本参数文件。一般情况下，只需要注意如下内容的调整：

```
-P installLocation="D:\sybase"                    ##可以改为你想要的目录
-W installSoftwareLicenseType.type=developer      ##改为真正想用的 license 类型
-W sysamLicense.proceedWithoutLicense=false
-W sysamLicense.useExistingLicenseServer=false
-W sysamLicense.licenseFile=                      ##如果有 license 文件，给出全路径
```

默认直接用 Developer 类型的 license 即可。可以一路安装下去。

静默安装的命令是：

```
setupConsole.exe –silent -options c:\ase1503_silentinstall.txt
-W SybaseLicense.agreeToLicense=true -G replaceExistingResponse="yesToALL"。
```

它最大的好处是可以自动执行，中间不需要人工干预，对于 DBA 来说非常便利。只要有了安装文件和静默安装参数文件，一个命令就可以完成 ASE 的安装。所以，从某种意义上来说，静默安装也是 DBA 应该掌握的一个基本技巧。

1.2.5　安装完成时 ASE 的目录结构

安装完以后，系统中将会在 Sybase ASE 的安装目录（这里是 D:\sybase）下生成如下子目录：

1. _jvm：存放 Java 虚拟机，ASE 的 GUI 管理都使用 Java。
2. ASE-15_0：这是 ASE 服务器的核心子目录，通常它对应于环境变量%SYBASE_ASE%。
3. ASEP：这是 ASE 的 GUI 管理器的 plugin 目录。
4. charsets：该目录用于存储 ASE 支持的所有字符集。
5. collate：用于存储排序相关信息。
6. data：ASE 数据库的系统库的存放位置，你会发现里边默认有 master.dat、sysproc.dat、sybmgmtdb.dat、sybsysdb.dat 等设备文件（其他数据库可能称之为数据文件，含义是一样的）。
7. Dataaccess：存储 ASE 的几个重要客户端访问驱动库，包括 ADO.NET、ODBC、OLEDB、

当然，在 Linux/UNIX 下，只会有 ODBC 目录（注意，ASE 是支持 Linux/UNIX 下的 ODBC 访问的）。

8. DBISQL：该目录是 dbisql 的工具目录。
9. ini：该目录在 Windows 下独有。这是 ASE 系统中配置文件的存储目录，其中有一个很重要的配置文件 sql.ini。在 Linux/UNIX 下，该配置文件直接对应于 $SYBASE/interfaces。
10. jConnect-6_0：这是 ASE 的 JDBC 驱动库目录。
11. jutils-2_0：这是很有用的小工具 jisql 和 ribo 的存储目录。
12. locales：用于存储语言和字符集的对应关系。
13. OCS-15_0：这是 OpenClient/OpenServer 库目录，它对应于环境变量%SYBASE_OCS%。
14. RPL-15_0：该目录用于存放复制服务工具。
15. Shared 目录：ASE 的 GUI 管理工具 Sybase Central 就存放在这里，它的运行要依赖于_jvm 目录中的 Java 虚拟机。
16. SYSAM-2_0：该目录存放 ASE 的相关 license 或 key 文件。
17. UAF-2_0：Sybase Unified Agent，它是一个可供 Web 访问的应用服务器。便于用户通过 Web 来管理 ASE 数据库。
18. WS-15_0：用于支持 Web Services 的目录，对应于环境变量%SYBASE_WS%。

我们在 Windows 下直接运行 set syb 命令（Linux 下使用命令 env|grep SYB），可以得到相关的 Sybase 环境变量：

```
D:\>set syb
SYBASE=D:\sybase
SYBASE_ASE=ASE-15_0
SYBASE_JRE=D:\sybase\Shared\Sun\jre142_013
SYBASE_OCS=OCS-15_0
SYBASE_SYSAM2=SYSAM-2_0
SYBASE_UA=D:\sybase\UAF-2_0
SYBASE_WS=WS-15_0
SYBROOT=D:\sybase

D:\>set ds
DSLISTEN=SEANLAPTOP
DSQUERY=SEANLAPTOP
```

这些环境变量对于 ASE 的运行来说相当重要。我们在%SYBASE%目录下边，还看到有一个批处理文件 ASE150.bat，里边有完整的环境变量信息。

在 Windows 下，系统会为 Adaptive Server、Backup Server、XP Server、Monitor Server 创建服务，服务的名称如图 1-16 所示。

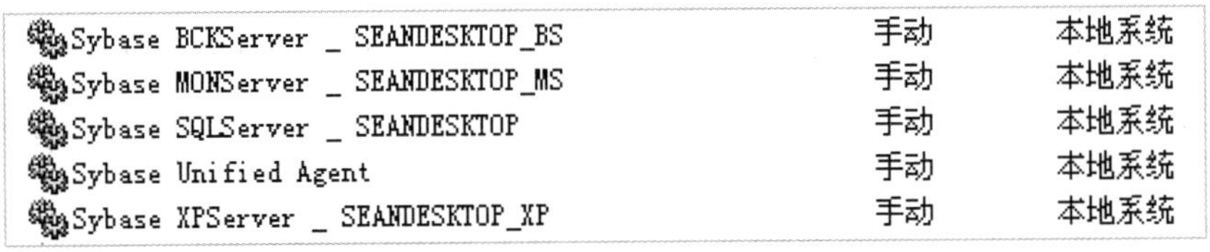

```
Sybase BCKServer _ SEANDESKTOP_BS        手动    本地系统
Sybase MONServer _ SEANDESKTOP_MS        手动    本地系统
Sybase SQLServer _ SEANDESKTOP           手动    本地系统
Sybase Unified Agent                     手动    本地系统
Sybase XPServer _ SEANDESKTOP_XP         手动    本地系统
```

图 1-16 ASE 数据库安装后创建的服务

在 UNIX/Linux 下，我们一定要把相关的 Sybase 环境变量添加到用户的 profile 当中。以 SuSE Linux 为例，我们先进入安装目录$SYBASE，看看环境变量文件 SYBASE.sh 的内容（以 Borne Shell 为例）。

```
PATH="/opt/sybase/OCS-15_0/bin":$PATH
export PATH
LD_LIBRARY_PATH="/opt/sybase/OCS-15_0/lib:/opt/sybase/OCS-15_0/lib3p":$LD_LIBRARY_PATH
export LD_LIBRARY_PATH
INCLUDE="/opt/sybase/OCS-15_0/include":$INCLUDE
export INCLUDE
LIB="/opt/sybase/OCS-15_0/lib":$LIB
export LIB
SYBASE_JRE="/opt/sybase/shared/jre142_013"
export SYBASE_JRE
SYBASE_SYSAM2="SYSAM-2_0"
export SYBASE_SYSAM2
PATH="/opt/sybase/UAF-2_0/bin":$PATH
export PATH
SYBASE_UA="/opt/sybase/UAF-2_0"
export SYBASE_UA
PATH="/opt/sybase/DBISQL/bin":$PATH
export PATH
SCROOT="/opt/sybase/shared/sybcentral43"
export SCROOT
PATH="/opt/sybase/ASEP/bin":$PATH
export PATH
PATH="/opt/sybase/RPL-15_0/bin":$PATH
export PATH
SYBASE_WS="WS-15_0"
export SYBASE_WS
PATH="/opt/sybase/ASE-15_0/jobscheduler/bin":$PATH
export PATH
LD_LIBRARY_PATH="/opt/sybase/DataAccess/ODBC/lib":$LD_LIBRARY_PATH
export LD_LIBRARY_PATH
SYBROOT="/opt/sybase"
export SYBROOT
```

我们要把上述内容复制到用户环境变量文件~/.bashrc 当中。同时，最好能同时添加环境变量 DSQUERY 和 DSLISTEN 的值：

```
export DSQUERY=SEANLINUX
export DSLISTEN=SEANLINUX
```

这两个环境变量的值的好处在于使用 isql 连接本机数据库时，不用再输入-s 参数及其值。

这些环境变量非常重要，Sybase ASE 的启动和运行，基本上都离不开这些环境变量。Sybase 的软件当中，所有环境变量在所有平台上全都是大写。UNIX 平台使用$前缀表示环境变量，Windows 平台使用%符号对来描述环境变量。几个通用的环境变量的作用和含义如下：

1. SYBASE：安装 Sybase 软件使用的根目录。
2. DSQUERY：本机上的客户端连接时用以连接的默认服务名。
3. DSLISTEN：本机上的服务器用以监听的服务名，如果在 Runserver 文件中没有指定-s 值时，就启用这个环境变量的值。
4. SYBASE_ASE：Adaptive Server Enterprise 服务器文件存放的子目录名。
5. SYBASE_EJB：EJB 服务器文件存放的子目录名。
6. SYBASE_OCS：Open Client/Open Server（简称 OCS）文件存放的子目录名。
7. SYBASE_SYSAM：Software Asset Manager（SAM）文件存放的子目录名。

在$SYBASE 目录下边有一系列子目录，有一些是产品相关的，有一些是公用的。某些公用的子目录，比如字符集（charsets）、排序（collate）、本地化（locales）、已安装（installed）可能包含多个 Sybase 产品（如 ASE、ASA（SQL Anywhere）、IQ（Adaptive Server IQ）、RS（Replication Server）以及其他）的相关信息。与具体产品相关的那些环境变量的值通常会带有产品的名字和相应的主版本号（如 ASE-15_0 或 REP-15_0）。下边这些子目录名就是安装 ASE 15.0 的时候产生的。

- ASE-15_0：它对应于$SYBASE_ASE，是 ASE 15.0 产品的顶级子目录，显然 ASE 12.5 将对应于目录 ASE-12_5。
- EJB-15_0：它对应于$SYBASE_EJB，是 EJB 15.0 安装的顶级子目录。
- jconnect-5_5：是 ASE jdbc 驱动的安装目录。
- jutils-2_0：是 jisql、Ribo 工具的安装目录。
- OCS-15_0：对应于$SYBASE_OCS 是 Open Client/Open Server 15.0 安装的顶级子目录，Open Client/Open Server 12.5.x 对应的目录应该是 OCS-12_5。
- SYSAM-1_0：是 SAM 的顶级子目录。

1.2.6　手动创建服务器

我们完全可以用命令行来创建 ASE 中的几个服务器程序。

以 Linux 为例，假设安装完 Sybase ASE 以后，SYBASE 指向目录：/testarea/ase1502/ase1502_esd1_c2。我们在目录$SYBASE/ASE-15_0/bin 下边创建一个配置文件，名为 sqlsrv.rs：

```
sybinit.release_directory: /testarea/ase1502/ase1502_esd1_c2
sybinit.product: sqlsrv
```

sqlsrv.server_name: BJEASLINUX1B
sqlsrv.new_config: yes
sqlsrv.do_add_server: yes
sqlsrv.network_protocol_list: tcp
sqlsrv.network_hostname_list: bjeaslinux1
sqlsrv.network_port_list: 5000
sqlsrv.server_page_size: USE_DEFAULT
sqlsrv.force_buildmaster: no
sqlsrv.master_device_physical_name: /testarea/ase1502/ase1502_esd1_c2/data/master.dat
sqlsrv.master_device_size: 120
sqlsrv.master_database_size: 50
sqlsrv.errorlog: USE_DEFAULT
sqlsrv.do_upgrade: no
sqlsrv.sybsystemprocs_device_physical_name: /testarea/ase1502/ase1502_esd1_c2/data/sybsystemprocs.dat
sqlsrv.sybsystemprocs_device_size: 195
sqlsrv.sybsystemprocs_database_size: 190
sqlsrv.sybsystemdb_device_physical_name: /testarea/ase1502/ase1502_esd1_c2/data/sybsystempdb.dat
sqlsrv.sybsystemdb_device_size: USE_DEFAULT
sqlsrv.sybsystemdb_database_size: USE_DEFAULT
sqlsrv.default_backup_server: BJEASLINUX1_BS

我们注意到这个文件是为 sqlsrv 服务而创建的，第二行 sybinit.product 就表明该文件所创建的服务器。

server_name 指的就是创建的服务名，它将被注册到 interfaces 文件当中。

在这里，我们依次指定了所需要的网络相关参数（支持的协议、主机名、端口号）、服务器逻辑页大小（这里采用默认大小，实际上为 2KB），然后是 master 数据库的设备物理位置、大小以及数据库本身大小（上述值比第 1 节中的推荐值要大一些），然后指定 sybsystemprocs 和 sybsystemdb 两个数据库的设备位置、设备文件大小以及数据库的大小。最后，还要指定 sqlsrv 服务对应的备份服务器 default_backup_server 的名字，这里是默认值 BJEASLINUX1_BS。

接着在此目录下，首先加载 SYBASE.sh 中定义的有关环境变量，然后通过可执行程序 srvbuild 来创建相应服务。

```
 [xionghe@bjeaslinux1 bin]$source ../../SYBASE.sh
[xionghe@bjeaslinux1 bin]$./srvbuild -r sqlserver.rs

Building Adaptive Server 'BJEASLINUX1B':
Writing entry into directory services...
Directory services entry complete.
Building master device...
Master device complete.
Writing RUN_BJEASLINUX1B file...
RUN_BJEASLINUX1B file complete.
Starting server...
Server started.
Building sysprocs device and sybsystemprocs database...
```

```
sysprocs device and sybsystemprocs database created.
Running installmaster script to install system stored procedures...
installmaster: 10% complete.
installmaster: 20% complete.
installmaster: 30% complete.
installmaster: 40% complete.
installmaster: 50% complete.
installmaster: 60% complete.
installmaster: 70% complete.
installmaster: 80% complete.
installmaster: 90% complete.
installmaster: 100% complete.
installmaster script complete.
Creating two-phase commit database...
Two phase commit database complete.
Installing common character sets (Code Page 437, Code Page 850, ISO Latin-1,
Macintosh and HP Roman-8)...
Character sets installed.
Setting server name in Adaptive Server...
Server name added.
Server 'BJEASLINUX1B' was successfully created.
[xionghe@bjeaslinux1 bin] isql –Usa
1> sp_password null, 'sybase1', sa
2> go
Password correctly set.
(return status = 0)
```

最终使用 sp_password，将 sa 用户的密码设置为 sybase1。

创建一个 XP Server，配置文件 xp.rs：

```
sybinit.release_directory: /testarea/ase1502/ase1502_esd1_c2
sybinit.product: xp
xp.server_name: BJEASLINUX1_XP
xp.new_config: yes
xp.do_add_xp_server: yes
xp.do_upgrade: no
xp.network_protocol_list: tcp
xp.network_hostname_list: bjeaslinux1
xp.network_port_list: 5003
sqlsrv.related_sqlsrvr: BJEASLINUX1B
sqlsrv.sa_login: sa
sqlsrv.sa_password: sybase1
```

这里需要提醒的是，为避免出问题，最好创建的服务器名都采用大写的形式，NT 平台也不例外，如这里将 XP 服务器名设为 BJEASLINUX1_XP，而不是小写形式 bjeaslinux1_xp。

创建备份服务器，其配置文件 backupsvr.rs 如下：

sybinit.boot_directory:	/linuxtea4_rel1/ase1502_esd1_c2
sybinit.release_directory:	/linuxtea4_rel1/ase1502_esd1_c2

sybinit.product:	bsrv
sqlsrv.sa_password:	
sqlsrv.sa_login:	sa
bsrv.do_add_backup_server:	yes
bsrv.server_name:	BJEASLINUX1_BS
bsrv.errorlog:	/linuxtea4_rel1/ase1502_esd1_c2/ASE-15_0/install/bsrv.log
bsrv.network_port_list:	5001
bsrv.network_hostname_list:	bjeaslinux1
bsrv.network_protocol_list:	tcp
bsrv.character_set:	cp850
bsrv.language:	us_english
bsrv.network_name_alias_list:	
bsrv.notes:	
bsrv.connect_retry_delay_time:	5
bsrv.connect_retry_count:	5
bsrv.addl_cmdline_parameters:	
bsrv.new_config:	yes
bsrv.do_upgrade:	no

在这里，我们还可以指定备份服务器用到的字符集和语言集。

创建监控服务器 Monitor Server，其配置文件 mon.rs 内容如下：

```
sybinit.release_directory: /testarea/ase1502/ase1502_esd1_c2
sybinit.product: msrv
msrv.server_name: BJEASLINUX1_MS
msrv.new_config: yes
msrv.do_add_monitor_server: yes
msrv.do_upgrade: no
msrv.network_protocol_list: tcp
msrv.network_hostname_list: bjeaslinux1
msrv.network_port_list: 5004
msrv.errorlog: USE_DEFAULT
sqlsrv.related_sqlsrvr: bjeaslinux1
sqlsrv.sa_login: sa
sqlsrv.sa_password: sybase1
```

创建过程如下：

```
 [xionghe@bjeaslinux1 bin]$ ./srvbuild -r mon.rs
Building Monitor Server 'BJEASLINUX1_MS':
Writing entry into directory services...
Directory services entry complete.
Writing RUN_BJEASLINUX1_MS file...
RUN_BJEASLINUX1_MS file complete.
Installing required script(s) in related Adaptive Server...
installmon: 10% complete.
installmon: 20% complete.
installmon: 30% complete.
```

```
installmon: 40% complete.
installmon: 50% complete.
installmon: 60% complete.
installmon: 70% complete.
installmon: 80% complete.
installmon: 90% complete.
installmon: 100% complete.
Script executed.
Starting server...
Server started.
Server 'BJEASLINUX1_MS' was successfully created.
```

ASE 中的四个服务器程序都可以通过 srvbuild 命令及对应的配置文件手工创建，这种创建过程还能加深对 ASE 服务器程序的理解，知道这些程序至少需要哪些有用的参数以及这些参数的意义。

还可以注意到，UNIX/Linux 下首次安装完 ASE 以后，在$SYBASE/$SYBASE_ASE 目录下，有下述几个资源文件：bsrv.res、js.res、msrv.res、sqlsrv.res、xp.res，它们分别对应 Backup Server、JSAgent Server、Monitor Server、Adaptive Server、XP Server。

这几个文件都可以作为参考。有时第一次安装，这几个资源文件对应的 Server 可能都没有创建成功，在修改了具体的系统参数或环境变量，并清除文件$SYBASE/$SYBASE_ASE/interfaces 中的内容以后，我们还可以通过 srvbuild 对它们进行重建。

在 Windows 下，没有 srvbuild 命令，但其实同样可以通过脚本和命令来进行手工创建。目前这个命令并没有见诸文档，使用的是命令 sybatch.exe。在%SYBASE%\%SYBASE_ASE%\sample\server 目录下边有几个示例资源文件，或者如同上边介绍的，第一次安装时装数据库服务，直接用$SYBASE/$SYBASE_ASE 下边的几个资源文件作参考。

取一个样例文件 sybase_ase.res，内容如下：

```
#
# --- This is a "sybatch.exe" sample resource file. ---
#
# This sample resource file will configure a new
# 2k pagesize Adaptive Server "SYBASE".
#
sybinit.boot_directory:                    C:\Sybase
sybinit.release_directory:                 C:\Sybase
sqlsrv.do_add_server:                      yes
sqlsrv.network_hostname_list:              localhost
sqlsrv.network_port_list:                  5000
sqlsrv.network_protocol_list:              tcp
sqlsrv.notes:
sqlsrv.connect_retry_delay_time:           5
sqlsrv.connect_retry_count:                5
sqlsrv.new_config:                         yes
```

```
#
sqlsrv.server_name:                              SYBASE
sqlsrv.sa_password:
sqlsrv.sa_login:                                 sa
sqlsrv.server_page_size:                         2k
#
# --- Set up master ----
#
sqlsrv.master_device_physical_name:              C:\Sybase\data\master.dat
sqlsrv.master_device_size:                       30
sqlsrv.master_db_size:                           13
sqlsrv.disk_mirror_name:
#
# --- Set up sybsystemprocs ----
#
sqlsrv.do_create_sybsystemprocs_device:          yes
sqlsrv.sybsystemprocs_device_physical_name:      C:\Sybase\data\sysprocs.dat
sqlsrv.sybsystemprocs_device_size:               132
sqlsrv.sybsystemprocs_db_size:                   132
sqlsrv.sybsystemprocs_device_logical_name:       sysprocsdev
#
# --- Set up sybsystemdb ----
#
sqlsrv.do_create_sybsystemdb:                    yes
sqlsrv.do_create_sybsystemdb_db_device:          yes
sqlsrv.sybsystemdb_db_device_physical_name:      C:\Sybase\data\sybsysdb.dat
sqlsrv.sybsystemdb_db_device_physical_size:      5
sqlsrv.sybsystemdb_db_size:                      5
sqlsrv.sybsystemdb_db_device_logical_name:       systemdbdev
#
sqlsrv.errorlog:                                 C:\Sybase\ASE-15_0\install\sybase.log
sqlsrv.sort_order:                               binary
sqlsrv.default_characterset:                     cp850
sqlsrv.default_language:                         us_english
#
sqlsrv.preupgrade_succeeded:                     no
sqlsrv.network_name_alias_list:
sqlsrv.resword_conflict:                         0
sqlsrv.resword_done:                             no
sqlsrv.do_upgrade:                               no
sqlsrv.characterset_install_list:
sqlsrv.characterset_remove_list:
sqlsrv.language_install_list:
sqlsrv.language_remove_list:
sqlsrv.shared_memory_directory:
sqlsrv.addl_cmdline_parameters:
sqlsrv.eventlog:                                 yes
sqlsrv.atr_name_shutdown_required:               yes
```

sqlsrv.atr_name_qinstall:	no
#	
sybinit.charset:	cp850
sybinit.language:	us_english
sybinit.resource_file:	
sybinit.log_file:	
sybinit.product:	sqlsrv
#	
sqlsrv.default_backup_server:	SYBASE_BS
#	
# optimize ASE	
#	
sqlsrv.do_optimize_config:	yes
sqlsrv.avail_physical_memory:	2048
sqlsrv.avail_cpu_num:	1
sqlsrv.tempdb_device_physical_name:	C:\Sybase\data\tempdb.dat
sqlsrv.tempdb_device_size:	100
sqlsrv.tempdb_database_size:	100
sqlsrv.application_type:	MIXED

通常，要想一次构建成功，我们要作如下修改：

（1）安装目录保持一致。

sybinit.boot_directory:	D:\Sybase
sybinit.release_directory:	D:\Sybase

所有的 C:\Sybase 目录都得调成现在安装的位置，包括数据文件。

（2）sysprocs.dat 文件原值 132 太小，改成至少 136。

sqlsrv.do_create_sybsystemprocs_device:	yes
sqlsrv.sybsystemprocs_device_physical_name:	D:\Sybase\data\sysprocs.dat
sqlsrv.sybsystemprocs_device_size:	150
sqlsrv.sybsystemprocs_db_size:	150
sqlsrv.sybsystemprocs_device_logical_name:	sysprocsdev

（3）字符集，这个是大家比较关心的，默认的是 iso_1, english。

sqlsrv.sort_order:	binary
sqlsrv.default_characterset:	**utf8**
sqlsrv.default_language:	us_english

字符集要一步调整到位，这样比图形界面安装要节省很多时间。以下是 Windows 下一次手动创建 ASE 服务的输出结果：

```
D:\sybase\ASE-15_0\bin>sybatch.exe -r d:\SOFT\sybatch_ase.res
CONNECTIVITY ERROR: Open Client message: 'ct_connect(): network packet layer: in
ternal net library error: Net-Lib protocol driver call to connect two endpoints
failed
```

```
Failed to connect to the server - Error is 10061    //由于目标机器积极拒绝，无法连接
。
'.
CONNECTIVITY ERROR: Initialization of auditinit connectivity module failed.
Running task: update Sybase Server entry in interfaces file.
Task succeeded: update Sybase Server entry in interfaces file.
Running task: create the master device.
Building the master device
......Done
Task succeeded: create the master device.
Running task: update Sybase Server entry in registry.
Task succeeded: update Sybase Server entry in registry.
Running task: start the Sybase Server.
waiting for server 'SYBASE' to boot...
waiting for server 'SYBASE' to boot...
waiting for server 'SYBASE' to boot...
waiting for server 'SYBASE' to boot...
waiting for server 'SYBASE' to boot...
Task succeeded: start the Sybase Server.
Running task: create the sybsystemprocs database.
sybsystemprocs database created.
Task succeeded: create the sybsystemprocs database.
Running task: install system stored procedures.
Installing system stored procedures : 10% complete...
Installing system stored procedures : 20% complete...
Installing system stored procedures : 30% complete...
Installing system stored procedures : 40% complete...
Installing system stored procedures : 50% complete...
Installing system stored procedures : 60% complete...
Installing system stored procedures : 70% complete...
Installing system stored procedures : 80% complete...
Installing system stored procedures : 90% complete...
Installing system stored procedures : 100% complete...
Task succeeded: install system stored procedures.
Running task: set permissions for the 'model' database.
Task succeeded: set permissions for the 'model' database.
Running task: set local Adaptive Server name.
Task succeeded: set local Adaptive Server name.
Running task: set the XP Server for the Adaptive Server.
Task succeeded: set the XP Server for the Adaptive Server.
```

前边的 auditinit 错误不影响结果，可以忽略。

当然我们还可以使用图形界面方式来创建全新的 Server。

Windows 下，通过菜单命令 Sybase→Adaptive Server Enterprise→Server Config 进入创建 Server 的界面。

下面给出 Linux 另一个创建新 Server 的实例。

UNIX/Linux 下，在拥有了合适的环境变量以后，进入 $SYBASE/$SYBASE_ASE/bin，运行 ./asecfg，如图 1-17 和图 1-18 所示。

图 1-17 配置 ASE 服务器　　　　图 1-18 asecfg 首页面

你会发现 asecfg 不仅可以配置一个新的 Server，还可以进行升级，设置安装好的 Server 的字符集相关信息，也可以编辑 interfaces 文件。

双击 Configure a new server 按钮，进入图 1-19。

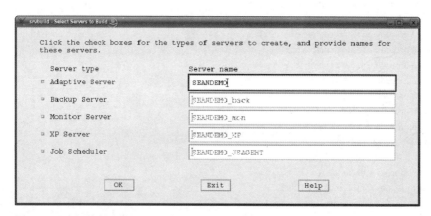

图 1-19 选择待创建的 Server 类型

在图 1-19 中，我们选择要创建一个 Adaptive Server，并输入 Server 名字 SEANDEMO。然后单击 OK 按钮，进入下一步，如图 1-20 所示。

我们在这里设定有关 Adaptive Server SEANDEMO 的一些参数：

- ASE 页大小默认为 2KB，也可以设定为 4KB、8KB、16KB。我们这里设定为 4KB，意思是 SEANDEMO 下各个数据库的每个物理页大小都为 4KB。
- Master 设备路径这里指定为 /opt/sybase/data2/master.dat。因为系统中已经安装了一个

Server——SEANLINUX，它占用了默认的设备路径/opt/Sybase/data/master.dat。
- Port number：这里指定一个不同于默认端口 5000 的值——7000，避免冲突。

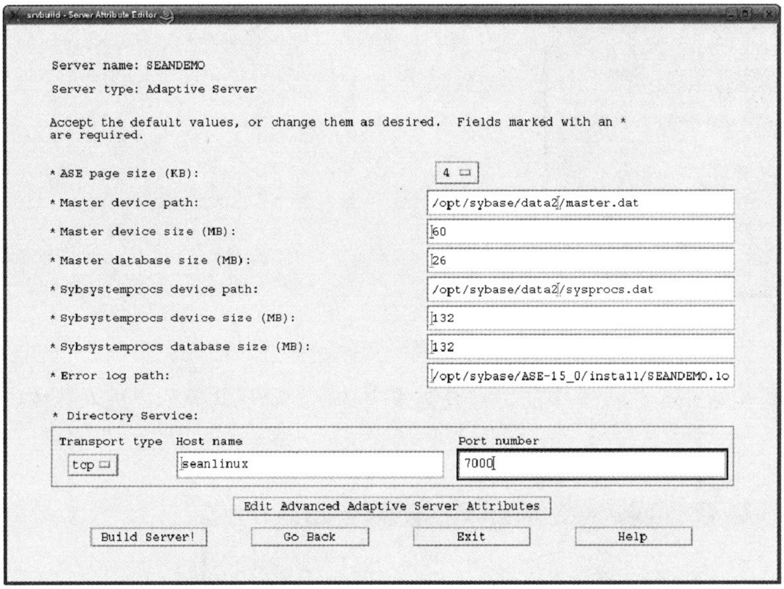

图 1-20 Adaptive Server 的相关参数

其他几个参数值都采用窗口中给定的默认值。然后单击"Build Server!"按钮，还会提示"端口号可能有冲突"，确认以后，即可以进入 Server 的创建过程，如图 1-21 所示。

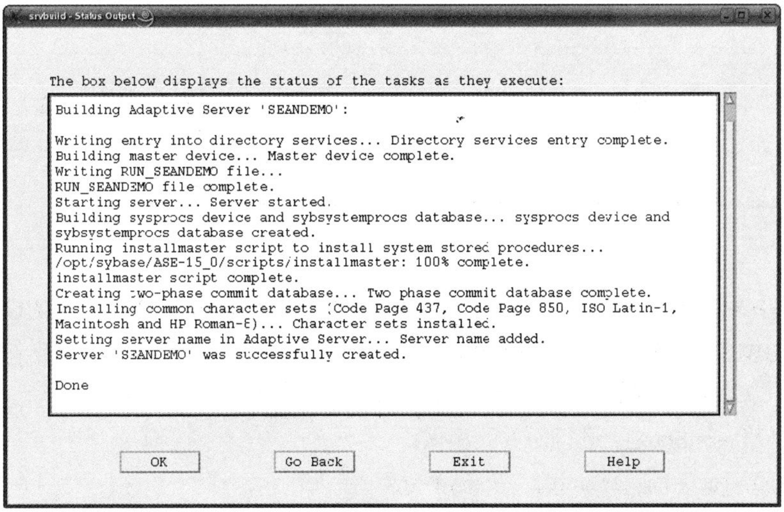

图 1-21 Server 创建成功

这时并没有完全创建完，我们会发现，系统默认安装的是通用字符，代码页 850、437 等，而不是我们通常所要的中文。单击 OK 按钮以后，系统会提示你是否要设置语言相关选项，如图 1-22 所示。

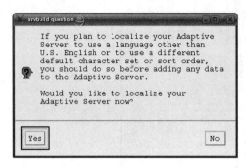

图 1-22　语言设置

假设我们这里将新建 Adaptive Server 中的语言设置成中文，则确认此提示，进入具体的语言设置，如图 1-23 所示。

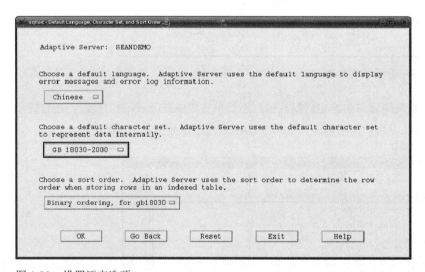

图 1-23　设置语言选项

我们将默认语言 English 改为 Chinese，将默认字符 cp850 改为 GB 18030-2000，对应的字符集排序规则也相应地改为 GB18030 的二进制排序方式。单击 OK 按钮以后，即可完成中文字符集的设置。当然，为了支持中文，也可将默认字符集改为 UTF-8。最后得到成功的提示。

我们可以在 Linux 机器 seanlinux 上进行简单的测试：

```
iihero@seanlinux:/opt/sybase/ASE-15_0/bin> isql -Usa -SSEANDEMO
保密字：
1> create table #t(id varchar(32))
```

```
2> go
1> insert into #t values('中文测试')
2> go
(1 row affected)
1> select * from #t
2> go
 id
 ------------------
 中文测试

(1 row affected)
1>
```

1.2.7　验证服务器是否在运行

一旦 ASE 服务器启动以后,管理员有必要确认 ASE 是否启动成功,服务器是否存在 ASE 服务进程。

在 Windows 各平台下,我们有多种方式可以判断 ASE 是否已经启动。

第一种方法是在服务的控制台管理器里查看 Sybase SQL Server _ <servername>的状态是否为已启动(这里 servername 是实际的 ASE 数据库服务名)。这种方法适用于通过服务方式来启动 ASE 的情形。

第二种方法是通过任务管理器,直接查看是否有我们需要的 sqlsrvr.exe 进程,如果 ASE 启动成功,则该进程肯定存在。

第三种方法是通过查询对应端口号的进程 ID,再查询对应 ID 的进程名,看是否是我们启动的 ASE 服务器进程 sqlsrvr.exe。请看下边的查询过程:

```
C:\Documents and Settings\hex>netstat -ano | grep 5000
  TCP    127.0.0.1:1369      127.0.0.1:5000     ESTABLISHED    5532
  TCP    127.0.0.1:1370      127.0.0.1:5000     ESTABLISHED    5532
  TCP    127.0.0.1:1371      127.0.0.1:5000     ESTABLISHED    5532
  TCP    127.0.0.1:1372      127.0.0.1:5000     ESTABLISHED    5532
  TCP    127.0.0.1:1373      127.0.0.1:5000     ESTABLISHED    5532
  TCP    127.0.0.1:1374      127.0.0.1:5000     ESTABLISHED    5532
  TCP    127.0.0.1:1375      127.0.0.1:5000     ESTABLISHED    5532
  TCP    127.0.0.1:1376      127.0.0.1:5000     ESTABLISHED    5532
  TCP    127.0.0.1:5000      0.0.0.0:0          LISTENING      7844
  TCP    127.0.0.1:5000      127.0.0.1:1369     ESTABLISHED    7844
  TCP    127.0.0.1:5000      127.0.0.1:1370     ESTABLISHED    7844
  TCP    127.0.0.1:5000      127.0.0.1:1371     ESTABLISHED    7844
  TCP    127.0.0.1:5000      127.0.0.1:1372     ESTABLISHED    7844
  TCP    127.0.0.1:5000      127.0.0.1:1373     ESTABLISHED    7844
  TCP    127.0.0.1:5000      127.0.0.1:1374     ESTABLISHED    7844
  TCP    127.0.0.1:5000      127.0.0.1:1375     ESTABLISHED    7844
  TCP    127.0.0.1:5000      127.0.0.1:1376     ESTABLISHED    7844

C:\Documents and Settings\hex>tasklist | grep 7844
```

sqlsrvr.exe	7844	0	3,692 K

C:\Documents and Settings\hex>tasklist | grep 5532

jsagent.exe	5532	0	400 K

我们先用 netstat -nao | grep 5000，查出所有与端口 5000 相关的进程。其中，ID 为 7844 的进程正在监听（listening），顺便我们还查出了 5532 进程正在建立到 5000 商品上的连接。再通过 tasklist，查询得到对应的进程名。

这里用到了 grep 命令，通过安装工具包 MKS Tookit，可以得到全套的仿 UNIX 命令集。

在我们的工作中，我还经常用到另外一个工具 Debugging tools for Windows，这个工具里边也有功能很强大的命令 tlist（列出所有正在运行的线程信息）。前边的 tasklist 命令完全可以使用 tlist 来代替，非常方便。

在 UNIX/Linux 平台下，执行 $SYBASE/$SYBASE_ASE/install 目录下的可执行程序 showserver，可以得到已经运行的 ASE 服务器程序的列表。

1.2.8 修改 sa 用户口令

刚安装完的 ASE 一点安全性也没有，因为超级用户 sa 的口令密码为空，这无论是在测试环境还是生产环境里都是无法接受的。

使用 isql 连接数据库：

```
D:\SybaseASE\ini>isql -Usa -Sxionghe
```

保密字为：

```
1>
```

然后直接运行以下代码：

```
1> sp_password null, 'sybase1', sa
2> go
Password correctly set.
(return status = 0)
```

这样即把密码从空改为 sybase1 了。

也许有人会问，想把密码再改回 NULL，如何操作？

运行以下代码：

```
1> sp_password 'sybase1',null,sa
2> go
Msg 10317, Level 14, State 1:
Server 'XIONGHE', Procedure 'sp_password', Line 118:
The specified password is too short. Passwords must be at least 6 character(s) long.
Msg 17720, Level 16, State 1:
Server 'XIONGHE', Procedure 'sp_password', Line 128:
```

```
Error:    Unable to set the Password.
(return status = 1)
1>
```

这里提示说密码太短。ASE 安装时默认的密码长度至少为 6（sa 默认密码为 NULL，岂不是有违此规定？有些自相矛盾）。解决的办法是修改此限制，首先将密码最短长度改为 0，再修改密码。

```
1> sp_configure minimum
2> go
Parameter Name              Default     Memory Used Config Value
-----------------------     -------     ------------------------
minimum password length     6           0

Run Value       Unit            Type
-------------   -------------   ---------
6               6 bytes         dynamic

(1 row affected)
(return status = 0)
> sp_configure "minimum password length", 0
2> go
Parameter Name              Default     Memory Used Config Value
-----------------------     -------     ------------------------
minimum password length     6           0

Run Value       Unit            Type
-------------   -------------   ---------
0               0 bytes         dynamic

(1 row affected)
Configuration option changed. ASE need not be rebooted since the option is
dynamic.
Changing the value of 'minimum password length' does not increase the amount of
memory Adaptive Server uses.
(return status = 0)
1> sp_password 'sybase1', null, sa
2> go
Password correctly set.
(return status = 0)
```

即又将密码恢复为 null 了。

1.2.9　Runserver 文件

服务器创建完毕，会生成相应的 Runserver 文件，它包含了主要的启动参数。该文件通过这些参数启动 dataserver 进程。这些参数主要有服务器名、master 设备的物理位置、errorlog 文件的物理位置、页大小、共享内存的位置等。Runserver 文件的默认名为 RUN_<SERVERNAME>，统一创建在物理目录$SYBASE/$SYBASE_ASE/install 下。

值得注意的是,在 NT 平台下虽然也创建了 Runserver 文件,但是实际的启动参数都保存在注册表中。我们从服务控制台里启动 ASE 的服务不会用到这个文件。这个批处理文件主要用于需要紧急重启的场合,改变该文件当中的设置不会影响正常的服务启动。

也就是说,启动 ASE 所使用的 Runserver 文件和启动 Sybase ASE 服务所使用的参数来源于不同的地方,前者的参数来源于文件本身,后者的参数来源于注册表,相应的注册表项是 HKEY_LOCAL_MACHINE\SOFTWARE\SYBASE\Server\<servername>\Parameters

NT 下的 Runserver 文件示例如下:

```
rem
rem Backup Server Information:
rem name:                    SEANLAPTOP
rem server page size:        2048
rem master device size:      120
rem errorlog:                D:\sybase\ASE-15_0\install\SEANLAPTOP.log
rem interfaces:              D:\sybase\ini\sql.ini
rem
D:\sybase\ASE-15_0\bin\sqlsrvr.exe -dD:\sybase\data\master.dat –sSEANLAPTOP
 -eD:\sybase\ASE-15_0\install\SEANLAPTOP.log -iD:\sybase\ini -MD:\sybase\ASE-15_0
```

UNIX 下的 Runserver 文件示例如下:

```
#!/bin/sh
#
# ASE page size (KB):    2k
# Master device path:    /opt/sybase/data/master.dat
# Error log path:        /opt/sybase/ASE-15_0/install/SEANLINUX.log
# Configuration file path: /opt/sybase/ASE-15_0/SEANLINUX.cfg
# Directory for shared memory files:    /opt/sybase/ASE-15_0
# Adaptive Server name:  SEANLINUX
#
/opt/sybase/ASE-15_0/bin/dataserver \
-d/opt/sybase/data/master.dat \
-e/opt/sybase/ASE-15_0/install/SEANLINUX.log \
-c/opt/sybase/ASE-15_0/SEANLINUX.cfg \
-M/opt/sybase/ASE-15_0 \
-sSEANLINUX \
```

需要注意的是,UNIX/Linux 对文件名是区分大小写的。

Runserver 中会用到下述参数:

- -c: 配置文件.cfg 的位置和名称。
- -d: master 设备的位置和名称。
- -e: 错误日志文件的位置和名称。

- -i：interfaces 文件的位置和名称。
- -M：共享内存目录的路径。
- -r：master 镜像的路径。
- -s：服务器名。
- -z：服务器逻辑页的大小，只作文档记录用，实际启动用的页大小是配置文件中的页大小。
- -m：master 恢复模式，用于单用户启动 Server，以恢复出错的 Master 数据库。
- -p：密码恢复模式，用于在启动时生成一个 SSO 密码，便于恢复你想要的 sa 用户密码。
- -u：登录解锁模式，在启动时对一个已经锁定的用户进行解锁。

1.3 如何卸载已经安装的 Sybase ASE

有时候，我们还需要卸载已经安装的 Sybase ASE 数据库，原因可能多种多样，比如：

1．数据库的安装位置不合要求。
2．数据库的有关参数配置不理想。
3．想抛弃旧版本的数据库，装一个新版本的数据库。

在 Windows 平台下，正确的使用方法是，先停掉所有的 ASE 相关服务：BCKServer、MONServer、SQLServer、Unified Agent 以及 XPServer。然后再通过添加/删除程序面板，来删除 Sybase ASE 数据库，最后重启机器即可。

如果是手动删除，那么在停掉所有的 ASE 相关服务以后，通过运行 regedt32 进入系统注册表，首先备份整个注册表。然后查找 Sybase 关键字，清除注册表中 Sybase ASE 数据库的相关注册项：HKEY_LOCAL_MACHINE\SOFTWARE\Sybase\SQLServer、HKEY_LOCAL_MACHINE\SYSTEM\CurrentControlSet\Services\SYB_<servername>、HKEY_LOCAL_MACHINE\SYSTEM\CurrentControlSet\Enum\Root\LEGACY_SYB_<servername>等。

尤其是相关服务的注册项，如 LEGACY_SYBSQL_<servername>，在注册表中无法直接删除，如图 1-24 所示。

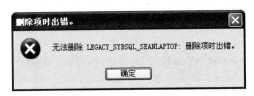

图 1-24　删除服务项

这时，我们通过修改注册项的权限，让 Everyone 用户组能"完全控制"该注册项，从而可以删除该项，如图 1-25 所示。其他服务相关注册项可采用同样的方法进行删除。

图 1-25　设置删除服务项时的用户权限

清理完注册表相关项之后，重启机器，就可以直接删除 ASE 原有目录，完成 ASE 的卸载。

在 UNIX/Linux 平台下，通过 showserver 查询得到所有的 ASE 相关进程，退出这些进程，然后手动删除 ASE 安装目录，即可卸载整个 ASE。

1.4　忘记了 sa 用户密码

这一节是专为粗心的 ASE 数据库管理员写的。对 ASE 数据库中的用户而言，最重要的用户是 sa，在安装完之后，它的密码一般要熟记起来。当然也有可能忘了 sa 的密码，ASE 为管理员提供了恢复和重置 sa 用户的密码的功能。这样不至于因为忘了 sa 的密码，就要重新安装整个数据库。

解决遗忘 sa 密码的办法是：进入%SYBASE%\%SYBASE_ASE%\install 目录，你会发现有一个文件，名为 RUN_<servername>.bat，这里<servername>为 SEANLAPTOP，这个文件就是启动 ASE Server 的批处理文件，内容如下：

```
rem
d:\ASE150\ASE-15_0\bin\sqlsrvr.exe    -dd:\ASE150\data\master.dat -sSEANLAPTOP -ed:\ASE150\ASE-15_0\install\SEANLAPTOP.log -id:\ASE150\ini -Md:\ASE150\ASE-15_0
```

我们在这个文件的末尾加上**-psa**：

```
rem
d:\ASE150\ASE-15_0\bin\sqlsrvr.exe    -dd:\ASE150\data\master.dat -sSEANLAPTOP -ed:\ASE150\ASE-15_0\install\SEANLAPTOP.log -id:\ASE150\ini -Md:\ASE150\ASE-15_0 -psa
```

搭建 Sybase ASE 环境   第 1 章

然后运行这个批处理文件，你会发现命令行窗口中有大量的提示信息：

```
……
ta cache for database 'sybsystemprocs'.
00:00000:00001:2008/08/29 21:42:12.90 kernel    Failed to log the current message
in the Windows NT event log
00:00000:00001:2008/08/29 21:42:12.90 server    Completed cleaning up the default
data cache for database 'sybsystemprocs'.
00:00000:00001:2008/08/29 21:42:12.90 kernel    Failed to log the current message
in the Windows NT event log
00:00000:00001:2008/08/29 21:42:12.90 server    Checking external objects.
00:00000:00001:2008/08/29 21:42:12.90 kernel    Failed to log the current message
in the Windows NT event log
00:00000:00001:2008/08/29 21:42:12.93 server    The transaction log in the databas
e 'sybsystemprocs' will use I/O size of 2 Kb.
00:00000:00001:2008/08/29 21:42:12.93 kernel    Failed to log the current message
in the Windows NT event log
00:00000:00001:2008/08/29 21:42:16.10 server    Database 'sybsystemprocs' is now o
nline.
00:00000:00001:2008/08/29 21:42:16.10 kernel    Failed to log the current message
in the Windows NT event log

New SSO password for sa:kdoatlovavedb7

00:00000:00010:2008/08/29 21:42:16.29 kernel    network name sean-laptop, interfac
e IPv4, address 192.168.1.3, type nlwnsck, port 5000, filter NONE
00:00000:00010:2008/08/29 21:42:16.29 kernel    Failed to log the current message
in the Windows NT event log
00:00000:00001:2008/08/29 21:42:17.00 server    The wash size of the 2K buffer poo
l in cache default data cache has been changed from 1638 Kb to 984 Kb due to a c
hange in the size of the pool.
00:00000:00001:2008/08/29 21:42:17.00 kernel    Failed to log the current message
in the Windows NT event log
00:00000:00001:2008/08/29 21:42:17.01 server    Recovery has tuned the size of '16
K' pool in 'default data cache' to benefit recovery performance. The original co
nfiguration will be restored at the end of recovery.
00:00000:00001:2008/08/29 21:42:17.01 kernel    Failed to log the current message
in the Windows NT event log
00:00000:00001:2008/08/29 21:42:17.01 server    Recovery has tuned the size of '2K
' pool in 'default data cache' to benefit recovery performance. The original con
figuration will be restored at the end of recovery.
00:00000:00001:2008/08/29 21:42:17.01 kernel    Failed to log the current message
in the Windows NT event log
……
```

你会发现中间带黑体的那一行（找里边的内容可能需要一点耐心，实在不行，可以将输出重定向到一个文本文件，在这个文件文件里查找 **New SSO** 的特定字符串），系统将 sa 用户的密码重置

31

为 kdoatlovavedb7，有了这个密码，就可以使用 isql 登录并修改 sa 的密码了，如要将 sa 的密码改为 sybase1，然后关闭 ASE 服务器，命令如下：

```
C:\Documents and Settings\hex>isql -Usa -Pkdoatlovavedb7
1> sp_password 'kdoatlovavedb7', 'sybase1', sa
2> go
口令设置正确
(return status = 0)
1> quit

C:\Documents and Settings\hex>isql -Usa -Psybase1
1> shutdown
2> go
服务器 SHUTDOWN 被请求
ASE 正在终止此进程
CT-LIBRARY error:
        ct_results(): 网络包层: 内部 net library 错误: 由于断开，使得 Net-Library 的操作中断
```

作了上述修改密码的操作以后，我们要将 RUN_<servername>.bat 的内容还原，即去掉末尾的 -psa，再运行它启动 ASE。这时我们发现使用 isql -Usa -Psybase1 很方便地就能连接到 ASE Server。

1.5 预装本书用到的 iihero 数据库

随书代码里有一个建库脚本 create_database_iihero.sql，它完整地创建了书中要用的 iihero 数据库。iihero 数据库创建在设备文件 C:\iihero.dat 上，占用空间 20MB。

执行 isql -Usa -iE:\MyDocument\MYBOOKS\ASE\code\create_database_iihero.sql 命令，按照提示输入 sa 用户的密码，即可完成 iihero 数据库的创建。

2 License 的使用

我们在上一章介绍安装 ASE 的时候，可以选择安装评估版（Evaluation），也可以选择基于已有的 License 执行安装。对于已经购买了 ASE 的正式用户而言，他们都可以得到正式的 License 文件（Sybase 的 License 基于 SySAM：Sybase 软件资产管理进行统一的许可管理，这个许可就是 License）。

2.1 评估版 License

对于初学者或者没有正式购买 Sybase 数据库的用户而言，使用评估版（Evaluation）License 是最合适的，它提供了 30 天的完全免费试用，支持当前版本的所有功能。一旦 License 过期，ASE 则不能正常启动，在错误日志文件%SYBASE%\%SYBASE_ASE%\install\errorlog 里会提示相关信息。

下边的输出显示的是一个 License 过期的示例：

```
00:00000:00000:2012/04/13 07:49:29.09 kernel  Registry keys for event logging are missing.
00:00000:00000:2012/04/13 07:49:29.45 kernel  SySAM: Using licenses from: D:\SybaseASE\\SYSAM-2_0\licenses
00:00000:00000:2012/04/13 07:49:29.67 kernel  SySAM: Failed to obtain 1 license(s) for ASE_CORE feature from license file(s) or server(s).
00:00000:00000:2012/04/13 07:49:29.67 kernel  SySAM: Feature has expired.
00:00000:00000:2012/04/13 07:49:29.67 kernel  SySAM: License feature name:   ASE_CORE
00:00000:00000:2012/04/13 07:49:29.67 kernel  SySAM: Expire date:    31-mar-2010
00:00000:00000:2012/04/13 07:49:29.67 kernel  SySAM: License search path:    D:\SybaseASE\\SYSAM-2_0\licenses\SYBASE.lic;D:\SybaseASE -
00:00000:00000:2012/04/13 07:49:29.67 kernel  SySAM:        \\SYSAM-2_0\licenses\SYBASE_ASE_DE.lic
00:00000:00000:2012/04/13 07:49:29.67 kernel  SySAM: FLEXnet Licensing error:-10,32
00:00000:00000:2012/04/13 07:49:29.67 kernel  SySAM: For further information, refer to the Sybase Software Asset
```

Management website at http://www.sybase.com/sysam
    00:00000:00000:2012/04/13 07:49:29.67 kernel   There is no valid license for ASE server product. Installation date is not found or installation grace period has expired. Server will not boot.

上边提示这个 License 已经在 2010 年 3 月 31 号过期了。对于初始学习 ASE 数据库的人来说，如果 License 过期，重装 ASE 是比较费时的一个工作。

其实 ASE 采用的 License 类型取决于一个一个的属性文件，它位于%SYBASE%\%SYBASE_ASE%\sysam\<hostname>.properties 里，它的头几行有以下内容：

PE=EE

LT=SR

它们分别表示 Product Edition 和 Licence Type，这里 EE 表示企业版（Enterprise Edition），如果没有有效的 License，最多只有 30 天的使用期。如果是评估版，对应的 LT 应该是 CP。

其实，ASE 对某些版本已经放开了，比如如果 PE 值为 DE（个人开发版），15.x 以后已经没有使用期限的限制了，这时 LT 值必须为 DT（Developing and Test）。当 PE 值为 XE（Express Edition，试验版），LT 为 CP。

所以，遇到这种情况，我们可以直接修改这个属性文件，使其成为开发版或者个人免费体验版。

（1）开发版。

PE=DE

LT=DT

（2）个人体验版。

PE=XE

LT=CP

这两种配置一般都能让 ASE 正常启动，只不过多了一些限制。

如果想一直使用企业版的评估版，有一种方法是安装一个与目标主机名相同的虚拟机，安装一个评估版的 ASE，然后复制上述属性文件到目标主机的相同位置，这样也是可行的。

2.2 License 的正式获取及使用

在安装和使用了一段时间评估版的 ASE 之后，如果购买了正式的 License，需要重新安装 ASE 吗？答案是不用重装。因为除了 License 信息以外，评估版和其他版本没有任何区别。

在购买完 Sybase 产品以后，会让你去官网注册一个账户，并登记一个邮箱。获取 License 的官网地址是 SPDC（Sybase Product Download Center）：https://sybase.subscribenet.com/。

申请成功之后，会收到具体的生成 License 文件的网址，按照那个网址上的提示，可以一步步生成想要的 License。

生成 License 的输入信息就是你机器的 hostid，如何得知自己机器的 hostid 呢？

运行目录%SYBASE%\SYSAM-2_0\bin 下的命令行 lmutil lmhostid，可以得到相应值。

该命令得到的 hostid 的具体值类型与具体的操作系统有关，它对应的操作系统底层命令如表 2-1 所示。

表 2-1 对应的操作系统底层命令

操作系统	HostID 类型	原始命令
Windows	网卡 Mac 地址	Ipconfig /all 查看 physical address 那一段
Solaris	32 位 hostid	hostid
Linux/Mac OS	网卡 Mac 地址	ifconfig echo0
IBM AIX	32 位 hostid	uname -m（返回 000276513100），这时要掉首尾的 0，保留中间的 8 位
SGI	32 位 hostid	/etc/sysinfo -s（转换成十六进制）
HP-UX PA-RISC	32 位 hostid	uname -I（转换成十六进制）
HP-UX Itanium	机器 ID	getconf CS_PARTITION_IDENT（加上前缀 "ID_STRING="）

比如在我的测试机器上，得到的是以下 hostid 信息：

D:\SybaseASE\SYSAM-2_0\bin>lmutil lmhostid
lmutil - Copyright (c) 1989-2006 Macrovision Europe Ltd. and/or Macrovision Corporation. All Rights Reserved.
The FLEXlm host ID of this machine is ""**00ff30cc1c8b 00234ed7891a 0024e891ec18 08002700582f**""
Only use ONE from the list of hostids.

这里出现了 4 个网卡地址。我们只能用其中的一个 MAC 地址去申请 License。一般来说，你用的 MAC 地址必须是你主机上提供对外网络访问的网卡的 MAC 地址，也就是主 IP 地址对应的 MAC 地址。

当然你也可以换成主机上硬盘的 ID 来代替 hostid。其命令是：

D:\SybaseASE\SYSAM-2_0\bin>**lmutil lmhostid -vsn**
lmutil - Copyright (c) 1989-2006 Macrovision Europe Ltd. and/or Macrovision Corporation. All Rights Reserved.
The FLEXlm host ID of this machine is "DISK_SERIAL_NUM=8060e304"

最终结果是在硬盘序列号前加了前缀：DISK_SERIAL_NUM=，在没有网卡的机器上或者网卡太多的机器上使用这种 hostid 还是有意义的。

在获取完 License 文件以后，将其放到%SYBASE%\SYSAM-2_0\licenses 目录下，找到该 License 对应的 PE=◇;LT=◇实际值，然后修改%SYBASE%\%SYBAS_ASE%\sysam\<host>.properties，将 PE 和 LT 的值改成与 License 文件中的值保持一样即可。

这就是使用 License 文件的基本过程。Sybase 产品获取 License 和最终使用 License 的过程基本上都是这样。

3

定义物理设备

本章主要介绍如何创建 ASE 中的物理设备和数据库。在 ASE 安装完毕以后，第一步就是要创建必要的物理设备。ASE 中的物理设备，指的是用于存放具体数据库的容器，大多数情况下，物理设备就是一个物理文件，当然也可以是一个裸设备（raw device：如未经格式化的磁盘分区，在 NT 平台下，倾向于使用物理文件）。一个物理设备上可以存放多个数据库，而一个数据库也可能跨多个物理设备。

3.1 物理设备管理

与其他数据库不同，ASE 使用逻辑设备而不是物理设备来管理具体的数据库空间。因为物理设备对应的是一个物理文件或裸设备的路径，而逻辑设备则是对该路径的一个映射，比较容易记忆和管理。

通常情况下，考虑到性能，物理设备应该尽量分散到多个磁盘分驱，以均衡 I/O，得到较好的性能。

我们通过一系列命令和存储过程来进行设备管理。设备信息以及与其相对应的数据库相关信息都会保存到 Master 主数据库当中，这些信息可以通过一些预定义的存储过程直接查询得到。需要注意的是，千万不要直接通过修改系统表来进行设备的创建和修改，这样做是非常危险的，ASE 提供了相应的命令和存储过程来完成设备管理。

作为一名 DBA，千万不要等到数据库空间用完的时候才开始创建新设备，而应该一开始就预算好总的大致空间开销，并且预先创建好一些设备作为预留，这样，当数据库空间用完时，可以很快将这些预留的空间分配给数据库。

在 ASE 里，master 主设备以及 sysprocsdev 设备（用于存储 sybsystemprocs 数据库，所有的系统存储过程以及一些重要的元信息表都存放在该数据库当中）都是在 Server 安装的过程中创建完成的。用于存储 sybscurity 数据库的设备则由 auditinit 工具创建完成。其他所有设备（文件）都使用 disk init 命令来完成。与设备本身的创建方法无关，每个创建完的设备在 master 数据库的 sybdevices 系统表里都有一行记录。

需要注意的是，每完成一次系统资源分配，都需要对 Master 数据库进行一次备份（执行 dump database 操作）。

3.1.1 创建设备

我们使用 disk init 命令来创建数据库设备。它的基本语法如下：

```
disk init
    name = "device_name",   physname = "physicalname",
    size = size_of_device
    [, vdevno = virtual_device_number ,]
    [, vstart = virtual_address ,]
    [, cntrltype = controller_number, ]
    [, dsync = {true | false}]
    [,directio = {true | false}]
```

第 1 个参数是设备的名称，也就是逻辑设备名。这个名称也是用于分配默认设备时参考的名称。该名称必须在 ASE 服务器上是唯一的，并且不能包含任何空格或者标点，并且它是区分大小写的，名字的最大长度不能超过 30 个字符。

对于设备的命名，最好是给出一个有意义的名字，像 d1,d2,…,dn 这样的命名，肯定不如 customer1_data1, products_data1, customer1_index1 这样有意义，使用后一种命令方式的好外是，当 customer 数据库的 index 空间用完时，你能快速的定位是哪个逻辑设备空间不够，可以迅速扩展它的索引段。

第 2 个参数是物理名，实际上它是对逻辑名的物理映射，是用以存储数据库的实际存储位置。取决于你使用裸设备还是常规的物理文件，系统管理员所在的操作系统用户必须拥有对裸设备或者物理文件所在目录的读写权限，否则 disk init 命令会执行失败。

> **注意**
> 物理文件必须使用全路径。当我们指定一个设备的物理名为 sean.dat，系统会在 Server 启动的目录里去创建这个文件，这是一个隐患。如果服务器的当前目录发生了改变，则该设备变得不可用。对于 UNIX 平台，同样会有这样的问题。

第 3 个参数是设备的大小。默认情况下，它使用数据库页的数量来表示，而不是以 M 为单位。从 12.5 开始，用户可以选择指定 disk init 时设备的大小单位，可以是 k 或 K（表示 KB）、m 或 M（表示 MB）、g 或 G（表示 GB）。在 15.0 中，甚至可以指定 t 或 T（表示 TB）。尽管是可选表示，这里还是强烈推荐使用带单位的表示法，这样更精确，不会随着数据库的页大小而发生改变。当带单位表示时，必须使用引号将其括起来。

ASE 数据库中默认的页大小是 2KB，因此，不带单位时，可以计算出实际要创建的设备的大小。如：

```
1> disk init name="sean",   physname="d:\sybase\sean.dat", size=5120
2> go
```

这里将为设备 sean 分配 5120 个页面，刚好是 5120*2K＝10MB。

与常规的物理文件不同，我们可以在一个目录下创建很多个物理文件，每个物理文件都可以对应一个逻辑设备。而使用裸设备时，这个裸设备目录只能创建一个逻辑设备，如果你创建了而不去使用，将会造成浪费。

第 4 个参数是 vdevno，虚拟设备号，它是为每个设备分配的唯一数字标识符。这个值必须确保没被其他设备所使用。如果总共有 10 个设备，那么 vdevno 的值不能超过 9。如果所有可用的设备号全部用完，则必须配置更多的设备。在 12.5 及后续版本里，我们可以完全忽略这个参数的值，服务器会自动为它创建一个可用的值。

要想确定 Server 当中配置的允许的设备数，可以运行命令 sp_configure "number of devices"或者在 ServerName.cfg 配置文件中查找参数即可。

一旦新创建的设备导致设备数量超标，则会创建失败。需要重新设置允许数量。如：

```
1> disk init name="sean2",   physname="d:\sybase\sean2.dat", size=5120
2> go
1> disk init name="sean3",   physname="d:\sybase\sean3.dat", size=5120
2> go
Msg 5162, Level 16, State 1:
Server 'SEANLAPTOP', Line 1:
已达到配置的设备的最大数 9。请重新配置 [number of devices
]，使之拥有一个较大的值，然后重试磁盘初始化
1>
```

要确定当前系统中使用了哪些设备号，可以查询系统表 sysusages 表获得：

```
1> select dbid, lstart, size, vstart, vdevno from sysusages
2> go
```

dbid	lstart	size	vstart	vdevno
1	0	6656	4	0
3	0	1536	6660	0
2	0	2048	8196	0
31513	0	1536	10244	0
31514	0	67584	0	1
31513	1536	512	0	2
31515	0	38400	0	3
4	0	1536	0	4
5	0	10240	0	5

5	10240	10240	0	6	
4	1536	3072	1536	4	
4	4608	2048	4608	4	

(12 rows affected)

第 5 及第 6 个参数，描述的是设备的起始虚拟地址和控制号，是可选参数，通常情况下不会被使用，默认值都是 0。不需要指定这两个参数的值。

第 7 个参数指定是否启用 dsync。我们在后边还会还会介绍这个参数的具体用法。

最后一个参数 directio，直接 IO，这个开关（在 disk reinit 及 sp_deviceattr 存储过程中也会用到它）用以指定是否允许 ASE 将数据直接写入磁盘，而不用经过操作系统的文件 cache 缓冲。directio 与裸设备执行 IO 的方式完全一样，但是它更易于使用和管理。由于 directio 是一个表态参数，因此需要 ASE 重启才能生效。默认情况下，directio 的值为 false，除了 Linux 平台下它的值是 true（on）。

directio 和 dsync 两个参数是互斥的，如果 dsync 设为 true，就不能将 directio 也设为 true；要想 directio 为 true，必须将 dsync 设为 false。

在 Server 安装完以后，master 设备是唯一的主设备，这可不是一件好事情。DBA 需要改变这个默认值，以避免将后续的用户数据库分配到这个主设备上，给维护带来不便。

创建默认设备的语法如下：

exec sp_diskdefault *logical_device_name*, 'defaultoff | defaulton'

比较理想的情况下，master 设备只应该包含系统数据库，而不要包含任何用户数据库。如果 master 数据库空间用完，并且 master 设备没有可用的剩余空间，那将是一个大问题，因为 master 数据库只能在 master 设备上分配可用空间。同时每个用户数据库都有数据页和日志页，最好空间上分开管理。

在安装完 ASE 之后，可以马上取消 master 设备作为默认设备，同时将临时数据库 tempdb 从 master 设备上分离出来。并创建一个或多个设备作为默认设备。下面是从 master 设备上分离 tempdb 的详细过程：

```
1> sp_diskdefault master, defaultoff
2> go
(return status = 0)
1> disk init name="tempdb_dev",physname="d:\sybase\data\tempdb.dat",size="50M"
2> go
1> alter database tempdb on tempdb_dev="30M"
2> go
将数据库在磁盘 tempdb_dev 上扩展 15360 页 (30.0 MB)
1> use tempdb
2> go
1> sp_dropsegment 'default',tempdb,master
```

```
2> go
1> sp_dropsegment logsegment,tempdb,master
2> go
DBCC 运行结束。如果 DBCC 打印出错误消息,请与有系统管理员角色的用户联系
数据库 tempdb 的最后机会阈值现在是 16 页
对设备的段引用被删除
(return status = 0)
1> sp_dropsegment system,tempdb,master
2> go
DBCC 运行结束。如果 DBCC 打印出错误消息,请与有系统管理员角色的用户联系
对设备的段引用被删除
```

执行上述操作之后,tempdb 将不再占用 master 设备了,而是直接使用设备 tempdb_dev。接着创建一个默认的设备 test_dev,后边创建一个测试数据库 test,如果不指定设备,则直接创建在该设备上,将日志存放到非默认设备 testlog_dev 上。

```
1> use master
2> go
1> disk init name="test_dev",physname="d:\sybase\test_dev.dat",size="50M"
2> go
1> sp_diskdefault test_dev, defaulton
2> go

1> disk init name="testlog_dev",physname="d:\sybase\test_devlog.dat",size="50M"
2> go
1> create database test on default=50 log on testlog_dev=50
2> go
CREATE DATABASE:分配磁盘 'test_dev' 上的 25600 逻辑页 (50.0 MB)
CREATE DATABASE:分配磁盘 'testlog_dev' 上的 25600 逻辑页 (50.0 MB)
数据库'test' 联机.
```

当然,我们随时可以对设备进行扩容。Sybase ASE 不允许减少设备空间的大小。使用命令 disk resize 即可对目标设备进行扩容。参数 size 的值指的是增加的空间大小。

```
1> disk resize name='demo', size="3M"
2> go
```

3.1.2 删除设备

删除设备的命令非常简单:

```
Exec sp_dropdevice 'logical_device_name'
```

什么情况下执行这类操作呢?有时候是因为创建设备时有误,定义了错误的大小、名字或者文件位置,或者需要重新分配一下各个设备的空间。ASE 提供了这个命令用以删除设备。设备的元信息存储到系统表 sysdevices 里,它位于 master 数据库当中。执行这个命令以后,sysdevices 表里会清除该逻辑设备对应的数据行,但是它并不删除对应的物理文件。如果不想重用这个物理文件,

可以手动删除这个物理文件。

3.1.3 裸设备与常规文件

上节只是从操作的角度提到了裸设备与普通文件的区别，从功能上比较，两者有什么区别呢？主要区别在于引用它们的方式、删除它们的方式不同，同时写入数据的方式和写入安全性方面也有区别。

常规文件使用操作系统自带的缓冲写操作。当你往数据库执行写操作时，ASE 将往磁盘上的设备文件执行写操作，当写操作完成时，操作系统会告诉 ASE 写操作已经完成，但实际上，这些写操作发生的改变只是缓存到操作系统的 cache 里，只有当写操作积累到一定的量时，才真正写回磁盘。如果服务器突然被停止（瞬间地），ASE 有可能会产生索引或者数据页链接错误。这意味着什么？意味着这种方式会有可能导致表数据的部分丢失，虽然这种可能性很小。如同正在运行的机器突然断点，结果操作系统中的某些系统文件丢失。

而裸设备并不缓冲写操作，ASE 发送一个写操作，相应的数据改变会在 ASE 收到写完成的确认之前立即写入磁盘。这是一种更安全的写操作。

安全性是一个区别，还有一个要考虑的是性能上的不同，常规文件加入了一个缓冲，而裸设备不用缓冲，因而在同等的硬件和操作系统条件下，常规文件比裸设备的效率要稍低一些。

管理的便捷性也是一个很重要的区别，要想管理好裸设备，需要投入更多的时间和精力去规划，限制条件也更多。如果使用常规文件只是简单地创建目录，当需要新设备的时候，只需简单地指向这些目录，并指定一个新文件名即可。而使用裸设备时，则受限于每块磁盘上有多少个分区，如果没有很好的磁盘管理工具，必须提前规划好每个裸设备的大小。

最终将由用户决定在数据库当中使用哪种设备。有些数据库要求没有那么严格，一旦发生数据损毁，从出错的前一天恢复数据是可以接受的，你可能会采用常规文件的方式。但是在生产环境的服务器中，你转而可能采用裸设备，在开发测试环境中采用常规文件作为设备。

3.1.4 Dsync 选项

创建设备中的 dsync 选项是从 12.0 版本开始加入的，它可以让你很方便地混合使用常规文件和裸设备。在使用常规文件时，ASE 使用 dsync 标志告诉 UNIX 减少缓冲行为，也就是说，所有的写操作都是有保障的，这样数据库的安全性有了保障，大大减少了数据损坏的可能性。

使用 dsync 选项的另一个好处是，虽然写的速度比常规文件快，但是它的安全性要高一些，但是读操作反而可能比常规文件要慢。Dsync 相当于是一个混合模式。

需要强调的是，具体的性能是需要细致测量的，无法确定到底是开启了 dsync 的文件设备快，还是关闭了 dsync 的裸设备快。通常情况下，tempdb（临时数据库）的设备最好使用高速缓冲的文件设备并且关闭 dsync 开关。如果没有在 disk init 命令里显示指定 dsync 的值，它的默认值为 true，即使是使用裸设备（对于裸设备而言，dsync 必须显式的设为 false）。dsync 标志在 NT 平台上会被忽略，在 NT 平台上要获取较好的安全性和性能，最好使用文件设备。

> **注意**
> master 数据库对服务器非常关键，在定义 master 设备文件时，其 dsync 必须为 true，你不能改变它的值，如果试图改变它的值，则会收到系统警告。

除了在使用 disk init 时指定 dsync 标志值，也可以在创建完设备以后再设定它的值，可以通过调用存储过程 sp_deviceattr 来改变它的值，必须重启 ASE 来使新值生效。

该存储过程的定义如下：

exec sp_deviceattr '*logical_name*', '*option_name*', '*option_value*'

如：

```
1> exec sp_deviceattr 'iihero', 'dsync', 'true'
2> go
Msg 18685, Level 16, State 1:
Server 'SEANLAPTOP', Procedure 'sp_deviceattr', Line 188:
属性 dsync 已变为 on
(return status = 1)
```

3.2 设备（文件）的限制条件

数据库设备有如下限制条件：
- 逻辑设备名：必须是唯一的，并且区分大小写，最多 30 个字符。
- 设备的大小：依赖于具体的系统，但在不改变操作系统相关设置的情况下，最大限制是 4TB。
- 设备的个数：最大 2GB。
- 虚拟设备号(vdevno)：从 1 到 2,147,483,647 (2G-1)，0 保留给 master 设备。
- dsync：裸分区和 NT 平台的文件系统都会忽略该标志；在 master 设备上永远为 true。
- 物理名：物理文件必须位于本地机器上，不能使用网络文件或者远程文件。全路径名最大长度不能超过 255 个字符。

3.3 创建 master 设备

master 设备是安装时创建的第一个设备，在 dsync 标志被推出之前，一直建议 master 设备创建到裸设备上。但推出 dsync 以后，我们完全可以把它创建到常规文件设备上。

一旦安装完成，将有 4 个数据库占用 master 设备上的空间。除了 master 数据库，还有 model 数据库、默认的 tempdb 数据库以及 sybsystempdb 数据库（从 12.0 版本开始，sybsystempdb 会创建在 master 设备上）。我们不能把用户数据库创建到 master 设备上，同时也有必要将 tempdb 从 master 设备上分离出来。在 3.1.1 节里有相关介绍。

创建 master 设备只能使用下面的 dataserver 命令行（不是在 isql 命令行里执行）：

```
dataserver -d master_device_name
    (-e, -c, -T, -I, etc.)
    [-b master_device_size [k|K|m|M|g|G|t|T]]
        [-z logical_page_size [k|K]]
```

- -d 选项：用于指定 master 设备的物理位置和物理名；
- -b 选项：用于指定为设备分配多少虚拟页。因为虚拟页总是以 2KB 大小计数的。所以，在不带单位（K、M、G、T）的情况下，将其与 512 相乘，即可得到实际有多少 MB 空间。
- -z 选项：指定要使用逻辑页的大小，它与实际的物理存储无关，因为所有的物理设备都是以 2K 的页大小进行存储的。它只是指出数据库将如何去组织使用空间，它的值可以是 2K、4K、8K、16K，如果在创建语句里不带 K，则使用一个逻辑页的实际字节数来表示，-z8192 与 -z8K 含义相同。

需要注意的是，dataserver 同时也用于 ASE 服务器的启动。在第 2 章我们讲到的 Runserver 文件中，都不带 -z 或 -b 选项，因为启动 ASE Server 不需要这两个选项，这两个选项仅用于 master 设备的创建阶段，而不是 ASE 服务器的启动阶段。

dataserver 命令对应于 NT 平台上的 sqlsrvr.exe，下面是一个创建 master 设备的例子：

```
sqlsrvr.exe -d "d:\sybase\master_demo.dat" -b50M -z4K
sqlsrvr.exe: master device size for this server: 50.0 Mb
sqlsrvr.exe: master database size for this server: 26.0 Mb
sqlsrvr.exe: model database size for this server: 6.0 Mb
    00:00000:00000:2010/04/11  10:09:17.82 kernel    SySAM: Using licenses from: D:\sybase\\SYSAM-2_0\licenses;
d:\SybaseASE125\SYSAM-1_0\licenses\license.dat
    00:00000:00000:2010/04/11  10:09:18.54 kernel    SySAM: Checked out license for 1 ASE_CORE (2009.1231/permanent/
19B3 47BC 0B7E 8DC3).
    00:00000:00000:2010/04/11  10:09:18.54 kernel    This product is licensed to: ASE Developer Edition - For Development and
Test use only
    00:00000:00000:2010/04/11  10:09:18.54 kernel    Checked out license ASE_CORE
    … …
```

3.4 设备镜像

ASE 提供了设备镜像的功能，当设备发生故障时，可以从镜像设备那里恢复。这样可以更好地保护数据库设备。镜像（Mirroring）的基本功能就是：每个写操作在主设备写完之后，都会写往从设备，这样会得到一个完整的数据拷贝。当然，从设备最好是定义在一个单独的物理磁盘上。设备镜像的语法如下：

```
disk mirror
    name = 'logical_device_name',
    mirror = 'physical_mirror_name'
     [, writes = {serial | noserial}]
```

disk mirror 语法与 disk init 基本相似,第一个参数是源逻辑设备名,第二个参数是物理镜像名,也就是要指定镜像的路径,不用指定它的大小和虚拟设备号,它的大小会跟源逻辑设备一样。同样,也不会在 sysdevices 表里记录它的信息,只会修改逻辑设备对应的相关信息。既然是镜像,对应路径必须有读写权限。

第三个参数告诉 ASE 对镜像设备是否使用同步写,这与常规文件和裸设备很类似。如果选择了 noserial(异步写),当主设备写操作完成时,会继续后边的操作,而不用等待镜像设备中的写操作。如果这时服务器宕机,则镜像设备中会有数据丢失。serial 方式可以更好地确保数据不会丢失,但同时也牺牲了部分性能。下边是一个使用镜像设备的示例。

首先要启用 ASE 的磁盘镜像选项。

```
1> sp_configure 'disable disk mirroring', 0
2> go
Parameter Name                  Default     Memory Used Config Value
-----------------------------   -------     ------------------------
disable disk mirroring             1             0
Run Value        Unit           Type
-----------      -------------  ---------
    0              1 switch     static

(1 row affected)
Configuration option changed. Since the option is static, Adaptive Server must
be rebooted in order for the change to take effect.
Changing the value of 'disable disk mirroring' does not increase the amount of
memory Adaptive Server uses.
(return status = 0)
```

这个选项的改变在 ASE 重启之后生效。

```
1> disk mirror name="demo", mirror="D:\sybase\demo_mirror.dat",writes=noserial
2> go
Creating the physical file for the mirror...
Starting Dynamic Mirroring of 25600 pages for logical device 'demo'.
The remaining 25600 pages are currently unallocated and will be mirrored as they
are allocated.
```

这时候如果我们查询设备 demo,会看到:

```
1> sp_helpdevice demo
2> go
demo
            D:\sybase\data\demo.dat
```

```
    file system device, special, MIRROR ENABLED, mirror = 'D:\sybase\demo_m
    irror.dat', nonserial writes, dsync on, directio off, reads mirrored,
    default disk, physical disk, 50.00 MB, Free: 50.00 MB
```

ASE 会创建一个与原来的设备 demo 一样大小的镜像文件 D:\sybase\demo_mirror.dat，并且开启异步写功能。在创建镜像的过程当中，ASE 会把主设备中的所有页都复制到镜像设备中，是一个 I/O 密集操作，因此注意一定要在 CPU 空闲的时候执行此操作。

我们要注意的是，这里的镜像是针对设备一级，而不是针对数据库。如果你的数据库跨多个设备，就需要对这些设备都执行镜像，否则只能部分地保护你的数据库。

具体要镜像哪些设备，或者选择完全镜像，取决于系统的实际情况。一个比较好的镜像是只针对事务日志设备进行镜像，因为事务日志对数据库恢复起非常重要的作用，不容有失。

最重要的数据库设备是 master 设备，对它采取镜像也是一种策略。对 master 设备执行镜像，除了使用 disk mirror 命令生成镜像以外，还需要修改 Runserver 文件，加入 -r 参数，指定镜像文件。

```
1> disk mirror name=master,mirror="D:\sybase\master_mirror.dat"
2> go
Creating the physical file for the mirror...
Starting Dynamic Mirroring of 25600 pages for logical device 'master'.
      512 pages mirrored...
     1024 pages mirrored...
     1536 pages mirrored...
     2048 pages mirrored...
     2560 pages mirrored...
     3072 pages mirrored...
     3584 pages mirrored...
     4096 pages mirrored...
     4608 pages mirrored...
     5120 pages mirrored...
     5632 pages mirrored...
     6144 pages mirrored...
     6656 pages mirrored...
     7168 pages mirrored...
     7680 pages mirrored...
     8192 pages mirrored...
     8704 pages mirrored...
     9216 pages mirrored...
     9728 pages mirrored...
    10240 pages mirrored...
    10752 pages mirrored...
    11264 pages mirrored...
    11776 pages mirrored...
    12288 pages mirrored...
    12800 pages mirrored...
    13312 pages mirrored...
```

```
            13824 pages mirrored...
            14336 pages mirrored...
            14848 pages mirrored...
            15360 pages mirrored...
            15872 pages mirrored...
            16384 pages mirrored...
            16896 pages mirrored...
            17408 pages mirrored...
            17920 pages mirrored...
            18432 pages mirrored...
            18944 pages mirrored...
            19456 pages mirrored...
            19968 pages mirrored...
            20480 pages mirrored...
            20992 pages mirrored...
            21504 pages mirrored...
            22016 pages mirrored...
            22528 pages mirrored...
            23040 pages mirrored...
            23552 pages mirrored...
            24064 pages mirrored...
            24576 pages mirrored...
            25088 pages mirrored...
            25344 pages mirrored...
The remaining 256 pages are currently unallocated and will be mirrored as they are allocated.
```

创建完 master_mirror.dat 以后，修改 RUN_SEANLAPTOP.bat，加入 -r D:\sybase\master_mirror.dat 即可。

镜像可以自动禁用和手动禁用。如果 ASE 发现在某个设备上有 I/O 错误，镜像会被禁，所有的读写操作都会指向仍能工作的那个设备上，同时相关错误信息会被记入错误日志中。

采用手动禁用时，可以通过命令 disk unmirror 来实现：

```
disk unmirror
name = 'logical_device_name'
[, side = {'primary' | 'secondary'}]
[, mode = {'retain' | 'remove'}]
```

第一个参数是逻辑设备名；第二个参数是 side，指出你要禁用的是哪个设备，可以是主设备，也可以是从设备，默认的是从设备；最后一个参数是 mode，表示是否保留禁用的设备（文件），如果是 retain（保留），则意味着以后还可以再次执行镜像（使用 disk remirror 命令），它只是将 sysdevices 表中的状态位重置，以表示该逻辑设备的镜像禁用，但是仍然保留着从设备的物理名。而如果是 remove（移除），则更新 sysdevices 中的状态位，表示没有镜像，并且删除对应的从设备的物理名，这个时候如果想再次镜像，则必须使用 disk mirror 命令，因为以前的设备镜像记录已经被删除。

disk mirror/unmirror 命令的一个有用的用法是，可以有效地将一个设备文件从一个地方移到另一个地方，下边的例子则是将设备 iihero 从原位置移到新位置 D:\sybase\data\iihero_mirror.dat。

```
1> disk mirror name=iihero,mirror="D:\sybase\data\iihero_mirror.dat"
2> go
Creating the physical file for the mirror...
Starting Dynamic Mirroring of 25600 pages for logical device 'iihero'.
        512 pages mirrored...
       1024 pages mirrored...
       1536 pages mirrored...
       2048 pages mirrored...
       2560 pages mirrored...
       3072 pages mirrored...
       3584 pages mirrored...
       4096 pages mirrored...
       4608 pages mirrored...
       5120 pages mirrored...
       5632 pages mirrored...
       6144 pages mirrored...
       6656 pages mirrored...
       7168 pages mirrored...
       7680 pages mirrored...
       8192 pages mirrored...
       8704 pages mirrored...
       9216 pages mirrored...
       9728 pages mirrored...
      10240 pages mirrored...
The remaining 15360 pages are currently unallocated and will be mirrored as they
are allocated.
1> disk unmirror name=iihero,side='primary',mode='remove'
2> go
```

在这以后，逻辑设备对应的原文件可以被物理删除了，只需保留新的物理文件 D:\sybase\data\iihero_mirror.dat。

```
1> sp_helpdevice iihero
2> go
 device_name          physical_name            description                status       cntrltype vdevno    vpn_low
vpn_high

    iihero            d:\sybase\data\iihero_mirror.dat        file system device, special, dsync on, directio off, physical disk,
50.00 MB, Free: 30.00 MB         16386         0                3        0          25599
    (1 row affected)
```

如果是以 retain 模式解除镜像的（默认模式），则可以使用 disk remirror 重新恢复镜像。

```
1> disk mirror name=demo, mirror="D:\sybase\demo_mirror.dat"
2> go
Creating the physical file for the mirror...
Starting Dynamic Mirroring of 25600 pages for logical device 'demo'.
The remaining 25600 pages are currently unallocated and will be mirrored as they
are allocated.
1> disk unmirror name=demo,side='secondary',mode='retain'
2> go
1> disk remirror name=demo
2> go
Starting Dynamic Mirroring of 25600 pages for logical device 'demo'.
The remaining 25600 pages are currently unallocated and will be mirrored as they
are allocated.
```

3.5 与设备信息相关的存储过程

最常用的几个与设备相关的存储过程如下：

Step 1 sp_helpdevice ['*logical_device_name*']
将返回指定逻辑设备的所有元信息

Step 2 sp_dropdevice '*logical_device_name*'
删除指定的逻辑设备，前提是该设备没有被任何数据库或存储过程引用

Step 3 sp_diskdefault '*logical_device_name*', {'*defaulton*' | '*defaultoff*'}
从默认的逻辑设备组里添加或者删除某逻辑设备
需要注意的是，当我们创建一个数据库时，如果指定的是 on default=<size>，则会依次从默认的逻辑设备组里提取空间，当第 n-1 个逻辑设备空间用完时，它会自动从第 n 个默认的逻辑设备里分配空间，直到指定大小的空间分配完毕。下边的例子是依次创建两个设备 demo 和 demo2，大小均为 10MB，然后将它们都设为默认设备，系统里的 master 设备在这之前已经被默认设备组里移除，在这之后，分别创建两个数据库 demo 和 demo2，数据库 demo 大小为 8MB，demo2 大小为 11MB，可以看到数据库 demo 将从设备 demo 里获取 8MB 空间，数据库 demo2 将从设备 demo 里获取余下的 2MB 空间，还有 9MB 将从设备 demo2 里获取。

```
1> disk init name=demo,physname="d:\sybase\data\demo.dat",size="10M"
2> go
1> sp_diskdefault 'demo', defaulton
2> go
(return status = 0)
```

```
1> disk init name=demo2,physname="d:\sybase\data\demo2.dat",size="10M"
2> go
1> sp_diskdefault demo2, defaulton
2> go
(return status = 0)
1> create database demo on default=8
2> go
CREATE DATABASE: allocating 4096 logical pages (8.0 megabytes) on disk 'demo'.
Database 'demo' is now online.
1> create database demo2 on default=11
2> go
CREATE DATABASE: allocating 1024 logical pages (2.0 megabytes) on disk 'demo'.
CREATE DATABASE: allocating 4608 logical pages (9.0 megabytes) on disk 'demo2'.
Database 'demo2' is now online.
1>
```

Step 4 sp_deviceattr '*logical_device_name*', '*option*', '*setting*'
改变逻辑设备的属性

Step 5 sp_configure '*number of devices*'
显示系统配置文件里最大允许的设备数

Step 6 sp_helpdb 'database_name'

```
1> sp_helpdb demo2
2> go
Warning: Row size (2619 bytes) could exceed row size limit, which is 1962
bytes.
```

name	db_size	owner	dbid	created	status
demo2	11.0 MB	sa	6	Apr 11, 2010	mixed log and data

(1 row affected)

device_fragments	size	usage	created	free kbytes
demo	2.0 MB	data and log	Apr 11 2010 11:31AM	320
demo2	9.0 MB	data and log	Apr 11 2010 11:31AM	9180

(return status = 0)

3.6 与设备相关的系统表

与物理设备定义相关的系统表有：sysdevices、sysusages、sysdatabases。

sysdevices：记录每个数据库或者 dump 设备的信息，也包括镜像设备的信息。

sysusages：每行都记录着数据库分段的相关信息，在 15.0 版本以前，这两张表通过条件（vstart

between low and high）关联，在 15.0 及以后版本中，这两张表都有一个 vdevno 字段。

sysdatabases：表中记录了每个数据库的相关信息，与 sysusages 表相关联。存储过程 sp_helpdb 可以查询每个数据库用到了哪些设备。

系统表的关联关系如图 3-1 所示。

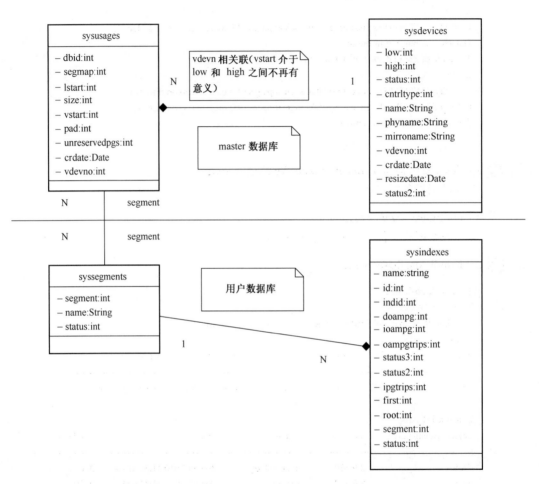

图 3-1　系统表的关联关系

对于灾难恢复而言，sysdevices、sysusages 及 sysdatabases 表中的数据非常重要，一旦系统信息发生改变，有必要对这几个表的数据进行备份。

表 sysdevices 完整地定义了设备的信息。通过该表的某些列，完全可以复原出 disk init 命令来创建该设备。下面的例子中，phyname 列对应的就是 disk init 命令的 physname 参数值。

```
1> select * from sysdevices
2> go
```

```
low          high         status     cntrltype     name          phyname
mirrorname   vdevno       created                  resizedate    status2
------------ ------------ ---------- ------------- ------------- ----------------------------------
------------ ------------ ---------- ------------- ------------- ----------------------------------
0            25599        16386      0             iihero        D:\sybase\data\iihero_mirror.dat
NULL         3                       Apr 11 2010  9:37AM NULL    0
0            5119         16387      0             demo          D:\sybase\data\demo.dat
NULL         4                       Apr 11 2010  9:45AM NULL    0
```

ASE 发展到 15.0 版本以后，low 和 high 值的行为发生了变化，我们也不太容易直接从 sysusages.vstart 列值找到对应的 sysdevices 对应的行，现在 low 和 high 两列的值允许支持更大的设备（4TB），在 15.0 以前，只支持最大 32GB。

在 15.0 版本以前，要查找没有包含任何数据库空间的空闲设备使用如下语句：

```
SELECT sd.name AS "Unused devices",
sd.phyname AS "Physical location",
((1. + (sd.high - sd.low)) * @@pagesize) / 1048576. AS "Device size in MB"
FROM master..sysdevices sd
WHERE
name not in (
SELECT sd.name
FROM master..sysdevices sd, master..sysusages su
WHERE
su.vstart >= sd.low
and su.vstart <= sd.high
)
and sd.name not like "tapedump%"
```

到了 15.0 版本以后，只能使用下述语句来进行查询：

```
1> SELECT sd.name AS "Unused devices",
2> sd.phyname AS "Physical location",
3> ((1. + (sd.high - sd.low)) * @@pagesize) / 1048576. AS "Device size in MB"
4> FROM master..sysdevices sd
5> WHERE
6> name not in (
7> SELECT sd.name
8> FROM master..sysdevices sd, master..sysusages su
9> WHERE
10> su.vdevno = sd.vdevno
11> )
12> and sd.name not like "tapedump%"
13> go
 Unused devices      Physical location          Device size in MB
 ------------------  -------------------------  ------------------
 demo2               D:\sybase\demo2.dat        10.00000000
```

这是一个很大的变化。

status 列是一个整数值，用于表示该设备具备哪些属性，这些值可以组合在一起：如果该设备是默认设备，则有状态值 1；如果是镜像设备，则有状态值 64；如果 dsync 开关开启，则有状态值 16384。

cntrltype 列则用于区分该设备是数据库设备还是备份（dump）设备，备份设备的 cntrltype 值要大于 0，并且 status 值为 16。

4 连接 ASE

本章主要介绍 ASE 安装完成以后，如何通过 ASE 客户端，远程访问管理 ASE 数据库，配置服务器，创建数据库及对应的登录用户，建立可用的开发应用环境。

4.1　ASE 客户端概述

ASE 数据库有如下几种形式的客户端：
- Sybase Central 图形管理工具：它是标准的 ASE 数据库管理工具，纯 Java 界面控制。
- Interactive SQL 客户端：它是 ASE 自带的图形界面 isql 客户端。
- isql 命令行工具：它是 ASE 中 Open Client 下的命令行工具，是 isql 控制台命令行交互的入口。
- jisql：这是 ASE 可选工具包 jutils 中自带的 Java 瘦客户端，可以执行一些简单的 sql 命令与 Server 端交互。

本章重点介绍第一种客户端。

4.2　网络连接

在使用客户端连接 ASE 数据库服务器之前，必须配置好到服务器端之间的数据库连接。客户端可以与一台或多台 ASE 服务器通信，ASE 服务器也可以与一台或多台其他 ASE 服务器通信。

ASE 使用自己特有的目录服务来定位有关服务器的网络位置信息。目录服务包含有网络中所有 Adaptive Server、Backup Server 和其他服务器的名称和网络地址信息，称之为地址记录。ASE 服务器或客户端都可以通过这些地址记录连接到远程的服务器。这些地址记录信息存放到 ASE 的 interfaces 文件当中。

4.2.1 interfaces 文件的内容

interfaces 文件在 Linux/UNIX 下的路径为 $SYBASE/interfaces；Windows 下，其路径为 %SYBASE%/ini/sql.ini。它包含网络上有关 ASE 所有服务器的网络信息。其中，典型的包括 Adaptive Server、Backup Server、XP Server 以及其他服务器应用程序，如 Monitor Server、Replication Server 和其他 Open Server。

interfaces 文件中的每个条目可以包括两种类型的记录信息：

- master 记录：服务器应用程序使用这条信息在网络上进行监听。此信息称为监听器服务。
- query 记录：客户端应用程序使用它们连接网络上的 ASE 各种服务器。此信息称为查询服务。

对于服务器来说，master 记录和 query 记录中包含的网络信息是一致的，因为服务器在客户端用来请求连接的同一端口上监听连接请求。

客户端因为不需要监听，所以它的 interfaces 文件不需要 master 记录。仅有 query 记录即可正常工作。

Linux/UNIX 下，interfaces 文件常用格式如下：

```
<ServerName>
    master tcp ether <hostname> <port>
    query tcp ether <hostname> <port>
```

这里<ServerName>指的是服务器名。它是服务器端监听的名字，同时也是作为客户端连接时要连的远程服务的名字。Linux/UNIX 下，ServerName 长度不能超过 30 个字符。第一个字符必须是字母（ASCII 的 a 到 z、A 到 Z），随后的字符必须是字母、数字或下划线（_）。需要注意的是：这里的 ServerName 是区分大小写的。

在 master 和 query 之前是一个<tab>制表符，后边每个标识符之间都以一个空格间隔。tcp 标识使用的网络协议，ether 指的是网络名称，意指"以太网"，ASE 当前并未真正使用这一项，只不过用它作为点位符。然后是目标主机名和端口号。主机名这一项也可以使用真实的 IP 地址来代替。

在 Windows 下，interfaces 文件（sql.ini）的常用格式与 UNIX 下还是有区别的，给出一个样例如下：

```
[SEANLAPTOP]
master=NLWNSCK,sean-laptop,5000
query=NLWNSCK,sean-laptop,5000
```

看起来就是传统的 ini 文件的标准格式。每个 Server 项都以"[]"括起来作标识。对于服务器名，有如下限制：

- 服务器名称长度不能超过 11 个字符。过去老式的 FAT 分区则不能超过 8 个字符。
- 服务器名称的首字符必须是字母（a-z、A-Z）。后面的字符可以是字母、数字、下划线（_）、井号（#）、"位于"符号（@）或美元符号（$）。

- Server 名称不区分大小写。

master 和 query 后以"="列出各项的值，NLWNSCK 指的是什么呢？其全称是 Net Library WinSocket。这个值指的是使用的网络驱动库的名称。它来源于另一个 Windows 下 ASE 的另一个配置文件 libtcl.cfg，它与 sql.ini 默认位于同一个目录（%SYBASE%/ini）下。我们来看一下 libtcl.cfg 的内容：

```
[DRIVERS]
NLWNSCK=TCP   Winsock TCP/IP Net-Library driver
NLMSNMP=NAMEPIPE   Named Pipes Net-Library driver
NLNWLINK=SPX   NT NWLINK SPX/IPX Net-Library driver
NLDECNET=DECNET   DecNET Net-Library driver
```

你会发现，ASE 支持上述四种网络驱动库。通常的 tcp 网络协议对应的就是 NLWNSCK，也是目前使用最广泛的网络驱动库。

4.2.2 interfaces 文件的工作原理

如果是 ASE 服务器，怎么根据 interfaces 文件来确定监听客户端的地址呢？就以 Adaptive Server 为例，如这里的 interfaces 文件有一个 Server 的记录信息：

```
[SEANLAPTOP]
master=NLWNSCK,sean-laptop,5000
query=NLWNSCK,sean-laptop,5000
```

当 Adaptive Server 启动时，会执行以下步骤：

Step 1 查找 -s 命令行参数中提供的服务器名，如果未提供服务器名，则进行下一步。

Step 2 通过检查 **DSLISTEN**（全称：**Data Server Listen**）环境变量的值来确定服务器名称，如果系统并未设定环境变量 DSLISTEN 的值，那么就假定服务器名为 SYBASE。

Step 3 在 interfaces 文件中查到与步骤 1 和 2 里得到的服务器名相匹配的记录。在我实验的机器上，DSLISTEN 环境变量的值刚好为 SEANLAPTOP，所以能找到上边对应的那条记录。

Step 4 使用找到的那条记录 master 条目中的信息进行监听。这表明 Adaptive Server 会在主机 sean-laptop 的 5000 端口上进行监听，并且使用的是 TCP 协议。

如果是 ASE 客户端，当它连接到服务器端时，会执行以下步骤：

Step 1 通过编程方式或引用 **DSQUERY**（全称：**Data Server Query**）环境变量来确定服务器的名称。如果程序中没有指定服务器名，则引用 DSQUERY 变量的值，如果 DSQUERY 环境变量没有设定，则服务器名称为 SYBASE。

Step 2 在 interfaces 文件查找到 1 中确定的服务器名对应的记录。

Step 3 使用 2 中查找到的记录里的 query 条目中的信息进行远程连接。如果没有找到对应的记录，则会提示出错。

仍以前边的 interfaces 文件为例，假设客户端现在有这个 interfaces 文件，并且想连接服务名为 SEANLAPTOP 的服务器，那么它就会通过 query=NLWNSCK,sean-laptop,5000 中的信息建立到服务器 sean-laptop 主机上 5000 端口的连接。

由此看来，当不指定 -s 参数的值时，环境变量 DSLISTEN 和 DSQUERY 的值非常重要。尤其是在连接本机数据库的时候。下边的 isql 命令证实了 DSQUERY 环境变量在客户端连接中的作用：

```
D:\ASE150>set DSQUERY=

D:\ASE150>isql -Usa -Psybase1
CT-LIBRARY error:
        ct_connect(): 目录服务层: 内部目录控制层错误: 要求的服务器名没找到

D:\ASE150>isql -Usa -Psybase1 -SSEANLAPTOP
1> quit

D:\ASE150>set DSQUERY=SEANLAPTOP

D:\ASE150>isql -Usa -Psybase1
1> quit
```

4.2.3 配置网络连接

配置网络连接，首先就是要对 interfaces 文件进行编辑和连接验证。对 interfaces 文件进行编辑，可以使用普通的文本编辑器按照上节中的格式要求进行编辑。也可以使用 ASE 自带的工具 dsedit 或者 UNIX/Linux 下的 asecfg 进行编辑。

如果使用 dsedit，可以直接进入 %SYBASE%\%SYBASE_OCS%\bin 目录，运行命令行 dsedit.exe，也可以执行系统菜单命令"程序"→Sybase→Connectivity→Open Client Directory Service Editor，即启动 dsedit 的图形界面，如图 4-1 所示。

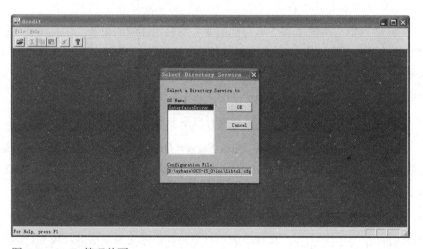

图 4-1 dsedit 管理首页

单击 OK 按钮，进入编辑管理界面，如图 4-2 所示。

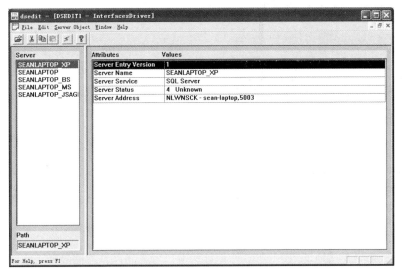

图 4-2　dsedit 编辑管理界面

在图 4-2 中，我们发现系统中已经默认生成了 5 个 interfaces 记录。它们是 ASE 数据库安装过程中生成的几个 Server 的网络连接信息。

通过 Server Object→Add 命令，输入服务名，这里输入一个示例名称 DemoServer，结果生成如图 4-3 所示的属性列表。

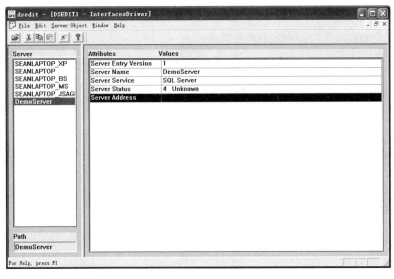

图 4-3　属性列表

这时我们需要填写 Server 的地址，双击 ServerAddress 选项，会弹出一个对话框，单击 Add 按钮，选定协议为 TCP，并输入网络地址，如图 4-4 所示。

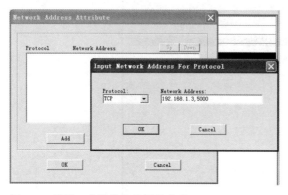

图 4-4 编辑一个服务名的网络信息

上图中的网络地址格式必须是（地址,端口号），即地址和端口号之间以","间隔开。地址可以是 IP 地址，也可以是主机名。这样单击 OK 按钮以后，即可生成一个 interfaces 记录。我们如果关闭 dsedit 管理工具，就会发现%SYBASE%\ini\sql.ini 里会有一条新记录：

```
[DemoServer]
master=TCP,192.168.1.3,5000
query=TCP,192.168.1.3,5000
```

最后还可以通过选中 Demo Server 选项，单击 Server Object→Ping Server 命令，来验证远程服务器是否可以连接，如图 4-5 所示。

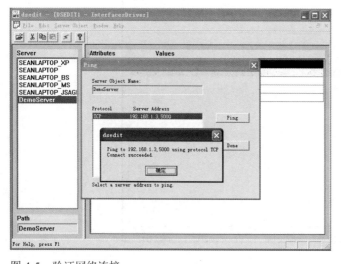

图 4-5 验证网络连接

需要记住的是，安装完 ASE 数据库，会生成默认的几个服务名及其 interfaces 网络信息记录。

```
[SEANLAPTOP]
master=NLWNSCK,sean-laptop,5000
query=NLWNSCK,sean-laptop,5000

[SEANLAPTOP_BS]
master=NLWNSCK,sean-laptop,5001
query=NLWNSCK,sean-laptop,5001

[SEANLAPTOP_MS]
master=NLWNSCK,sean-laptop,5002
query=NLWNSCK,sean-laptop,5002

[SEANLAPTOP_XP]
master=NLWNSCK,sean-laptop,5003
query=NLWNSCK,sean-laptop,5003

[SEANLAPTOP_JSAGENT]
master=NLWNSCK,sean-laptop,4900
query=NLWNSCK,sean-laptop,4900
```

我们看到 Adaptive Server 主服务名 SEANLAPTOP 对应的端口号为 5000；Backup Server 服务名 SEANLAPTOP_BS 的端口号为 5001；Monitor Server 服务名 SEANLAPTOP_MS 的端口号为 5002；XP Server 服务名 SEANLAPTOP_XP 的端口号为 5003；JSAGENT Server 对应的端口号为 4900。

在通过 Open Client 或者 JDBC 编程时，需要指定的端口指的就是 Adaptive Server 的端口 5000。

4.3 使用 ASE 客户端

4.3.1 连接 ASE

当配置好网络连接以后，我们就可以使用 Sybase Central 客户端来对 ASE 数据库作一些日常的管理了。

通过"程序"→Sybase→Sybase Central 命令进入该客户端的主界面（在 UNIX/Linux 下进入目录 $SYBASE/shared/sybcentral43/，然后运行 **./scjview.sh** 即可打开 Sybase Central，需要注意的是，当前用户一定要导入 Sybase 的环境变量，详见第 1 章的 ASE 在 Linux 下的安装）。通过选择"文件"→"连接"命令，打开"连接"对话框，如图 4-6 所示。

从服务器名的列表里，可以找到上一节中我们创建的 Server 名 DemoServer 并选中，然后单击"确定"按钮，连接成功，即可进入 DemoServer 服务名对应的 ASE 服务器。这个列表里所列的服务器名称全部来自于 interfaces[①]（或者 NT 下的 sql.ini 文件）中登记了 query 值的那些服务名。如图 4-7 所示。

图 4-6　"连接"对话框

图 4-7　Sybase Central 管理界面

---

① 如果特殊说明，后文的 interfaces 文件既指 UNIX 下的 interfaces 文件，也指 NT 下的 sql.ini 文件。

通过 Sybase Central，可以对如下各项进行管理：
- 数据库；
- 登录信息；
- 数据库设备；
- 转储设备；
- 远程服务器；
- ASE 中的进程；
- 引擎组；
- 执行类；
- 角色；
- 事务；
- 锁。

通过 sa 账户连接上 Adaptive Server 以后，非常重要的一个步骤是修改 sa 账户的密码。在 Sybase Central 中，通过 DemoServer→"登录"命令，然后双击 sa 账户，打开"sa 登录 属性"对话框，选择"参数"选项卡，可以更改 sa 的口令，如图 4-8 所示。

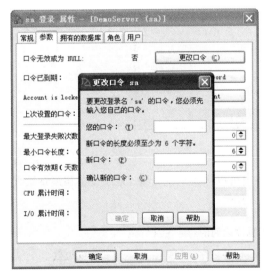

图 4-8　更改 sa 账户的口令

4.3.2　创建数据库设备

数据库设备是存储组成数据库的对象的物理资源。"设备"不一定指特定的物理设备，它可以是一块盘，如磁盘的一个分区，也可以是操作系统的文件。

数据库用于存储一套互相关联的对象（如表）的有关信息（数据）。当用户建立自己的数据库时，需要定义如何组织自己的数据。

ASE 数据库对象包括：表、规则、默认、存储过程、触发器、视图等。

在 ASE 服务器中，默认有多个设备文件，每个设备文件都有一个设备名。这些设备与数据库之间存在着多对多的关系，即一个设备上可以创建多个用户数据库；同时一个用户数据库也可以从多个设备上使用其中的数据空间。设备是基于物理意义上的文件，数据库则是逻辑意义上的表的集合，这些表创建在数据库所用的各设备上，是逻辑意义上的容器。

我们可以看到，默认情况下，ASE 设备如图 4-9 所示。

图 4-9　默认设备列表

这些设备全部用来创建系统数据库，这些设备也可以称为系统设备。系统设备和系统数据库之间的对应关系如表 4-1 所示。

表 4-1　系统设备和数据库之间的对应关系

数据库名及用途	设备名	存储位置
master，包含一些系统表，存储管理 Adaptive Server 所用的数据	master	%sybase%\data\master.dat
model，是用于创建新数据库的模版数据库	master	%sybase%\data\master.dat
sybsystemdb，包含用于分布式事务管理特性的数据	master，systemdbdev	%sybase%\data\master.dat %sybase%\data\systemdbdev.dat
sybsystemprocs，主要用于存储系统存储过程	sysprocsdev	%sybase%\data\sysprocsdev.dat
tempdb，主要用于存储临时表	master	%sybase%\data\master.dat

我们可以为用户数据库创建独立的设备。其过程如下：

单击 DemoServer→"数据库设备"命令，右击"新建"→"数据库设备"命令，然后指定"设备名"和"设备路径"，如图 4-10 所示。

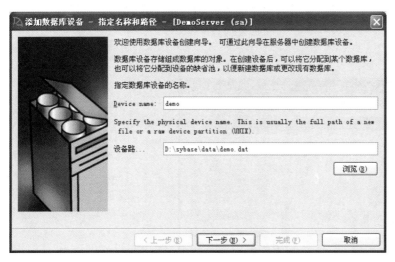

图 4-10 指定设备名及其路径

单击"下一步"按钮，指定设备大小，这里设定为 25MB，设备号可以直接使用提示的值。

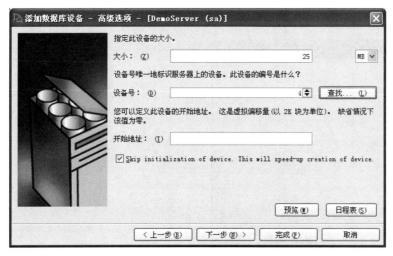

图 4-11 创建设备高级选项

然后不断单击"下一步"按钮，直至完成。

创建单独的数据库设备的好处是，将系统数据库的存储空间和用户数据库的存储空间分开处理,这样便于维护。一般情况下,不建议使用系统设备来创建用户数据库。

4.3.3 创建数据库

使用 Sybase Central 可以很方便地创建数据库。登录进入 Sybase Central 管理界面,在 DemoServer 下的管理项"数据库"下右击,然后执行"新建"→"数据库"命令,输入数据库名称,如图 4-12 所示。

图 4-12 创建数据库

这时,还可以单击"预览"按钮,看到创建数据库的即时脚本,如图 4-13 所示。

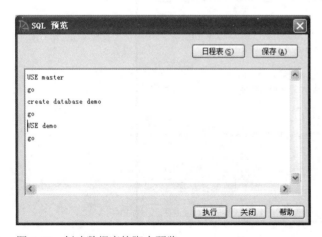

图 4-13 创建数据库的脚本预览

接着单击"下一步"按钮,进入"添加数据库-设备信息"对话框,再单击"添加"按钮,如图 4-14 所示。

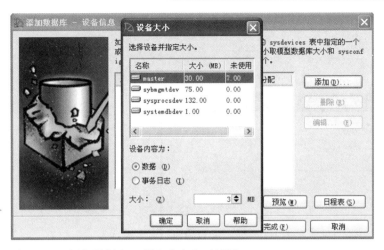

图 4-14 "添加数据库 - 设备信息"对话框

我们选择其中一个设备名 master,并指定数据部分大小为 30MB,然后单击"确定"按钮。同样,也可以指定日志部分的设备名和大小。然后单击"下一步"按钮,如图 4-15 所示。

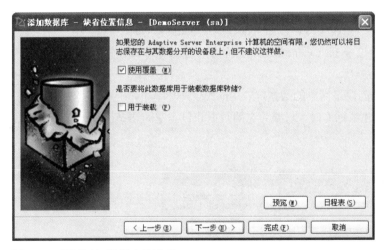

图 4-15 设置添加数据库的选项

我们勾选"使用覆盖"复选项,允许日志部分被覆盖。接着单击"下一步"按钮,如图 4-16 所示。指定是否为该数据库创建 Guest 用户。如果勾选该复选项,则允许普通的 Guest 用户访问该数据库。这时我们可以"预览"一下建库的脚本:

```
USE master
go
create database demo    on master = 3 with override
go
```

```
USE demo
go
exec sp_adduser 'guest'
go
```

图 4-16　其他数据库选项

最后，单击"下一步"→"完成"按钮，即可创建完成 demo 数据库。

以上步骤只是示例如何通过 Sybase Central 简单地创建一个数据库。在实际使用过程中，并不建议使用 master 设备来创建数据库。

当然，上边只是演示如何创建最简单的数据库。在正式生产环境中，用户数据库通常都创建在单独的设备（文件）上，在创建数据库时，只要为它的数据和日志空间选择已经创建好的数据库设备即可，如图 4-17 所示。

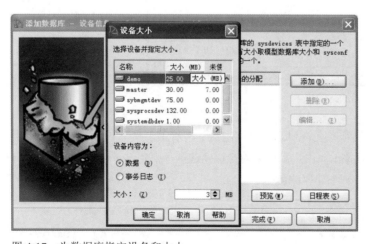

图 4-17　为数据库指定设备和大小

建完数据库以后，在开发应用测试中，有几个重要的选项需要设置，如图 4-18 所示。

图 4-18　几个重要的数据库选项

- abort tran on log full：当日志满时，放弃当前事务。
- ddl in tran：支持在事务里支持 DDL 操作，即在一个事务里可以创建表，甚至可以在异常的时候进行回滚，此时这个表就会删除。这个特性非常有用。
- select into/bulkcopy/pllsort：这个操作允许数据库支持 select into、bulkcopy、pllsort 操作，可以直接通过 select into 操作完成建表操作。
- trunk log on chkpt：在检查点操作以后，截断日志。

4.3.4　创建登录账户和数据库用户

刚开始学习 ASE 的时候，很多人容易把登录账户和数据库用户弄混淆，以为二者是同一个概念。其实，登录是基于 Server 级别的概念，即整个 Server 范围内的用户。而数据库用户则是针对具体的数据库的。当一个登录账户没有指定给任何用户数据库或系统数据库时，它并不能使用具体的数据库，但是它能连接到 Server。要登录 Server，用户必须是登录账户。要进入数据库，必须是数据库的有效用户。

所以，要想建立一个全新的数据库用户，首先得创建一个登录账户。创建登录账户的过程如下：

单击 DemoServer→"登录"命令，右击"新建"→"登录"命令，打开"添加登录 - 指定名称和口令"对话框，如图 4-19 所示。

图 4-19　指定登录名和口令

然后单击"下一步"按钮，指定登录名对应的默认数据库，如图 4-20 所示。

图 4-20　指定默认数据库

接着单击"下一步"按钮，指定对特定数据库的访问权限，如图 4-21 所示。

我们还可以通过预览查看完整的创建登录账户的 T-SQL 脚本：

```
exec sp_addlogin 'spring', 'spring1', @defdb = 'demo', @deflanguage = 'us_english', @auth_mech = 'ANY', @fullname='Spring.Myself'
```

```
go
USE demo
go
exec sp_adduser 'spring' , 'spring' , 'public'
go
USE master
go
exec sp_adduser 'spring' , 'spring' , 'public'
go
USE master
go
```

图 4-21　指定数据库访问权限

由于这里只能选择 public 权限，账户 spring 只能对 demo 数据库执行 select 操作，并不能执行 CUD（Create、Update、Delete）操作。

在这个脚本中，sp_adduser 本身就蕴含着添加用户的功能，只不过这个用户是只读用户，这可以通过 isql 命令行创建简单的表来印证：

```
D:\>isql -Sseanlaptop -Uspring
Password:
1> create table ttt(id int)
2> go
Msg 10331, Level 14, State 1:
Server 'SEANLAPTOP', Line 1:
CREATE TABLE permission denied, database demo, owner dbo
1>
```

因此，我们还需要对 demo 数据库中的用户 spring 的权限重新进行设置，才能使用户 spring 合理地使用 demo 进行 CRUD（Create、Read、Update、Delete）操作，其过程如下：

单击 DemoServer→"数据库"→Demo→"用户"命令，然后选择用户 spring，双击打开"spring 用户 属性"对话框，如图 4-22 所示。

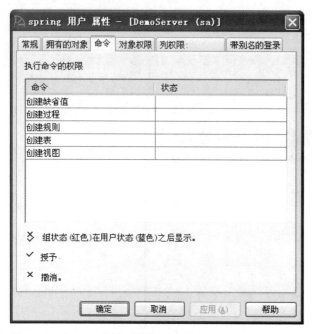

图 4-22　"spring 用户 属性"对话框

选择"命令"选项卡，在"状态"栏中，将这 5 个创建的权限全部勾选上，然后"确定"按钮，即可完成对 spring 用户的权限设置，它将有这 5 个权限。由于 Update、Delete 权限都隐含在 Create 权限里边，所以这时 spring 用户拥有完整的 CRUD 权限。这时我们发现简单的 isql 操作可以顺利完成。

```
D:\>isql -Sseanlaptop -Uspring
Password:
1> create table ttt(id int primary key, col2 varchar(32))
2> go
1> insert into ttt values(1, '创建用户')
2> go
(1 row affected)
1> select * from ttt
2> go
 id          col2
 ----------- ------------------
   1 创        建用户
```

```
(1 row affected)
1> delete from ttt
2> go
(1 row affected)
```

4.3.5　使用 Interactive SQL 客户端

在连接到具体的数据库以后，可以选择指定的数据库，如 Demo，单击右键，选择"打开交互式 SQL"命令，即进入 Interactive SQL 客户端，如图 4-23 所示。

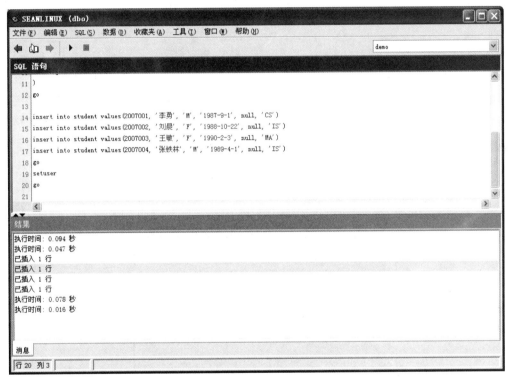

图 4-23　交互式 SQL

在上述 SQL 语句输入框里输入下述 SQL 语句：

```
setuser 'spring'
go
create table student
(
    sno int not null primary key,
    sname varchar(32) not null,
    sgender char(1) not null,
```

```
    sbirth datetime not null,
    sage numeric(2) null,
    sdept varchar(128) null
)
go

insert into student values(2007001, '李勇', 'M', '1987-9-1', null, 'CS')
insert into student values(2007002, '刘晨', 'F', '1988-10-22', null, 'IS')
insert into student values(2007003, '王敏', 'F', '1990-2-3', null, 'MA')
insert into student values(2007004, '张铁林', 'M', '1989-4-1', null, 'IS')
go
setuser
go
```

即可以用 spring 用户的身份创建 student 表。在结果一栏里还可以看到 SQL 命令执行的结果。

除了通过 Sybase Central 中的数据库,间接进入交互式 SQL 图形界面以外,也可以直接通过菜单命令 Sybase→Adaptive Server Enterprise→Interactive SQL 进入。首次进入时,要求选择服务器名,输入用户/密码。这同 Sybase Central 完全一样,其过程如图 4-24 所示。

图 4-24　交互式 SQL 登录

4.4 启动和关闭服务器

4.4.1 启动 Adaptive Server

启动 Adaptive Server，可以使用命令行方式启动。在 Windows 下还可以通过服务管理器来启动。

Windows 下，进入目录 %SYBASE%\%SYBASE_ASE%\install，执行批处理脚本 RUN_<ServerName>.bat。其中的<ServerName>为系统安装时生成的默认 Adaptive Server 名。

也可以通过运行 services.msc 或者通过"控制面板"→"管理工具"命令进入服务管理器，然后启动服务 Sybase SQLServer _ <ServerName>。服务器启动以后，生成的日志文件位于相同目录<ServerName>.log 中，许多启动信息都记录在这里。将该服务设置自动启动，还可以让该服务在操作系统启动以后自动启动。

当然，也可以使用命令行方式启动后台服务：

```
D:\>net start "sybase sqlserver _ SEANLAPTOP"
Sybase SQLServer _ SEANLAPTOP 服务正在启动 ...
Sybase SQLServer _ SEANLAPTOP 服务已经启动成功。
```

UNIX/Linux 下，进入目录 $SYBASE/$SYBASE_ASE/install，运行命令：

```
startserver  -f  RUN_<ServerName>
```

运行完，可以立即运行 showserver 命令，确认 Adaptive Server 是否成功启动。

```
iihero@seanlinux:/opt/sybase/ASE-15_0/install/ > ./showserver
F S UID       PID  PPID  C PRI  NI ADDR SZ WCHAN  STIME TTY          TIME CMD
0 S iihero   9824  9823  3  75   0 - 24594 -      17:33 ?        00:00:25 /opt/sybase/ASE-15_0/bin/dataserver
-d/opt/sybase/
     data/master.dat     -e/opt/sybase/ASE-15_0/install/SEANLINUX.log       -c/opt/sybase/ASE-15_0/SEANLINUX.cfg
-M/opt/sybase/ASE-15_0 -sSE
     ANLINUX
iihero@seanlinux:/opt/sybase/ASE-15_0/install>
```

以上命令显示 Adaptive Server 在运行。这里需要注意一下 XP Server 的启动，你不需要手动去启动 XP Server。

一般情况下，XP Server 是被动启动，它主要用于扩展的存储过程。当第一个扩展存储过程调用时，会激发 Adaptive Server 启动 XP Server。XP Server 会一直运行到 Adaptive Server 关闭为止。

> **注意**
>
> 有一个选项对 XP Server 非常重要，那就是 sp_configure 'xp_cmdshell context',0，默认情况下，这个选项的值为 1。xp_cmdshell 会将权限限制为在操作系统级具有系统管理特权的用户。如果此参数设置为 0，xp_cmdshell 将使用 Adaptive Server 运行所用的操作系统账号（也就是 ASE 数据库的操作系统账户）的安全性环境。

所以，在 UNIX/Linux 下，在修改该值之前，运行 xp_cmdshell 会出现用户上下文无法切换的错误。因为我们运行 ASE 数据库的操作系统用户通常都是非 root 用户。所以，该值必须为 0 才有效。

```
1> xp_cmdshell 'whoami'
2> go
09/02/15 05:54:50 PM    User access denied. Failed to change the user context.

User access denied. Failed to change the user context.
 xp_cmdshell

(0 rows affected)
(return status = 1)

1> sp_configure 'xp_cmdshell context', 0
2> go
(1 row affected)
Configuration option changed. ASE need not be rebooted since the option is
dynamic.
Changing the value of 'xp_cmdshell context' does not increase the amount of
memory Adaptive Server uses.
(return status = 0)

1> xp_cmdshell 'whoami'
2> go
 xp_cmdshell
 -------------
 iihero

(1 row affected)
(return status = 0)
1>
```

4.4.2 关闭服务器

关闭服务器可以使用 isql 命方式，在 Windows 下，可以使用服务管理器停止相应的服务。

```
D:\Sybase\ASE-15_0\install>isql -Usa -Psybase1 -SSEANLINUX
1> shutdown SEANLINUX with wait
2> go
1> quit
```

超级用户使用 isql 登录到服务器，运行 shutdown [服务名] [with wait]即可关闭服务器。

服务名和 with wait 都是可选项，如果没有提供服务名，则默认关闭 Adaptive Server。如果没有提供 with wait 选项，则立即关闭服务器进程，这会加长下次 ASE Adaptive Server 启动时的检查时间。因此，一般都推荐添加 with wait 选项。

ASE 除了主服务名以外，还有 XP Server、Backup Server 及 Monitor Server 都可以通过命令行停止。

还有一种关闭 Adaptive Server 的方法是，打开 Sybase Central，连接到服务器，然后选择具体的服务器名 SEANLINUX 并右击选择"关闭"选项，即可关闭相应的服务器。

如果不想通过查看 errorlog 来获取上次启动 ASE 的具体时间，在启动 ASE 之后，可以查询全局变量@@boottime 来获取此信息。这个全局变量是从 ASE 12.5 起开始引入的。

5

ASE 的交互命令行工具

5.1 SQL 交互命令 isql

在 ASE 数据库当中,每个用户都拥有唯一的登录名和口令所标识的登录账号,该账号由系统管理员建立,基本内容包括:
1. 登录名:在服务器中是唯一的。
2. 口令。
3. 默认数据库(可选):如果定义了默认数据库,用户在定义的数据库中启动每个 ASE 会话的时候,无须再发出 use <database> 命令;如果未定义默认数据库,则会在 master 数据库中启动每个会话。
4. 默认语言(可选):它指定提示和消息所使用的语言,如果未定义,则使用安装时设置的 ASE 的默认语言。
5. 全名(可选):这是用户的全名,对文档记录和身份识别非常有用。

下面我们来看看 isql 的使用方法。

5.1.1 启动和停止 isql

启动 isql,只需发出 isql 命令即可:

```
#001 D:\>isql
#002 保密字:
#003 Msg 4002, Level 14, State 1:
#004 Server 'SEANLAPTOP':
#005 Login failed.
```

```
#006 CT-LIBRARY error:
#007            ct_connect(): 协议特定层: 外部错误: 试图连接服务器失败
```

由以上命令可以看到，由于系统默认连接出错，所以 isql 执行不成功。

```
#001 D:\>isql -U spring -S seanlaptop
#002 保密字:
#003 1>quit
#004
```

使用参数选项-U 指定用户名，-S 指定数据库服务器名，再输入指令，即可成功进入，退出时直接发出 quit 或者 exit 命令（两者等效）即可。

进入 isql 以后可以输入任意有效的 SQL 语句进行查询或者 DDL/DML 操作。

5.1.2　isql 的命令选项

使用如下命令，可以看到 isql 命令的参数选项。

```
#001 D:\>isql --help
#002 Syntax Error in '--help'.
#003 usage: isql [-b] [-e] [-F] [-p] [-n] [-v] [-X] [-Y] [-Q]
#004        [-a display_charset] [-A packet_size] [-c cmdend] [-D database]
#005        [-E editor [-h header -H hostname -i inputfile]
#006        [-I interfaces_file] [-J client_charset] [-K keytab_file]
#007        [-l login_timeout] [-m errorlevel] [-M labelname labelvalue]
#008        [-o outputfile] [-P password] [-R remote_server_principal]
#009        [-s col_separator] [-S server_name] [-t timeout] [-U username]
#010        [-V [security_options]] [-w column_width] [-z localename]
#011        [-Z security_mechanism]
```

相关的格式选项如下：

1．-h header：列标题之间显示的行数，默认为 1。

2．-s col_separatoer：更改列的分隔符，默认是空格。

3．-w column_width：显示列宽，默认为 80。

4．-e：在输出中包含发送给 isql 的每个命令，默认是不包含。

5．-n：删除编号和提示符，默认是不删除。

6．-o outputfile：将结果转储到输出文件，当结果较多时比较有用。

这些选项名称都区分大小写，看看下边的示例：

```
D:\>isql -Uspring -h 1 -s "|" -w 200 -e -n
保密字:
use iihero
go
use iihero
select cno, cpno, ccredit from course
```

```
go
select cno, cpno, ccredit from course
 |cno        |cpno       |ccredit    |
 |-----------|-----------|-----------|
 |          1|          5|          4|
 |cno        |cpno       |ccredit    |
 |-----------|-----------|-----------|
 |          2|       NULL|          2|
 |cno        |cpno       |ccredit    |
 |-----------|-----------|-----------|
 |          3|          1|          4|
 |cno        |cpno       |ccredit    |
 |-----------|-----------|-----------|
 |          4|          6|          3|
 |cno        |cpno       |ccredit    |
 |-----------|-----------|-----------|
 |          5|          7|          4|
 |cno        |cpno       |ccredit    |
 |-----------|-----------|-----------|
 |          6|       NULL|          2|
 |cno        |cpno       |ccredit    |
 |-----------|-----------|-----------|
 |          7|          6|          4|

(7 rows affected)
```

我们看到，各列都使用"|"分隔开，并且由于使用了-e 选项，已经发出的 SQL 命令 use iihero 以及 select cno、cpno、ccredit from course 都回显一次；使用了-n 选项，命令行的提示符及行号都删除了。对比下边默认的 isql，就可以看到不同的效果。

```
D:\>isql -Uspring
保密字：
1> use iihero
2> go
1> select cno, cpno, ccredit from course
2> go
     cno         cpno        ccredit
 ----------- ----------- -----------
           1           5           4
           2        NULL           2
           3           1           4
           4           6           3
           5           7           4
           6        NULL           2
           7           6           4

(7 rows affected)
1>
```

5.1.3 指定 interface 文件、语言、字符集、数据库名

有时候，我们可以需要指定自定义的 interface 文件、语言和字符集，使用选项-I 指定 isql 要用的 interface 文件，使用-a 则指定显示使用的语言，而使用-J 则用于指定 isql 要使用的字符集（isql 作为客户端，有它默认的字符集）。使用-D 则可以用于指定登录用户的默认数据库名，指定以后，就无须在命令行里重新发起 use <dbname>的命令了。

这里，在 G:\下创建了一个 interface 文件 testase.ini，内容如下：

```
[SEAN_ASE]
master=NLWNSCK,sean-laptop,5000
query=NLWNSCK,sean-laptop,5000


[SEANLAPTOP_XP]
master=NLWNSCK,sean-laptop,5004
query=NLWNSCK,sean-laptop,5004
```

现在，要想在 isql 里使用自定义的 interface 文件 testase.ini，并且使用 iso_1 客户端字符集：

```
D:\>isql -Uspring -Ig:\testase.ini -Jiso_1 -SSEAN_ASE -Diihero
???:
??! ??????????????????????????????('?')?
1> quit
```

由于数据库安装时采用的是 eucgb（实质上是 gbk）字符集，所以，客户端采用 iso_1 字符集必然会导致乱码，将-J 设置为 eucgb 或者 cp936 字符集，则会正确显示：

```
D:\>isql -Uspring -Ig:\testase.ini -Jcp936 -SSEAN_ASE -Diihero
保密字:
1>
```

上例中，我们还指定了 spring 用户默认的数据库为 iihero。

5.1.4 改正输入

在输入命令内容的时候，难免中途会出错，但这时可以通过输入 reset 或者按 Ctrl+C 组合键，清除当前缓冲区中的命令。

```
#001 1> select * from course
#002 2> abcd
#003 3> reset
#004 1> go
#005 1> select * from course
#006 2> abcd
#007 1> go
#008 1> select cno from course
```

```
#009 2> go
#010     cno
#011     -----------
#012        1
#013        2
#014        3
#015        4
#016        5
#017        6
#018        7
#019
#020 (7 rows affected)
#021 1>
```

上例中，在第 2 行和第 6 行都有错误输入，紧接着下一行输入 reset 或者按 Ctrl+C 组合键都可以清除掉命令缓冲。

5.1.5　性能统计信息收集与更改命令终结符

isql 命令还提供了性能统计的参数选项-p，通过-c 还可以指命令终结符。

```
#001 D:\>isql -Uspring -Diihero -p
#002 保密字:
#003 Execution Time (ms.):       0           Clock Time (ms.):    0
#004 1> select count(*) from ULCustomer
#005 2> go
#006
#007    -----------
#008        30
#009
#010 (1 row affected)
#011 Execution Time (ms.):      62           Clock Time (ms.):   62
#012 1>
```

我们看到在加了参数选项-p 以后，每个命令执行完以后，都有相应的性能统计结果（第 3 行和第 11 行）。改变终结符的示例如下：

```
#001 D:\>isql -Uspring -Diihero -p -c .
#002 保密字:
#003 Execution Time (ms.):       0           Clock Time (ms.):    0
#004 1> select count(*) from ULCustomer
#005 2> .
#006
#007    -----------
#008        30
#009
```

```
#010 (1 row affected)
#011 Execution Time (ms.):        16           Clock Time (ms.):        16
#012 1>
```

这里指定终结符为".",用以代替 isql 默认的终结符 go,在输入完一条或多条 SQL 命令以后,输入"."即可终结并执行这些命令。

5.1.6 设置 isql 的网络包大小

通过正确设置网络包大小,可以极大地提高 Adaptive Server 的性能。

-A size 选项指定 isql 会话使用的网络包大小,默认值为 2048 个字节。若要将当前 isql 会话的包大小设置为 4096 字节,可以输入:

```
isql -A 4096
```

若要检查网络包大小,请键入:

```
select * from sysprocesses
```

此 isql 会话的值在 sysprocesses 表中 network_pktsz 标题下面显示。

5.1.7 设置输入和输出文件

isql 可以使用-i 和-o 分别指定输入和输出文件,正如第 2 节里所言,-e 选项会使 isql 在输出中显示输入的命令,这样,输出的文件会同时包含输入的命令和输出的结果。

当执行大批量的 SQL 查询处理时,建议采用-i 指定输入文件,有大量的输出结果,我们也可以使用-o 指定输出文件。

例如,我们要建立 iihero 数据库,本书提供了建库的 sql 文件,名为 E:\MyDocument\MYBOOKS\ASE\code\create_database_iihero.sql。

要执行此文件,我们只需要运行以下语句:

```
isql -Usa -P<password> -i
E:\MyDocument\MYBOOKS\ASE\code\create_database_iihero.sql
```

以下是一个指定输出文件的示例:
首先定义一个输入文件 test.sql,内容如下:

```
Select * from sysprocesses
go
```

接着执行:

```
G:\>Isql -Usa -itest.sql -og:\isql.out
保密字:
```

这时我们能看到 g:\isql.out 当中记录着上边的 sql 执行的结果。

值得注意的是，如果有-o 选项，必须前边带有-i 选项，否则 isql 会一直处于挂起状态。

在 UNIX/Linux 操作系统下，可以采用 UNIX 重定向符号"<"和">"提供与-i 和-o 选项类似的机制，如下所示：

```
isql -Usa < input > output
```

可以指示 isql 接受来自终端的输入，如下例所示：

```
xionghe@bjeaslinux2:~/ase150> isql -Usa -Psybase1 -Sbjeaslinux2 <<EOF>out.txt
> select * from sysprocesses
> go
> EOF
xionghe@bjeaslinux2:~/ase150> cat out.txt | more
 spid    kpid       enginenum    status              suid
           hostname
           program_name
           hostprocess
           cmd
           cpu        physical_io memusage    blocked dbid     uid
           gid
           tran_name

           time_blocked network_pktsz fid
           execlass
           priority
           affinity
           id         stmtnum      linenum      origsuid    block_xloid
           clientname
           clienthostname
           clientapplname                                          sys_id
           ses_id     loggedindatetime
……
```

"<<EOF"指示 isql 接受来自终端的输入，直到遇到字符串 EOF。out.txt 里记录着 EOF 之前所有 SQL 语句的执行结果。

可以使用任何字符串替换 EOF。同样，以下示例使用 Ctrl+D 组合键表示输入结束：

```
isql -Usa << > output
```

5.2 导入/导出数据 bcp

bcp 命令是 Sybase ASE 数据库当中最常用的数据导入/导出命令。下面介绍该命令的基本用法和使用技巧。

5.2.1 使用 bcp 导出数据

现在数据库 iihero 当中有 course 表（课程表），其创建的初始 SQL 语句为：

```
create table course
(
    cno int identity not null primary key,
    cname varchar(128) not null,
    cpno int null,
    ccredit int not null
)
go
```

基本数据如下：

```
#001 insert into course(cname, cpno, ccredit) values('数据库', 5, 4)
#002 insert into course(cname, cpno, ccredit) values('数学', null, 2)
#003 insert into course(cname, cpno, ccredit) values('信息系统', 1, 4)
#004 insert into course(cname, cpno, ccredit) values('操作系统', 6, 3)
#005 insert into course(cname, cpno, ccredit) values('数据结构', 7, 4)
#006 insert into course(cname, cpno, ccredit) values('数据处理', null, 2)
#007 insert into course(cname, cpno, ccredit) values('C 语言', 6, 4)
#008 go
```

现在要把这个表的数据导出成一个外部文件，使用如下命令：

```
#001 D:\>bcp iihero..course out g:\couse.out -Sseanlaptop -Uspring -Pspring1 -c

启动拷贝...

已拷贝 7 行.
时钟时间(ms.): 共计= 47  平均 = 6 (148.94 行每秒.)
```

我们看到第一行中的 -c 表示默认采用字符格式进行导出。导出的文件 course.out 中的内容如下：

```
1    数据库      5    4
2    数学        2
3    信息系统    1    4
4    操作系统    6    3
5    数据结构    7    4
6    数据处理         2
7    C 语言      6    4
```

5.2.2 使用 bcp 导入数据

我们接上一节，将导出的数据重新导入到表 course 当中，先将该表的数据全部删除，然后再将数据导入。其过程如下：

```
C:\Documents and Settings\hex>isql -Uspring -Pspring1 iihero
1> truncate table course
2> go
1> select count(*) from course
2> go

 -----------
           0

(1 row affected)

C:\Documents and Settings\hex>bcp iihero..course in f:\course.out -Uspring -Pspring1 -c

启动拷贝...

已拷贝 7 行
时钟时间(ms.): 共计= 547    平均 = 78 (12.80 行每秒)

C:\Documents and Settings\hex>isql -Uspring -Pspring1 iihero
1> select count(*) from course
2> go

 -----------
           7
```

6 使用 Transact-SQL

SQL（结构化查询语言）是在关系数据库系统中使用的高级语言。SQL 最初由 IBM 的 San Jose Research Laboratory 在 20 世纪 70 年代末开发成功，现经改编后已用于很多关系数据库管理系统。美国国家标准协会（ANSI）和国际标准化组织（ISO）已将其批准为正式关系查询语言标准。

而 Transact-SQL 则是 ASE 数据库使用的 SQL 查询语言，它改进了 SQL 的功能，并尽量避免用户借助于编程语言来实现某些功能。

Transact-SQL 主要有如下几个扩展：

1. 使用 compute 子句来执行集合函数（sum、max、min、avg、count、count_big）。
2. 控制流语言：支持 begin...end、break、continue、declare、if...else、print、raiseerror 等多个控制流标识。
3. 存储过程：通过 Transact-SQL 能够创建存储过程，从而极大地提高了 SQL 数据库语言的功能、效率和灵活性。
4. 扩展存储过程：具有存储过程的接口，但它不包含 SQL 语句和控制流语句，而是执行已编译成动态链接库（DLL）的过程语言代码。
5. 触发器：是一个存储过程，在尝试进行特定更改时，指示系统采取一项或多项操作。触发器能防止对数据进行错误的、未授权的或不一致的更改，有助于保持数据库完整性。
6. 默认值和规则：Transact-SQL 提供维护实体完整性（确保为要求值的每列都提供值）和域完整性（确保列中的每个值都属于该列的合法值集合）的关键字。默认值和规则定义了在数据输入和修改过程中起作用的完整性约束。
7. 错误处理和 set 选项：为 Transact-SQL 程序员提供了许多错误处理技术，包括捕获存储过程的返回状态、定义存储过程的自定义返回值、从过程向其调用方传递参数以及从全局变量（如@@error）获得报告等功能。

同时，在运算符上，Transact-SQL 支持否定比较（!=、!>、!<）、按位运算（--、^、|和&）、连接运算符（*=和=*）、通配符（[]和-）以及 not 运算符（^）。以下将 Transact-SQL 简称为 T-SQL。

SQL 语句主要分为以下 3 个类别：

- DDL（Data Definition Language，数据定义语言）：这类语句将定义不同的数据库对象（设备、数据库、段、表、列、索引、视图等）。常用的 DDL 语句关键字有 create、drop、alter 等。
- DML（Data Manipulation Language，数据操纵语言）：用于添加、删除、更新和查询数据库记录，并检查数据完整性。常用的语句关键字主要包括 insert、delete、update 和 select 等。
- DCL（Data Control Language，数据控制语言）：用于控制用户访问数据库对象的权限和访问级别。这些语句主要用于定义数据库、表、用户的访问权限和访问级别。主要的语句关键字为 grant、revoke 等。

其实，DDL 和 DML 两类操作也常被人统称为 CRUD（即 create、read、upsert、delete）操作。create 对应于 DDL，read 对应于 select 查询，upsert 对应于 update 和 insert。这种称呼似乎更贴近于实际的开发应用。

6.1 数据库对象

从逻辑上划分，数据库是由一系列对象组合而成。SQL 是数据库产品中最常用的查询数据的手段，它是一种语言，通常包括语句、函数或过程、数据类型。在使用 SQL 之前，我们最好能够了解一下每个 ASE 数据库中最基本的数据库对象。这些对象包括：

- 表；
- 视图；
- 规则；
- 索引；
- 锁；
- 触发器；
- 存储过程；
- 日志文件。

6.1.1 T-SQL 中的数据类型

T-SQL 中的数据类型主要用于指定表字段、存储过程参数以及局部变量的信息类型、大小及存储格式。ASE 的表中每个字段都会关联着对应的数据类型。ASE 支持两种重要的数据类型：一种是系统内置数据类型，简称系统数据类型；另一种是用户自定义数据类型。

6.1.2 系统数据类型

T-SQL 中主要会用到表 6-1 中的系统数据类型，ASE 允许每种数据类型不区分大小写，但默认情况下，都是以小写形式输出。

表 6-1 ASE 中的系统数据类型

数据类型	别名（同义词）	取值范围	占用字节数
精确数值整数			
bigint		-2^{63} 到 $2^{63}-1$ 之间的整数	8
int	integer	-2^{31} 到 $2^{31}-1$ 之间	4
smallint		-2^{15} 到 $2^{15}-1$ 之间（即-32768 到 32767）	2
tinyint (字节)		0 到 255	1
unsigned bigint		0 到 $2^{64}-1$ 之间	8
unsigned int		0 到 $2^{32}-1$ 之间	4
unsigned smallint		0 到 $2^{16}-1$ 之间	2
精确数值小数			
numeric(p, s)		-1038 到 1038-1	2 到 17
decimal(p, s)	dec	-1038 到 1038-1	2 到 17
近似数值			
float(p)		与计算机有关	p<16 时为 4 p>=16 时为 8
double precision		与计算机有关	8
real		与计算机有关	4
货币			
smallmoney		-2^{31} 到 $2^{31}-1$ 之间	4
money		-2^{63} 到 $2^{63}-1$ 之间	8
日期/时间			
smalldatetime		1900 年 1 月 1 日到 2079 年 6 月 6 日	4
datetime		1753 年 1 月 1 日到 9999 年 12 月 31 日	8
date		0001 年 1 月 1 日至 9999 年 12 月 31 日	4
time		12:00:00AM 到 11:59:59:999PM	4
字符			
char(n)	character	页大小	n
varchar(n)	character varying, char varying	页大小	实际串长度
unichar	unicode char	页大小	n*2 (@@unicharsize = 2)
univarchar	unicode char varying	页大小	实际字符数*@@unicharsize

续表

数据类型	别名（同义词）	取值范围	占用字节数
nchar(n)		页大小	n*@@ncharsize
nvarchar(n)		页大小	n*@@ncharsize
text		最多 2^{31}-1 个字节	初始化后为 2K 的倍数
unitext		最多 2^{30}-1 个字符	初始化后为 2K 的倍数
二进制值			
binary(n)		页大小	n
varbinary(n)		页大小	实际字节数
image		最多 2^{31}-1 个字节	初始化后为 2K 的倍数
位			
bit		0 或 1	1（1 个字节最多可保存 8 个 bit）

6.2 数据库对象的创建

6.2.1 使用和创建数据库

数据库是相关表和其他数据库对象的集合。安装 ASE 时，它会包含以下系统数据库：

- **master**：从整体上控制用户数据库和 ASE 服务器的操作。
- **sybsystemprocs**：包含系统存储过程。
- **sybsystemdb**：包含有关分布式事务的信息。
- **tempdb**：存储临时的数据库对象，包含创建的具有名称前缀 tempdb.. 的临时表。
- **model**：ASE 将它用作创建新用户数据库的模板。

用户的全部数据存储在用户数据库当中。而 ASE 则通过系统表来管理每一个用户数据库，master 数据库和其他系统数据库中的数据字典表均被视为系统表。

1. 使用数据库

可以使用 use 命令来访问现有数据库，其语法为：

use <db_name>

如要访问 iihero 数据库，输入：

use iihero

只有当你是 iihero 数据库中的已知用户时，此命令才允许你访问 iihero 数据库。否则，会出现错误信息。

2. 创建数据库

当系统管理员授予你创建 create database 命令的权限时，你就可以创建数据库。创建数据库时，必须确定当前使用的数据库是 master 数据库。在实际开发应用环境中，许多数据库都是由系统管理员直接创建全部数据库，这时，数据库的创建者也成了数据库的所有者。要想变更数据库的所有者，可以通过 sp_changedbowner 来改变。

数据库所有者负责授予用户访问数据库的权限，并负责授予或撤消其他某些权限。有时需要临时获得其他用户的某些权限，可以使用 setuser 命令。

创建数据库的命令语法如下：

```
create [temporary] database database_name
    [on {default | database_device} [=size]
        [, database_device[=size]]…]
    [log on database_device} [=size]
        [, database_device[=size]]…]
    [with {override | default_location = "pathname"}]
    [for {load | proxy_update}]
```

最简单的命令就是 create database <dbname>，系统将会在默认的设备文件上创建该数据库。

使用 temporary 选项，表明要创建一个临时数据库。

我们要创建 test 数据库，首先确保使用的是 master 数据库。然后输入以下命令：

```
use master
create database test on default = 5 log on testlog = 3
go
```

这里使用 on 子句，可以指定要创建的数据库的大小，不带具体单位，则默认以 MB 为单位，这里创建的 test 数据库大小为 5MB。Default 这里是指默认的设备名。也可以在多个设备名上指定相应的大小。

log on 子句则指定事务日志的存储位置及大小。除非要创建极小且不太重要的数据库，否则务必要指定 log on 选项，这样可以确保系统将事务日志放到单独的数据库设备之上。上例中事务日志的大小为 3MB。这样做的原因如下：

（1）它允许 dump transaction 而不是 dump database，从而节省了时间和空间。

（2）允许建立固定大小的日志，避免与其他数据库争用空间。

事务日志所需的设备大小因更新活动量和事务日志转储频率不同而有所不同。作为一般准则，为日志分配的空间应为分配的数据库空间的 25%左右。

For load 选项形式主要用于装载数据库转储。

3. 变更数据库

变更数据库常用于扩大数据库的数据大小和日志大小。在实际环境中，有时数据存储空间全部用满，则不能再向其添加新数据。这时，必须扩大数据库的存储空间以容纳更多的数据。

默认情况下，alter database 权限授予给数据库的所有者，不能将其转让给其他用户。同样，也必须使用 master 数据库运行此命令。

完整的 alter database 语法允许将数据库扩展指定的兆字节数（最小 1MB），并允许指定增加存储空间的位置：

```
alter database database_name
    [on {default | database_device} [= size]
    [, database_device [= size]]...]
    [log on {default | database_device} [ = size ]
    [, database_device [= size]]...]
    [with override]
    [for load]
```

alter database 命令中的 on 子句与 create database 中的 on 子句相似。for load 子句与 create database 中的 for load 子句相似，并只能用于使用 for load 子句创建的数据库。

要在数据库设备 testdata 上为 test 增加 2MB 的分配空间，并在数据库设备 newdata 上为其增加 3MB 的分配空间，请输入：

```
alter database test
on testdata = 2, newdata = 3
```

4. 删除数据库

可以使用 drop database 命令来删除数据库。drop database 从 Adaptive Server 中删除数据库及其所有内容，释放为其分配的存储空间，并从 master 数据库中删除对它的引用。

语法如下：

```
drop database database_name [, database_name]...
```

不能删除正在使用中的数据库，即由其他用户打开进行读写操作的数据库。

如前述，可用一个命令删除多个数据库。例如：

```
drop database newpubs, newdb
```

不能使用 drop database 来删除已经损坏的数据库。

5. 数据库选项

ASE 支持很多数据库一级的选项，这些选项只对单个数据库起作用，ASE 15.0 支持的一些选项如下：

- Abort tran on log full：为 true 时，在日志满时会放弃当前事务。
- Allow nulls by default：为 true 时，在创建表时，默认情况下，列值将允许为 null。该选项默认为 false。
- Auto identity：强制所有表都创建一个 identity 列 SYB_IDENTITY_COL，该列为 10 个数字宽度，并且不能用于 select * 查询。
- Dbo use only：只有该数据库的属主用户才能使用该数据库，该选项经常用于系统维护，阻止其他非属主用户访问该数据库。

- Ddl in tran：为 true 时，允许在一个事务中执行 DDL 语句。
- Identity in nonunique index：该选项会让 identity 列成为非 unique 索引的一部分，提高索引效率。
- No chept on recovery：为 true 时，Server 一旦重启，执行完恢复进程以后，数据库将不执行 checkpoint 操作。
- No free space acctg：用于增量恢复，通过延迟计算所有非日志段的可用空间，并禁调用设置可用空间的阈值的存储过程来实现。
- Read only：为 true 时，该数据库不能被修改。
- Select into/bulkcopy/pllsort：为 true 时，会让下述操作产生最少量的 log：select into 语句、使用 writetext、使用 bcp 工具、并行排序。
- Single user：数据库将只允许一个用户能访问该数据库，最常见的情况下，当 SA 执行维护任务时，通常要其他用户访问数据库。该选项不能用于 tempdb 数据库。
- Trunk log on chkpt：当后台 check-point 任务激活时，ASE 将大约每分钟截断一次日志。该选项对 dbo 触发的 check point 不起作用。如果日志非常小，则不会被截断。该选项经常用于开发环境，因为在开发环境当中，没有必要进行增量备份。
- Unique auto_identity index：在所有的 unique 非 cluster 索引里包含一个 IDENTITY 列。
- Disable alias access：禁止使用跨数据库的别名访问。

使用存储过程 sp_dboption 设置数据库选项，例如：

```
1> sp_dboption demo, "select into/bulkcopy/pllsort", true
2> go
Database option 'select into/bulkcopy/pllsort' turned ON for database 'demo'.
Running CHECKPOINT on database 'demo' for option 'select into/bulkcopy/pllsort'
to take effect.
(return status = 0)
1> checkpoint
2> go
```

6.2.2 使用和创建表

表是数据库中含有特殊列的记录的集合。在 ASE 当中，一个数据库最多可以支持 2000,000,000 张表。

1. 创建表

创建表是 SQL 中 DDL（数据定义语言）的最常见语句。

创建表的常用语法格式如下：

```
create table [database.[owner].]table_name (column_name data_type [default *]
    [[constraint constraint_name] {{unique|primary key}
```

```
            | references ref_table [ ref_column]}, …
    [lock {datarows | datapages | allpages}]
    [on segment_name]
)
```

ASE 中的表名或者列名所用的标识符都有长度限制：常规标识符不超过 255 个字节。此限制适用于大多数用户定义的标识符，包括表名、列名、索引名等。

对于变量名，由于"@"前缀占用了 1 字节，因此它的最大长度为 254 个字节。

创建表时，默认情况下，在当前数据库和当前用户下创建表。也可以显示地在表名前加上 database.owner.前缀以指定该表创建给哪个数据库下的哪个用户，这样创建的前提是，当前用户拥有在指定数据库下创建表的权限。

如果运行 set quoted_identifier on 命令，则表名和列名都可以是分隔标识符。对于每个数据库下的同一个用户，表名必须是唯一的。

例如：

```
use master
create table iihero.spring.test(id int primary key, col2 varchar(32) null default null，col3 int references test_ref(id))
go
```

该语句首先是在 master 数据库下，接着想在 iihero 数据库中，以 spring 用户的身份创建表 test。主键为 id 列，列数据类型为 int 型，col2 列的默认值为 null，列数据类型为 varchar，即变长字符串。col3 列为 int 型，同时它也是表 test 的外键，依赖于表 test_ref 的 id 列。外键的意思是 col3 列的值必须来源于表 test_ref(id)列的值。

默认情况下，主键使用的是聚簇索引，而 unique 约束则使用的是非聚簇索引。关于聚簇索引和非聚簇索引的区别，我们在后续章节中会逐步提到。

前边的语法中，还有 lock {datarows | datapages | allpages}标识，它用于指定该表使用什么样的锁定方案，其默认值是对应于 lock sheme 的全服务器范围的设置。

on segment_name 用于指定在其上放置表的名称。使用 on segment_name 时，必须已经用 create database 或 alter database 将逻辑设备指派给了数据库，且必须已经用 sp_addsegment 在数据库中创建了段。有关数据库中可用段名的列表，可以使用 sp_helpsegment 得到。

关于列定义，ASE 有一种特殊的列类型，叫做 identity 列，通过在列定义后边加上 identity 标识来设置。数据库的每个表都可以具有一个以下数据类型的 identity 列：

- 精确 numeric 类型：小数位数为 0。
- 任意整数数据类型：包括有符号或无符号 bigint、int、smallint 或 tinyint。

identity 列不能更新，也不允许有空值。

identity 列用于存储由 Adaptive Server 自动生成的序列号，如发票编号或职员编号。identity 列的值唯一地标识表的每一行。

Default 用于指定列的默认值。如果指定了默认值，而用户在插入数据时没有为列提供值，

Adaptive Server 就会插入默认值。默认值可以是常量表达式、内置的（插入执行插入操作的用户的名称）或者是 null（插入 NULL 值）。Adaptive Server 以 tabname_colname_objid 为格式生成默认名称，其中 tabname 是表名的前 10 个字符，colname 是列名的前 5 个字符，而 objid 是默认对象 ID 号。使用 identity 属性为列定义的，可以在不引用数据库对象的 create table 语句的 default 部分引用全局变量。

可通过在 create table 语句中指定关键字 identity，而不是 null 或 not null 来定义 identity 列。identity 列的数据类型必须为 numeric 且标度为零，或者为任何整数类型。请在新表中将 identity 列定义为任何所需精度（1 到 38 位）：

Create table table_name (column_name numeric(precision, 0) identity)，则可能的最大列值为 $10^{precision} - 1$。如创建一个表，其 identity 列的最大值为 $10^5 - 1$（即 9999）：

```
Create table test_identity
    (test_id numeric(5, 0) identity,
    item_name varchar(32) null)
```

列定义的后边可以跟随 null 或 not null，指定在无默认值的情况下， Adaptive Server 在数据插入过程中的行为。

null 指定如果用户没有提供值，则 Adaptive Server 分配一个空值。not null 指定如果没有默认值，则用户必须提供一个非空值。

如果不指定 null 或 not null，默认情况下，Adaptive Server 将使用 not null。不过，为了使此默认值与 SQL 标准兼容，可以使用 sp_dboption 对它进行切换。在标准 SQL 中，默认情况将使用 null（Oracle 数据库默认情况使用的就是 null）。

该命令如下：

```
sp_dboption <dbname>, "allow nulls by default", true
```

在 ASE 数据库中，还有一种特殊的表叫**临时表**，临时表的数据没有持久性。临时表分为以下两种：

（1）可以在 ASE Server 会话之间共享的表，通过将 tempdb 指定为 create table 语句中表名的一部分，可以创建**可共享临时表**。例如，下列语句创建可在 Adaptive Server 会话之间共享的临时表：

```
create table tempdb..authors
(au_id char(11))
drop table tempdb..authors
```

ASE Server 不更改以此方式创建的临时表名。该表会一直存在，直至当前会话结束或者表的所有者使用 drop table 将其删除。

（2）临时表，是只能由当前 ASE Server 会话或过程访问的表。

通过在 create table 语句中的表名前面指定井号（#），可以创建非共享临时表。例如：

create table #authors(au_id char (11))

该表会一直存在，直至当前会话或过程结束，或者其所有者用 drop table 将其删除。

> 注意
> 可将规则、默认值和索引与临时表建立关联，但不能在临时表上创建视图，或将触发器与之建立关联。

总结起来，**临时表**的创建方法与普通表有一些区别。它有两种创建方法：
1）使用 "#" 作为表名前缀，以表明该表是临时表，这种用法最常见。
2）使用 tempdb…来指定数据库名。这就和前边的语法格式完全吻合。

关于创建表，还有一种特殊的命令来创建，那就是 select into，其语法如下：

select <column, column, …> into <新表> from table_name ……

例如：

1> select * into iihero..new_students from iihero.spring.student
2> go
(4 rows affected)

上述语句将创建基于表 spring.student 创建表 dbo.new_students，它们的表结构完全相同。

借用这个特性，加上合适的 where 条件，我们还可以创建一个新表，与原表保持同样的结构，但不插入数据。

例如：

1> drop table iihero..new_students
2> go
1> select * into iihero..new_students from iihero.spring.student where 0 = 1
2> go
(0 rows affected)
1>

能成功使用该命令的一个前提条件是，目标数据库必须将 select into/bulkcopy/pllsort 数据库选项设置为 on。在一般开发和应用环境中，大多会将此选项设为 on，其命令如下：

sp_dboption iihero, "select into/bulkcopy/pllsort", true

当该选项为 on 时，则可使用 select into 子句来建立新的永久表，而不必使用 create table 语句。可用 select into 直接生成一个临时表，即使 select into/bulkcopy/pllsort 选项不是为 on。

使用 sp_helpdb <database_name>命令可以看到目标数据库相关选项值。例如：

```
1> sp_helpdb iihero
2> go
 name        db_size        owner        dbid        created
status
 ----------- -------------- ------------ ----------- --------------------
 ---------------------------------------------------------------------------------
 iihero      20.0 MB        sa           5           Sep 14, 2008
select into/bulkcopy/pllsort, trunc log on chkpt, ddl in tran, abort tran on log full, mixed log and data
```

2. 变更表

使用 alter table 命令可以对表结构作如下修改：

- 添加列和约束
- 更改列默认值
- 添加 null 列或 not null 列
- 删除列和约束
- 更改锁定方案
- 分区或未分区表
- 改变列数据类型
- 转换现有列的 null 默认值
- 增加或减小列长度

变更表的常用语法如下：

```
alter table table_name
    [add column_name datatype [identity | null |
    not null] [, column_name datatype [identity
    |null | not null]]]
    [drop column_name [, column_name]
    [modify column_name {[data_type]
    [[null] | [not null]]}
    [, colum_name datatype [null | not null]]]
```

其中：table_name 是变更的表名，datatype 是变更的列的数据类型。

例如：

- 添加列。

    alter table t add address varchar(32) null, col5 int null, col6 int default 99

将向表 t 添加两列（address varchar(32) null, col5 int null, col6 int default 99）。

需要说明的是，我们不能直接添加列（col_name data_type not null），这时必须采用带有默认值的方式（col_name data_type default <value>）来代替。

- 更改列。

    ```
    alter table t modify col5 int not null
    ```

    将表 t 中的列 col5 由 null 改为 not null。

- 更改列中的默认值。

    ```
    alter table t add col7 int default 1000
    alter table t replace col7 default 99
    ```

    上述语句先添加一列（col7 int），默认值为 1000，接着又将其默认值改为 99。

- 删除列。

    ```
    alter table t drop col5, col6
    ```

    将 col5 和 col6 列从表 t 中删除。

- 删除列中的主键。

    设 t 表结构为：create table t(id int primary key, col2 varchar(32))

    要删除主键 primary key(id)，在 ASE 中不能直接用命令：

    ```
    Alter table t drop primary key(id)
    ```

    需要先查出完整性约束名：

    ```
    1> sp_helpconstraint t
    2> go
     name                    definition                            created
     ----------------------  ------------------------------------  -------------------------
     t_id_3511562411         PRIMARY KEY INDEX ( id ) : CLUSTERED   四月  23 2012   7:5

    (1 row affected)
    (return status = 0)
    ```

    然后删除这个约束：

    ```
    1> alter table t drop constraint t_id_3511562411
    2> go
    ```

    同样，unique 约束也是这样进行删除：

    ```
    1> alter table t add unique(id)
    2> go
    1> sp_helpconstraint t
    2> go
     name                definition                              created
     ------------------  --------------------------------------  -------------------------
     t_3511562412        UNIQUE INDEX ( id ) : NONCLUSTERED      四月  23 2012   8:0
    ```

```
(1 row affected)
(return status = 0)
1> alter table t drop constraint t_3511562412
2> go
```

这是通过 constraint 名字进行删除。另外，ASE 提供了系统存储过程 sp_dropkey 来删除主键。调用命令 sp_dropkey t, id 即可。

- 重命名列。

  对某一列进行重命名，没有直接的 SQL 语法进行支持。但 ASE 提供了系统存储过程 sp_rename 来实现。其语法如下：

```
sp_rename objname, newname [,"index" | "column"]
```

例如：将表 t 的列 address 改名为 new_address。这时 objname 必须是"表名.列名"的形式。

```
1> sp_rename "t.address", new_address, "column"
2> go
Column name has been changed.
(return status = 0)
```

本例中末尾的 column 也可以不要，只要前边的名字是唯一的即可。

3. 重命名表

表的重命名要借助于系统过程 sp_rename。

例如：将表 t 重命名为 new_t，语法如下：

```
1> sp_rename "t", new_t
2> go
Object name has been changed.
(return status = 0)
1>
```

4. 删除表

删除表的 SQL 语法相对简单。其语法如下：

drop table *[[database.]owner.] table_name*
[, [[database.]owner.] table_name]...

我们可以通过列出多个表名，一次删除多张表。

此命令从数据库中删除指定的表，连同表中的内容以及相关的所有索引与特权。原来绑定到该表的规则和默认值将不再绑定，其他方面均无变化。

只有表的所有者才能删除它。然而，在用户或前端程序对表进行读写操作时，任何人都不能将它删除。不能对任何系统表使用 drop table 命令，无论是在 master 数据库还是在用户数据库中。

drop table 和 truncate table 权限不能移交给其他用户。

例如：删除当前数据库中的表 new_t，语法如下：

```
1> drop table new_t
2> go
1>
```

6.3 操纵数据库对象（DML）

DML 操作是指对数据库中表记录的操作，主要包括对表记录的插入（insert）、更新（update）、删除（delete）和查询（select），是开发人员最常使用的 SQL 操作。

6.3.1 插入记录

表创建完以后，就可以往表里插入记录，插入记录的基本语法如下：

INSERT INTO <tablename> (field1, field2, ……, fieldn) VALUES(val1, val2, ……,valn)

如数据库 iihero 有如下一张表：

```
create table emp
(
    empno int not null,
    ename varchar(32) null,
    job varchar(32) null,
    mgr int null,
    hiredate datetime null,
    sal decimal(8, 2) null,
    comm decimal(8, 2) null,
    deptno decimal(2) null,
    comments text null,
    photo image null,
    gender char(1) not null
)
go
alter table emp add constraint emp_fk_dept foreign key(deptno) references dept
go

/* 表内循环依赖 */
alter table emp add constraint emp_fk_emp foreign key(mgr) references emp
go
```

向它插入一条记录如下：

```
1> insert into spring.emp(empno, ename, job, mgr, hiredate, sal, deptno, gender)
 values(8000, 'Johannes', 'DEVELOPER', 7839, '2002-05-17', 1000.00, 20, 'M')
2> go
(1 row affected)
```

当然，也可以不用指定字段名，这样就必须设置每个字段的值，并且值的顺序与表中字段的排列顺序保持一致。

例如：

```
1> insert into spring.emp values(8001, 'Jothy', 'MANAGER', 7839, '2001-01-09', 3
000, 200, 20, null, null, 'F')
2> go
(1 row affected)
```

指定字段名的好处是：①可以任意调整字段的顺序；②对于某些可以为空（null）的字段、带有默认值的字段、自增字段，完全可以不用出现在 insert 后的字段列表里边，values 后只需要写对应字段名称的值，而那些没有出现的字段自动设置为 null、默认值或者自增的下一个数值。

例如，这里只对表 emp 中的 empno、mgr、deptno、gender 设值：

```
1> insert into spring.emp(empno, mgr, deptno, gender) values(8002, 7839, 20, 'M'
)
2> go
(1 row affected)
```

插入以后的实际值如下：

```
1> select * from spring.emp where empno = 8002
2> go
empno,ename,job,mgr,hiredate,sal,comm,deptno,comments,photo,gender
8002,NULL,NULL,7839,NULL,NULL,NULL,20,NULL,NULL,'M'
```

你会发现，那些没有设值的字段全部为 NULL。

不过，如果表中有字段为自增字段，使用时需要小心。

```
1> create table t1(id int identity not null primary key, b image null)
2> go
1> insert into t1 values(1, 0xabc)
2> go
Msg 7743, Level 16, State 1:
Server 'SEANLAPTOP', Line 1:
当字段列表被使用时，表't1'的标识字段的显式值只能在插入语句中被指定
1>
```

在默认情况下，自增字段值是不能显示插入的，请看下例：

```
1> set identity_insert t1 on
2> go
```

```
1> insert into t1(id, b) values(1, 0xabcd)
2> go
(1 row affected)
1> insert into t1(id, b) values(2, 0xef02)
2> go
(1 row affected)
1> set identity_insert t1 off
2> go
1> insert into t1(id, b) values(3, 0x12)
2> go
Msg 584, Level 16, State 1:
Server 'SEANLAPTOP', Line 1:
当 'SET IDENTITY_INSERT' 是 OFF 时,在表 't1' 中为标识字段指定的显式值
1> insert into t1(b) values(0x12)
2> go
(1 row affected)
1> select id from t1
2> go
 id
 -----------
 1
 2
 3

(3 rows affected)
1>
```

通过设置 identity_insert 开关可以很好地控制 identity 列的插入。其语法如下：

set identity_insert <table_name> on | off

上边的例子中，当该值设为 on 以后，我们可以连带字段 id 插入数据，当关闭该开关以后，则不能显示设置 id 值，但是 id 的新值会基于表中最大的 id 值递增。

该选项开关对数据库中涉及到 identity 表数据的导入和导出非常有用。当我们在另一个库中创建相同的表结构时，要想导进相同的 identity 值，必须先将目标表的 identity_insert 设置为 on，待数据插入结束以后，再将其设置为 off，以保证后续插入使用的 identity 值是从最大的 identity 值开始的。

6.3.2 更新操作

表里的记录值可以通过 update 命令进行更改。其语法如下：

UPDATE tablename SET field1=value1, field2=value2, …… field$_n$=value$_n$ [WHERE CONDITION]

例如，将表 st_course 中学号为 2007002，课程号为 2 的成绩从 90 改为 95，语法如下：

```
1> select * from st_course
2> go
   sno           cno       grade
   ------------- --------- ---------
   2007001       1         92
   2007001       2         85
   2007001       3         88
   2007002       2         90
   2007002       3         80

(5 rows affected)
1> update st_course set grade=95 where sno=2007002 and cno=2
2> go
(1 row affected)
1> select * from st_course
2> go
   sno           cno       grade
   ------------- --------- ---------
   2007001       1         92
   2007001       2         85
   2007001       3         88
   2007002       2         95
   2007002       3         80

(5 rows affected)
```

当然，where 条件可以更复杂一些，在下例中，通过更新表 st_course，将学生李勇的各科成绩都加上一分：

```
1> select a.sno,sname, cno, grade from student a, st_course b where a.sno=b.sno
2> go
   sno           sname      cno       grade
   ------------- ---------- --------- --------
   2007001       李勇        1         92
   2007001       李勇        2         85
   2007001       李勇        3         88
   2007002       刘晨        2         95
   2007002       刘晨        3         80

(5 rows affected)
1> update st_course set grade=grade+1 where sno=(select sno from student where sname='李勇')
2> go
(3 rows affected)
1> select * from st_course
2> go
   sno           cno       grade
   ------------- --------- ---------
   2007001       1         93
   2007001       2         86
   2007001       3         89
```

```
2007002    2    95
2007002    3    80

(5 rows affected)
```

6.3.3 删除操作

如果记录不再需要，则可以通过 delete 命令进行删除。语法如下：

```
DELETE FROM tablename [WHERE CONDITION]
```

例如，将表 emp 中名为 JAMES 的记录都删除，其命令为：

```
1> delete from emp where ename='JAMES'
2> go
(1 row affected)        (受影响行数为 1 行，表示有一行记录被删除)
```

不过，在 Sybase ASE 数据库中，也有不太方便的地方，它不支持 cascade delete 操作，即如果父表有对应的子表，那么通过命令直接删除父表会产生错误。必须先删除子表中对应的记录，才能删除父表中的相应记录行。

例如，将表 student 中的名为"李勇"的记录都删除，其命令为：

```
1> delete from student where sname='李勇'
2> go
Msg 547, Level 16, State 1:
Server 'SEANLAPTOP', Line 1:
在一个引用完整性约束中相关外键违约。数据库名 = 'iihero'，表名 =
'student'，约束名 = 'st_course_880003135'.
命令已异常终止
(0 rows affected)
```

删除失败，这是因为子表 st_course 有一个字段 sno 引用了外键依赖于父表 student，要想删除"李勇"，必须先删除 st_course 中对应于"李勇"的 sno 的所有记录：

```
1> delete from st_course where sno=(select sno from student where sname='李勇')
2> go
(3 rows affected)
1>    delete from student where sname='李勇'
2> go
(1 row affected)
1>
1> select a.sno,sname, cno, grade from student a, st_course b where a.sno=b.sno
2> go
 sno         sname        cno       grade
 -----------  -----------  --------  --------
 2007002      刘晨          2         95
 2007002      刘晨          3         80
```

我们可以看到，李勇的所有记录都已经被删除。

> **注意**
> 如果不带任何 where 条件，那么表中的所有记录将被删除。当然，删除表中所有记录，还可以使用命令 TRUNCATE TABLE tablename，唯一的区别是它没有事务处理。而 DELETE FROM tablename 基于事务处理，已经删除的记录是可以回滚的。同时，TRUNCATE TABLE 只能由数据库的属主用户发出。

```
1> create table t1(id int primary key, col2 varchar(32))
2> go
1> begin trans
2> go
1> begin tran
2> insert into t1 values(1, 'sean')
3> insert into t1 values(2, 'james')
4> commit
5> go
(1 row affected)
(1 row affected)
1> begin tran
2> delete from t1
3> go
(2 rows affected)
1> rollback tran
2> go
1> select * from t1
2> go
 id          col2
 ----------- ----------------
    1        sean
    2        james

(2 rows affected)
1> begin tran
2> truncate table t1
3> rollback tran
4> go
Msg 226, Level 16, State 1:
Server 'SEANLAPTOP', Line 2:
在多语句事务中不允许使用 TRUNCATE TABLE 命令
1> truncate table t1
2> go
1> select * from t1
2> go
 id          col2
 ----------- ----------------
```

(0 rows affected)
1>

6.4 SQL 查询操作（DQL）

6.4.1 使用"*"查询所有记录

最简单的查询操作是查询一个表中的所有记录。使用"*"可以不用显示指定要查询的字段名。

```
1> select * from salgrade
2> go
 grade       losal       hisal
 ----------- ----------- -----------
     1          700        1200
     2         1201        1400
     3         1401        2000
     4         2001        3000
     5         3001        9999

(5 rows affected)
1>
1> select * from salgrade where grade >3
2> go
 grade       losal       hisal
 ----------- ----------- -----------
     4         2001        3000
     5         3001        9999

(2 rows affected)
1>
```

当然也可以带上一定的限定条件。如上例中，查询级别大于 3 的薪水记录。

6.4.2 TOP 限定记录及 distinct 消重

ASE 15.0 支持使用 TOP 来限定记录的条数，TOP 同时还可用于 update 和 delete 语句当中。distinct 标识符可以用于在查询中消除重复的记录。

distinct 消重的应用示例如下：

```
#001 C:\Documents and Settings\hex>isql -Uspring -Pspring1 -Diihero
#002 1> select * from st_course
#003 2> go
#004  sno         cno         grade
#005  ----------- ----------- -----------
#006  2007001          1          92
```

```
#007    2007001    2    85
#008    2007001    3    88
#009    2007002    2    90
#010    2007002    3    80
#011
#012 (5 rows affected)
#013 1> select distinct sno from st_course
#014 2> go
#015    sno
#016    -----------
#017    2007001
#018    2007002
#019
#020 (2 rows affected)
#021 1> select sno from student
#022 2> go
#023    sno
#024    -----------
#025    2007001
#026    2007002
#027    2007003
#028    2007004
#029
#030 (4 rows affected)
#031 1>
```

第 13 行就表示从 st_course（学生-课程）表中查询得到选了课程的学生的学号，有些学生选了多门课，有的学生基本上没有选课。使用 distinct 可以只从各批重复的学号中各取一个，从而得到真正选了课的学生的学号。

以下是 TOP 应用的示例，从雇员表（emp）中取出雇员号排在前 5 的五名雇员记录。

```
1> select top 5 empno, ename from emp order by empno
2> go
 empno       ename
 ----------- ------------------------------
       7369  SMITH
       7499  ALLEN
       7521  WARD
       7566  JONES
       7654  MARTIN

(5 rows affected)
1>
```

如果要查询雇员号排在第 6 到第 10 之间的五名雇员记录，查询方法如下：

```
1> select top 5 empno, ename from (select top 10 empno, ename from emp order by
empno desc) order by empno
2> go
```

```
Msg 154, Level 15, State 20:
Server 'SEANLAPTOP', Line 1:
在派生表.中不允许使用一条 ORDER BY 子句。
1>
```

ASE 在这里并不支持在派生表中使用 order by 子句,一种可替代的方法是借助于临时表,先将记录排序放于临时表,并生成一个主键,然后再从临时表里取 id 为 6 到 10 的 5 条记录。在生成临时表的过程中加入"top 10"限定,是为了减少生成记录的条数,提高临时表的生成效率。这也是用于 ASE 数据库记录分页的一个基本方法。

```
1> select top 10 empno, ename, id=identity(10) into #tmp_tb from emp order by empno
2> go
(14 rows affected)
1> select empno, ename from #tmp_tb where id between 6 and 10
2> go
 empno       ename
 ----------- ----------------
       7698  BLAKE
       7782  CLARK
       7788  SCOTT
       7839  KING
       7844  TURNER

(5 rows affected)
```

6.4.3 Like 通配符模糊查询

SQL 查询的基本应用中,比较值得注意的是 Like 匹配字符串。Like 关键字检索与某一模式相匹配的字符串,可以用于 varchar、nchar、nvarchar、unichar、unitext、univarchar binary、varbinary、text 和 date/time 数据。

Like 的基本语法为:

{where|having} [not] column_name [not] like "match_str"

match_str 通常使用的符号如表 6-2 所示。

表 6-2 匹配串用到的符号

符号	含义
%	与 0 个或多个字符的任意字符串相匹配
_	与单个字符相匹配
[string]	中括号将范围或集合括起来,如[a-f] 或[abcdef]。这里有两种形式,形如 a-f 表示 a 到 f 之间的任一字符,形如 abcdef 表示这 6 个字符中任一字符。里边的字符区分大小写
[^string]	尖号(^)位于分类符之前,表示不包括。[^a - f] 表示"不在 a~f 的范围内";[^a2bR] 表示"不是 a、2、b 或 R"

假设有一张学生表：

```
#001 create table student
#002 (
#003     sno int not null primary key,
#004     sname varchar(32) not null,
#005     sgender char(1) not null,
#006     sbirth datetime not null,
#007     sage numeric(2) null,
#008     sdept varchar(128) null
#009 )
#010 go
#011
#012 select * from student where name like "李%"
#013 select * from student where name like "H[^c]%"
#014 select * from student where name like "%李%"
#015 select * from student where name like "H[abc]o"
```

第 12 行表示名字以"李"打头的所有学生；

第 13 行表示名字以 H 打头，第二个字符不是 c 的所有学生；

第 14 行表示名字中带有"李"字的所有学生；

第 15 行表示名字以 H 打头，后面紧跟的字符必须为 abc 中的一个，然后是字母 o 的所有学生。

通常，where name not like "<pattern>"与 where not name like "<pattern>"完全等价。

那么，如果匹配串本身就含有通配符，该如何处理呢？

ASE 提供了两种方法将通配符转义并将其作为文字搜索来搜索通配符：使用方括号和 escape 子句。匹配字符串也可以是包含通配符的表中的一个变量或一个值。

可将方括号中可以是百分号、下划线、左括号和右括号。若要搜索破折号（不是用它来指定范围），应在一组括号中将破折号用作第一个字符，表 6-3 描述了方括号的使用方法。

表 6-3 使用方括号检索通配符

like 子句	检索
like "a%"	a 以后跟包含 0 或多个字符的任意字符串
like "a[%]"	a%精确串
like "_n"	an、bn、cn 等，第一个字符为任意字符
like "[_]n"	_n
like "[a-cdf]"	a、b、c、d 或 f
like "[-acdf]"	-、a、c、d 或 f
like "[[]"	[
like "[]]"	]

我们也可以 escape 子句来指定转义字符，转义字符必须是单个字符串，其用法如表 6-4 所示。

表 6-4　使用 escape 子句

Like 子句	检索
like "a@%" escape "@"	a%
like "*_n" escape "*"	_n
like "%80@%%" escape "@"	包含 80%的字符串
like "*_sql**%" escape "*"	包含_sql*的字符串
like "%#####_#%%" escape "#"	包含##_%的字符串

如果字符在某一模式中作为转义字符出现两次，则该字符串必须包含四个连续的转义字符，参看表 6-4 的最后一个示例。

在 ASE 中，Like 查询可以将列名作为通配符，例如：

```
#001 1> select empno, ename from emp
#002 2> go
#003    empno       ename
#004    -----------  ----------------
#005    7369        SMITH
#006    7499        ALLEN
#007    7521        WARD
#008    7566        JONES
#009    7654        MARTIN
#010    7698        BLAKE
#011    7782        CLARK
#012    7788        SCOTT
#013    7839        KING
#014    7844        TURNER
#015    7876        ADAMS
#016    7900        JAMES
#017    7902        FORD
#018    7934        MILLER
#019
#020 (14 rows affected)
#021
#022 1> create table #ttt(ename varchar(32))
#023 2> go
#024
#025 1> insert into #ttt values('SM%')
#026 2> insert into #ttt values('J%')
#027 3> insert into #ttt values('SC%')
#028 4> go
#029 (1 row affected)
#030 (1 row affected)
#031 (1 row affected)
#032
#033 1> select empno, emp.ename from emp, #ttt where emp.ename like #ttt.ename
```

```
#034 2> go
#035   empno      ename
#036   ----------  ----------------
#037   7369       SMITH
#038   7566       JONES
#039   7900       JAMES
#040   7788       SCOTT
#041
#042 (4 rows affected)
#043 1>
#044
```

在这里，我们要查询 emp 表中以 SM、J 或 SC 打头的所有员工的员工号和姓名，第 33 行直接使用 emp.ename like #ttt.ename 进行匹配，而不用执行三个带有 like 的 or 子句。当匹配条件很多时，这种方式就非常有利。

6.4.4 NULL 值及其含义

在 ASE 中，NULL 值的含义就是列的数据值为 "未知" 或者 "不确定"、"不可用"。它与 0 或者 " " 的含义不同。使用 NULL 值可以区分是特意在某列置 0 或者 " "，还是该列没有值。

在允许空值的列中：

- 如果未输入任何数据，则自动置为 NULL。
- 用户可显示输入**不带引号**的单词 NULL 或者 null，如果带上引号，系统则认为这是一个字符串。

关于 NULL 值的精确定义，目前各种关系数据库尚无完全统一的标准，当 NULL 值用于比较时，最终结果还要依赖于它们自身的具体实现。通常情况下，NULL 值的比较是未知的，即我们无法确定 NULL 值是等于（或不等于）给定值或另一个 NULL。在 ASE 中，如果表达式 a 是求值结果为 NULL 的任意列、变量或文字，则下列情况返回 TRUE：

- a is NULL。
- a = NULL。
- a = @x，其中，@x 是包含 NULL 的变量或参数。
- a != n，其中，n 是不包含 NULL 的文字，a 的值为 NULL。

如果采用 Like 运算符，则会得到相反的结果，a like null 返回 false，a not like null 则返回 true，如下例所示：

```
#001 1> create table #t (id varchar(5) null)
#002 2> go
#003 1> insert into #t values(null)
#004 2> insert into #t values(null)
#005 3> select * from #t where id=null
#006 4> go
```

```
#007 (1 row affected)
#008 (1 row affected)
#009    id
#010    -----
#011    NULL
#012    NULL
#013
#014 (2 rows affected)
#015 1> select * from #t where id like null
#016 2> go
#017    id
#018    -----
#019
#020 (0 rows affected)
#021 1> select * from #t where id not like null
#022 2> go
#023    id
#024    -----
#025    NULL
#026    NULL
#027
#028 (2 rows affected)
```

值得注意的是，空列与空列无法连接，发生这种情况时，无论比较运算符是什么，始终返回未知，且结果不包括该行。例如：

```
#001 1> create table #t(id int null)
#002 2> insert into #t values(null)
#003 3> go
#004 (1 row affected)
#005 1> create table #a(id int null)
#006 2> insert into #a values(null)
#007 3> go
#008 (1 row affected)
#009 1> select #a.id from #a, #t where #a.id=#t.id
#010 2> go
#011    id
#012    -----------
#013
#014 (0 rows affected)
#015 1> select * from #a
#016 2> go
#017    id
#018    -----------
#019    NULL
#020
#021 (1 row affected)
#022 1> select * from #t
```

```
#023 2> go
#024    id
#025    -----------
#026    NULL
#027
#028 (1 row affected)
#029 1>
```

表#t 和表#a 都有一行空值记录,但做了连接查询以后,返回 0 条记录。也许这时有人会尝试使用!=进行连接,其实这么做以后,结果还是 0 条记录,因为 NULL 值和 NULL 值进行连接没有任何意义。

```
1> select #a.id from #a, #t where #a.id!=#t.id
2> go
 id
 -----------

(0 rows affected)
1>
```

Null 值的替代可以使用 isnull(a, value),其中 a 为表达式,value 为替代值。NULL 与任何数字表达式的所有算术运算都会返回 NULL。字符串与 NULL 相加时,则能返回字符串本身。例如:

```
#001 1> select isnull(null, 'unknown')
#002 2> go
#003
#004    -------
#005    unknown
#006
#007 (1 row affected)
#008 1> select 1+null
#009 2> go
#010
#011    -----------
#012    NULL
#013
#014 (1 row affected)
#015 1> select 'abc' + null + ' is abc'
#016 2> go
#017
#018    ----------
#019    abc is abc
#020
#021 (1 row affected)
```

需要额外值得注意的是,在 T-SQL 中,系统生成的 NULL 与用户分配的 NULL 具有不同的行为,例如:

```
#001 1> if (1 != NULL) print "yes" else print "no"
#002 2> go
#003 yes
#004 1> if (1 != convert(integer, null)) print "yes" else print "no"
#005 2> go
#006 no
```

要避免混淆，取得完全一致的结果，可以使用 set ansinull on：

```
#001 1> set ansinull on
#002 2> go
#003 1> if (1 != NULL) print "yes" else print "no"
#004 2> go
#005 no
#006 1> if (1 != convert(integer, null)) print "yes" else print "no"
#007 2> go
#008 no
```

此时系统生成的 NULL 和用户生成的 NULL 都是返回"未知"值。

6.4.5 SQL 查询的标准格式

常见的 SQL 查询的标准格式是：

```
SELECT [ALL | DISTINCT]
{[<qualifier>.]<column_name> | * | <expression>}
    [AS <column_alias>],...
FROM <tablg_or_view_name> | <inline_view> [<table_alias>]
[WHERE <predicate>]
[GROUP BY [<qualifier>.]<column_name>,...
    [HAVING <predicate>]]
[ORDER_BY [<qualifier>.]<column_name> | <column_number>
    [ASC | DESC],...];
```

前边提到了 DISTINCT、TOP、Like 及 NULL 值查询。更复杂一点的查询有：对查询结果进行排序，聚合以及分组。

1. 分组

我们会经常遇到这样的应用需求：对查询的结果按某个字段的值进行排序，这就要用到 order by 子句。其格式如下：

```
Select field1, field2, ... fieldn FROM tablename [WHERE CONDITION] [ORDER BY field1 [DESC | ASC], field2 [DESC|ASC], ... fieldn [DESC | ASC],
```

其中，DESC 和 ASC 分别是按照字段进行降序和升序排列（DESC：descend；ASC：ascend），ORDER BY 后可以跟多个不同的字段，每个字段都可以分别指定升序或者降序，如果不指定升序或降序，则默认为升序。

例如，对 emp 表中的雇员按照薪水从高到低进行排序：

```
1> select empno, ename, sal from emp order by sal desc
2> go
 empno       ename                  sal
 ----------- ---------------------- ---------------
        7839 KING                          5000.00
        8001 Jothy                         3000.00
        7788 SCOTT                         3000.00
        7902 FORD                          3000.00
        7566 JONES                         2975.00
        7698 BLAKE                         2850.00
        7782 CLARK                         2450.00
        7499 ALLEN                         1600.00
        7844 TURNER                        1500.00
        7934 MILLER                        1300.00
        7521 WARD                          1250.00
        7654 MARTIN                        1250.00
        7876 ADAMS                         1100.00
        8000 Johannes                      1000.00
        7369 SMITH                          800.00
        8002 NULL                             NULL

(16 rows affected)
```

对 emp 中薪水降序，在薪水相同的情况下，按照雇员号升序排列：

```
1> select empno, ename, sal from emp order by sal desc, empno asc
2> go
 empno       ename                  sal
 ----------- ---------------------- -----------
        7839 KING                          5000.00
        7788 SCOTT                         3000.00
        7902 FORD                          3000.00
        8001 Jothy                         3000.00
        7566 JONES                         2975.00
        7698 BLAKE                         2850.00
        7782 CLARK                         2450.00
        7499 ALLEN                         1600.00
        7844 TURNER                        1500.00
        7934 MILLER                        1300.00
        7521 WARD                          1250.00
        7654 MARTIN                        1250.00
        7876 ADAMS                         1100.00
        8000 Johannes                      1000.00
        7369 SMITH                          800.00
        8002 NULL                             NULL
```

(16 rows affected)

2. 聚合分组查询

这类查询主要用于对表中的数据进行统计和汇总。它通常会使用 where、group by、having 子句。在 Sybase ASE 中，还可能会用到计算列 compute 子句。这是它不同于其他数据库的地方。

```
1> select empno, sum(sal), deptno from emp group by deptno
2> go
 empno       deptno           sal
 ----------- ---------------- --------
 7782          8750.00         10
 7839          8750.00         10
 7934          8750.00         10
 7788         10875.00         20
 7566         10875.00         20
 7876         10875.00         20
 7369         10875.00         20
 7902         10875.00         20
 7698          9400.00         30
 7499          9400.00         30
 7521          9400.00         30
 7844          9400.00         30
 7900          9400.00         30
 7654          9400.00         30

(14 rows affected)

1> select empno, deptno, sal from emp order by deptno desc compute sum(sal) by deptno
2> go
 empno       deptno        sal
 ----------- ------------- -----------
 7698          30          2850.00
 7499          30          1600.00
 7521          30          1250.00
 7844          30          1500.00
 7900          30           950.00
 7654          30          1250.00

 Compute Result:
 -----------------------
 9400.00

 empno       deptno        sal
 ----------- ------------- -----------
 7788          20          3000.00
 7566          20          2975.00
 7876          20          1100.00
 7369          20           800.00
```

```
    7902      20     3000.00

Compute Result:
------------------------
0875.00
    empno    deptno    sal
    -------  --------  --------
    7782     10        2450.00
    7839     10        5000.00
    7934     10        1300.00

Compute Result:
------------------------
8750.00

(17 rows affected)
```

6.5　创建表的索引

6.5.1　索引简介

索引是提高 SQL 查询效率的重要手段，它有助于帮助 ASE 定位数据的物理位置。当表中的数据量比较少时，全表扫描并不会有多大的时间消耗，但是一旦数据达到一定规模，几十万行甚至更多时，全表扫描将会消耗大量的时间。这时就要借助于索引。

ASE 支持以下类型的索引：

- 组合索引：这些索引涉及多个列。当两个或更多列因其逻辑关系而最好作为一个单位进行搜索时，推荐使用此类型的索引。
- 唯一索引：这些索引不允许指定列中任意两行的值相同。Adaptive Server 在创建索引（如果数据已经存在）和每次添加数据时检查重复值。
- 聚簇索引或非聚簇索引：聚簇索引强制 Adaptive Server 持续对表中的行进行排序和重排序，以使其物理顺序始终与其逻辑（或索引）顺序相同。每个表只能有一个聚簇索引。非聚簇索引不要求行的物理顺序与其索引顺序相同。可以使用每个非聚簇索引以各种排序顺序访问数据。
- 本地索引：本地索引是一个索引子目录树，它仅编制一个数据分区的索引。可以对本地索引进行分区，并且对于所有类型的分区表都支持本地索引。
- 全局索引：全局索引编制表中所有数据分区的索引。对循环分区表支持非分区全局聚簇索引，而对所有类型的分区表支持非聚簇全局索引。不能对全局索引进行分区。

6.5.2 创建索引

索引有两种创建方式：一种方式是通过 create index 语句进行显式的创建；另一种方式是通过建表语句中的 unique 或 primary key 的完整性约束来隐式地创建索引。隐式创建索引会受到如下限制：

1. 不能创建非唯一索引。
2. 不能使用 create index 命令提供的选项来定制索引的工作方式。
3. 只能通过 alter table 语句将这些索引作为一个约束进行删除。

创建索引需要额外的索引页，占用额外的存储空间。当聚簇索引（cluster index）重建时，非聚簇索引也将自动重建。

插入、更新及删除索引列中的数据所花费的时间通常比未索引列中的时间要长，主要是因为索引列的数据变更时，在更新数据页的同时还要更新索引页。而更新未索引列时，只会更新数据页，时间开销要小。

创建索引的基本原则：

1. 如果打算在 identity 列进行手动插入，则创建一个唯一索引，以保证不会插入重复值。
2. 对于经常需要进行排序的列（由 order by 发起），推荐创建索引。
3. 对于经常与其他表进行连接运算的列，最好创建索引，创建索引的好处是，通过排序能快速执行连接运算。
4. 存储表的主键列通常有聚簇索引，当它与其他表中的列连接时更是如此。每个表只有一个聚簇索引。聚簇索引能确保数据的存储顺序与索引的存储顺序完全一致。
5. 对聚簇索引而言，最大的优势在于范围搜索，找到符合搜索范围内的第一个值，即可保证后续的数据行物理相邻。聚簇索引对于单值搜索并不占太大优势。

以下情况不应该创建索引：

1. 在查询中很少或者从不引用的列不需要创建索引。
2. 相对于表行数而言有很多重复值，但几乎没有唯一值，这样的列创建索引的用处不会太大。

有一点必须明白，如果系统确实需要搜索未被索引的列，则搜索将逐行进行，执行这种搜索所花费的时间将与表中的行数成正比，因为它执行的是全表扫描。

为数据库中的表创建索引之前，确保数据库已经打开 SELECT INTO 选项。

Sp_dboption db0, "SELECT INTO", true

创建索引的最简单形式：

create index index_name on table_name (column_name)，如要对 student 表的 sname 列创建索引，可以执行：

create index ind_student_sname on student (sname)

并不是所有字段类型都能创建索引，对于 bit、text 或 image 类型的列，不能创建索引。

稍复杂一点的索引是对表中的多列创建索引。例如：

Create index ind_student_2 on student(sdept asc, sname desc),对 student 表的 sdept（系别）和 sname（姓名）进行索引，前者升序，后者降序，升降序直接决定了列值的排序顺序。

6.5.3 聚簇索引和非聚簇索引

如果不指定聚簇（clustered）关键字，默认情况下都是创建非聚簇（nonclustered）索引。利用聚簇索引，ASE 可以随时对数据行进行排序，使其物理顺序与其逻辑顺序相同。聚簇索引的叶节点包含表的实际数据。创建聚簇索引的时候会自动重建非聚簇索引。

每个表只有一个聚簇索引，通常对主键创建聚簇索引。

对于非聚簇索引，行的物理顺序与其索引顺序不同。非聚簇索引的叶节点包含指向数据页上的行的指针。更确切地说，每个叶节点包含索引值以及指向该值所在行的指针。换句话说，非聚簇索引在索引结构与数据本身之间有一个额外的级别。

对 student 表中的 sname 创建聚簇索引：

Create clustered index student_idx1 on student(sname)

如果 student 表有主键，那么要先删除这个主键约束，才能创建上述聚簇索引，因为同一张表最多只能有一个聚簇索引。

删除一个索引的语法比较简单：

Drop index <表名>.<索引名>

6.5.4 创建索引的几个选项

创建索引有几个可用的选项，它们的使用情形如表 6-5 所示。

表 6-5 索引类型及选项

索引类型	选项
聚簇索引	ignore_dup_row | allow_dup_row
唯一聚簇索引	ignorfe_dup_key
非聚簇索引	无
唯一非聚簇索引	ignore_dup_key
唯一非聚簇索引	ignore_dup_row

语法为：create index index_name on table_name(colum_name, …) with option

Step 1 ignore_dup_key 选项。

如果在已经具有唯一索引的列中插入重复值，则该操作会被取消。如果创建该唯一索引时设置了该选项，则可避免此情况出现。

唯一索引包括聚簇和非聚簇的。开始数据输入时，重复值的插入都会被取消，并提示错误消息。无论是否设置 ignore_dup_key，已经包含了重复值的列不能创建唯一索引。创建唯一索引之前，必须消除重复值。

例如：

```
1> create table t0(id int null)
2> go
1> create unique clustered index ind on t0(id) with ignore_dup_key
2> go
1> insert into t0 values(1)
2> go
(1 row affected)
1> insert into t0 values(1)
2> go
重复键被忽略
(0 rows affected)
1> insert into t0 values(null)
2> go
(1 row affected)
1> insert into t0 values(null)
2> go
重复键被忽略
(0 rows affected)
插入时重复的 null 值也被忽略
1> drop index t0.ind
2> go
1> insert into t0 values(null)
2> go
(1 row affected)
1> select * from t0
2> go
 id
 -----------
 NULL
 1
 NULL

(3 rows affected)
1> create unique clustered index ind on t0(id) with ignore_dup_key
2> go
Msg 1505, Level 16, State 2:
Server 'XIONGHE', Line 1:
```

异常终止在重复键上创建唯一的索引。主键是'<NULL>'。

我们注意到，NULL 值也不允许重复，这是 ASE 数据库与其他数据库不同的一个地方。

Step 2 ignore_dup_row 和 allow_dup_row 选项。

在创建普通的非唯一聚簇索引时，这两者是完全互斥的选项，首先，非唯一聚簇索引允许键值重复，但是不允许有重复的行，除非指定了 allow_dup_row 选项（注意，非聚簇索引中，ASE 会为每一行附加一个唯一的行标识，因此永远不会出现重复行，但聚簇索引中并没有这样做）。如果设置了 allow_dup_row，则可以对包含重复行的表创建非唯一的聚簇索引，随后也可以插入重复行。如果表中有任何一列创建了唯一聚簇索引，则不能使用此选项。

ignore_dup_row 选项将从这一批数据中消除重复。输入重复行时，ASE 将忽略该行并取消操作。允许正常插入非重复行。该选项只适用于具有非唯一索引的表，如果表中存在唯一索引，则不能使用此选项。

例如：

```
1> create table t0(id int)
2> go
1> insert into t0 values(null)
2> go
(1 row affected)
1> insert into t0 values(1)
2> go
(1 row affected)
1> insert into t0 values(1)
2> go
(1 row affected)
1> create clustered index idx on t0(id) with ignore_dup_row
2> go
警告：删除重复行。主键是'1'
1> select * from t0
2> go
 id
 -----------
 NULL
 1

(2 rows affected)
```

创建聚簇索引以后，发现重复行被删除了。利用这一特性可以删除表中已经存在的重复行。

Step 3 sorted_data 选项。

例如：

```
1> create table t0(id int, name varchar(32))
2> go
1> insert into t0 values(1, 'wang')
2> go
(1 row affected)
1> insert into t0 values(2, 'wang2')
2> go
(1 row affected)
```

```
1> insert into t0 select id + (select max(id) from t0), name from t0
2> go
(2 rows affected)
1> insert into t0 select id + (select max(id) from t0), name from t0
2> go 3
(16 rows affected)
3 xacts:
```

这样插入以后，表 t0 的 id 列都是有序的。

```
1> create clustered index idx on t0(id) with sorted_data
2> go
```

当表中的数据已处于排序顺序时（例如，当使用 bcp 将已排序的数据复制到空表），create index 的 sorted_data 选项会加速索引的创建。速度提高对大型表尤其明显，对 1GB 以上的表，速度将提高许多倍。如果指定 sorted_data，但数据未处于排序顺序，则将显示错误消息并中止命令。此选项只提高编制聚簇索引或唯一非聚簇索引的速度。

Step 4 on segment 选项。

该选项主要用于在指定的段上创建索引。此选项非常有利于将索引和数据表的存储进行分离。如将表 t0 的 name 列创建索引 idx2 到段 mydata 上：

```
1> create index idx2 on t0(name) on mydata
2> go
```

6.5.5 索引删除与索引统计信息的更新

删除索引的语法为：drop index *table_name.index_name*[, *table_name.index_name*]...

使用此命令，ASE 将从数据库中删除指定的索引，并回收其存储空间。一次可以指定多个索引进行删除。要注意，系统表或者系统数据库中的索引是不允许删除的。

要了解一个表中具有哪些索引，可以调用存储过程 sp_helpindex table_name 或者 sp_statistics table_name。

例如：

```
1> sp_helpindex t0
2> go
```

具有下列索引：

index_name	index_keys	index_description	index_max_rows_per_page	index_fillfactor	index_reservepagegap
index_created		index_local			
idx	id	clustered	0	0	0
四月 24 2012 9:08PM		Global Index			
idx2	name	nonclustered	0	0	0

执行 sp_statistics，会把表的索引信息给列出来。如图 6-1 所示，执行的是 sp_statistics t0 的结果。

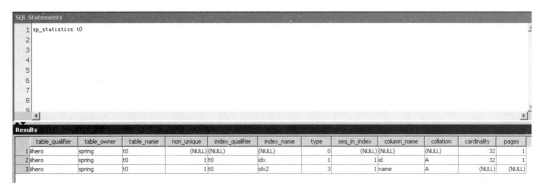

图 6-1　查询结果

索引统计信息的更新，通过使用语句：update statistics table_name index_name 完成。

如果不指定 index_name，则对表中所有索引统计信息进行更新。

6.6　ASE Transact–SQL 中的内置函数

内置的 SQL 函数可以简化 SQL 语句的编写，提高效率。ASE 中提供了常见的 SQL 内置函数。这里将结合具体的系统数据类型进行相关介绍。

6.6.1　获取数据库系统信息的系统函数

语法格式：select function_name(args)，主要有以下函数：

1. user_id (user_name)，带用户名作参数，返回该用户名对应的 user_id，不带参数时，返回当前用户的 user_id，例如：

   ```
   1> select user_id()
   2> go
   -----------
   ```

```
    3
(1 row affected)
1> select user_id('spring')
2> go
 -----------
    3
(1 row affected)
```

2. user_name (user_id)，返回 user_id 对应的用户名，不带参数时返回当前用户名，例如：

```
1> select user_name()
2> go
 -----------
  spring
(1 row affected)
1> select user_name(3)
2> go
 -----------
  spring
(1 row affected)
```

3. col_length (*object_name, column_name*)，获取目标列的列长度，这是列定义的长度，数据的实际长度用另一个函数 datalength。

   col_name (object_id, column_id[, database_id])返回指定列 ID 和表 ID 对应的列名。

   例如：

```
1> select datalength('abc ')
2> go
 -----------
    4
(1 row affected)
1> select max(datalength(name)) from t0     //取表 t0 的列 name 最大实际长度
2> go
 -----------
    5
(1 row affected)
1> select col_length('t0', 'name')     //取表 t0 的列 name 定义的长度
2> go
 -----------
   32
(1 row affected)
1> select col_name(object_id('t0'), 1)     //取表 t0 的第一列的列名
2> go
 -----------
  id
```

4. curunreservedpgs(*dbid, lstart, unreservedpgs*)，返回磁盘区段中的可用页数。
5. datalength(*expression*)，返回指定列或字符串的实际长度，单位为字节。
6. data_pages ([*dbid*,] *object_id, {doampg | ioampg }*)，返回表（doampg）或索引（ioampg）所使用的页数，其结果不包括内部结构的页，通常结合 sysindexes 表来查询，例如：
   查询当前数据库中所有数据对象占用的空间：

   ```
   1> select count( data_pages(db_id(),id,doampg) + data_pages(db_id(),id,ioampg)) from sysindexes where id <> 8
   2> go

    -----------
         94

   (1 row affected)
   ```

   查询所有日志对象占用的空间：

   ```
   1> select count( data_pages(db_id(),id,doampg) + data_pages(db_id(),id,ioampg)) from sysindexes where id = 8
   2> go

    -----------
          1
   ```

7. db_id ([*database_name*])，返回数据库的 ID 号，数据库名必须是字符表达式，并且要用引号引起来。不带参数时，返回的是当前数据库的 ID。反过来，对应的函数是 db_name([*db_id*])。例如：

   ```
   1> select db_id('iihero')
   2> go

    -----------
          4
   (1 row affected)
   1> select db_name(4)
   2> go
    -----------
    iihero
   (1 row affected)
   ```

8. host_id()和 host_name()，返回客户端的主机进程 ID，客户端的主机名，例如：

   ```
   1> select host_id(), host_name()
   2> go
    ------------------------ -------------------------
    13276                    XIONGHEXP
   (1 row affected)
   ```

9. index_col (*object_name, index_id, key_#[, user_id]*)，返回索引列的名称，当 object_name 不是表或视图时，则返回 NULL。主要用于系统过程 sp_helpindex。

10. isnull (exp1, exp2)，当求得的表达式 exp1 值为 NULL 时，用 exp2 的值替代。例如：

    ```
    1> select isnull(null, 'abcd')
    2> go
     -----------
     abcd
    (1 row affected)
    1> select isnull(123, 'abcd')
    2> go
     -----------
     123
    (1 row affected)
    ```

11. identity_burn_max (table_name)，跟踪给定表的最大 identity_burn 值，它只返回值。

12. is_quiesced(db_id)，指示数据库是否处于静默停止状态（quiesced database），如果是，则返回 1，否则返回 0。

13. lockscheme (*object_id [, db_id]*)，返回指定对象的锁定方式。例如：

    ```
    1> select lockscheme(object_id('t1'))
    2> go
     -----------
     allpages
    ```

14. next_identity(table_name)，返回 insert 的下一个 identity 值，只是查询，并不改变值。例如：

    ```
    1> select next_identity('t1')
    2> go
     -----------
     3
    (1 row affected)
    1> insert into t1(id) values(8)
    2> go
    (1 row affected)
    1> select next_identity('t1')
    2> go
     -----------
     9
    (1 row affected)
    ```

15. pagesize (*object_name [, index_name]*)，返回指定对象的页大小，单位为字节。
    如获取表 t1 的页大小：

    ```
    1> select pagesize('t1')
    2> go
    ```

```
      -----------
       4096
    (1 row affected)
```

16. object_id (object_name),返回对象的 ID,这个函数使用非常普遍。反过来,从对象 ID 得到对象名,函数为:

    object_name(*object_id[, database_id]*)

17. proc_role("role_name"),返回有关是否已授予用户指定角色的信息。
18. reserved_pgs(object_id, {doampg | ioampg}),返回分配给表或索引的页数,包括用于内部结构的报告页。
19. suser_id ([server_user_name]),返回登录用户的 ID。
20. suser_name([server_user_id]),返回服务器用户名。

    如:
    ```
    1> select suser_id()
    2> go
     -----------
       1
    (1 row affected)
    1> select suser_name()
    2> go
     -----------
       sa
    ```

21. tsequal (timestamp, timestamp2),比较两个 timestamp 值,timetamp 是被浏览的行的时间戳,timestamp2 是所存储行的时间戳。
22. tran_dumpable_status ("database_name"),返回 true/false,用于指示是否允许执行 dump trans。
23. user_id([user_name]),返回用户 ID。
24. user,返回用户名。
25. user_name([user_id]),返回用户名,无参数时,等效于 user。

6.6.2 字符串相关函数

ASE 中的字符串相关函数比较丰富,主要有以下函数:

1. ascii(char_exp),返回表达式中第一个字符的 ASCII 码值。
2. char(integer_exp),返回一个整数对应的字符值,参数值范围在 0~255 之间。
3. charindex (expression1, expression2),在 expression2 中搜索首次出现的 expression1 并返回表示其起始位置的整数。如果未找到 expression1,将返回 0。如果 expression1 包含通配符,charindex 会将它们视为文字。

4. char_length (char_expr)，返回一个表示字符表达式中字符数的整数，或返回一个 text 或 unitext 值。对于表列中的可变长度数据，char_length 将返回字符数。对于固定长度数据，将返回列的已定义长度。对于多字节字符集，表达式中的字符数通常小于字节数；使用 datalength 系统函数可确定字节数。

5. difference (char_expr1, char_expr2)，返回一个整数，表示两个 soundex 值之间的差值。参见下面的 soundex。

6. len (string_expression)，返回指定字符串表达式（不包括尾随空白）的字符数（而不是字节数）。

7. lower (char_expr)，将大写转换为小写。返回一个字符值。

8. ltrim (char_expr)，删除字符表达式中的前导空白，而且只删除 SQL 特殊字符说明中等于空格字符的值。

9. patindex ("%pattern%",char_expr [using {bytes |chars | characters}])，返回一个整数，表示指定字符表达式中首次出现的 pattern 的起始位置；如果未找到 pattern，则返回 0。默认情况下，patindex 返回的偏移以字符表示。若要返回以字节表示的偏移（多字节字符串），请指定 using bytes。pattern 的前后必须添加通配符"%"（搜索第一个或最后一个字符的情况除外）。patindex 可用于 text 和 unitext 数据。

10. replicate (char_expr, integer_expr)，返回与 char_expr 具有相同数据类型的字符串。该字符串是通过将同一表达式重复指定的次数后得到的，其长度为指定次数下的字符串长度或 255 字节的空间（取两者的最小值）。

11. reverse (expression)，返回字符或二进制表达式的逆序形式；如果 expression 为 abcd，它将返回 dcba；如果 expression 为 0x12345000，它将返回 0x00503412。

12. right (expression, integer_expr)，返回字符或二进制表达式中从右端起具有指定字符数的部分。返回值的数据类型与字符表达式的数据类型相同。

13. rtrim (char_expr)，删除尾随空白。而且只删除 SQL 特殊字符定义中等于空格字符的值。

14. soundex (char_expr)，对于由有效单字节或双字节罗马字母的邻接序列所组成的字符串，它将返回四个字符的 soundex 代码。

15. space (integer_expr)，返回由指定数目的单字节空格所组成的字符串。

16. str (approx_numeric[, length [, decimal]])，返回表示浮点数的字符。length 设置要返回的字符数（包括小数点、小数点左右两侧的所有位数及空白）；decimal 设置要返回的小数位数。length 和 decimal 是可选参数。如果给出这两个参数，就必须使用非负数。length 的默认值为 10；decimal 的默认值为 0。str 将舍入数字的小数部分，使结果不超出指定的 length。

17. str_replace ("string_expression1","string_expression2","string_expression3")，将第一个字符串表达式（string_expression1）中出现的第二个字符串表达式（string_expression2）的所有

实例替换为第三个表达式（string_expression3）。

18. stuff (char_expr1, start, length, char_expr2)，在 char_expr1 的 start 处删除 length 字符，然后将 char_expr2 插入到 char_expr1 的 start 处。若要只删除字符而不插入其他字符，char_expr2 应为 NULL，而不是表示单个空格的""。

19. substring (expression, start, length)，返回字符或二进制字符串的一部分。start 指定子字符串的字符起始位置，length 指定子字符串中的字符数。

20. to_unichar (integer_expr)，返回具有整数表达式值的 unichar 表达式。如果整数表达式介于 0xD800 和 0xDFFF 之间，将会引发 sqlstate 例外、输出错误并中止操作。如果整数表达式介于 0 和 0xFFFF 之间，将返回一个 Unicode 值。如果整数表达式介于 0x10000 和 0x10FFFF 之间，将返回一个代理对。

21. upper (char_expr)，将小写转换为大写。返回一个字符值。

22. uhighsurr (uchar_expr, start)，如果 start 处的 Unicode 值是代理对的上半部分（应首先出现在对中），则返回 1，否则返回 0。该函数允许编写用于处理代理的显式代码。应特别指出，如果子字符串以 uhighsurr() 为真的 Unicode 字符开头，则提取至少具有 2 个 Unicode 值的子字符串，因为 substr() 不能只提取具有 1 个 Unicode 值的子字符串，且不提取半个代理对。

23. ulowsurr (uchar_expr, start)，如果 start 处的 Unicode 值是代理对的下半部（应出现在对中的第二部分），则返回 1，否则返回 0。该函数允许在 substr()、stuff() 和 right() 所执行的调整周围显式代码。应特别指出，如果子字符串以 ulowsurr() 为真的 Unicode 值结尾，将提取少 1 个或更多字符的子字符串，这是因为 substr() 不会提取包含不匹配代理对的字符串。

24. uscalar (uchar_expr)，返回表达式中第一个 Unicode 字符的 Unicode 标量值。如果第一个字符不是代理对的高阶半部，则值的范围将在 0 和 0xFFFF 之间。如果第一个字符是代理对的高阶半部，则第二个值必须是低阶半部，并且返回值的范围在 0x10000 和 0x10FFFF 之间。如果在包含不匹配代理对半部的 uchar_expr 上调用该函数，将引发 sqlstate 例外、输出错误并中止操作。

应用示例：Ascii 和 char 函数：

```
1> select ascii('abc')
2> go
-----------
 97
(1 row affected)
1> select ascii('置 ab 置 c')
2> go
-----------
 231
```

```
(1 row affected)
1> select char(97)
2> go

 -
 a
(1 row affected)
```

Charindex 和 patindex：

前者只能用于字符串，不能使用模式匹配，后才则可以。例如：

```
1> select charindex("abc", "1234abc"), patindex("%abc%", "1234ababc"), patindex("abc", "1234abc")
2> go
 ----------- ----------- -----------
      5           7           0
(1 row affected)
1>
```

stuff, right, left：

```
1> select right('abc', 2)
2> go

 ---
 bc
(1 row affected)
1> select left('abc', 2)
2> go

 ---
 ab
(1 row affected)
1> select stuff('abc', 2, 3, 'defg')    //填充替换
2> go

 -------
 adefg
(1 row affected)
1> select 'abc' + space(5) + 'abc'    //中间填充 5 个空格，字符串拼接，直接使用"+"运算符
2> go
 -----------
 abc     abc

(1 row affected)
```

6.6.3 操作 TEXT/IMAGE 的文本函数

操作 TEXT/IMAGE 的文本函数如下：

1. patindex (ì%pattern%î, char_expr [using {bytes |chars | characters}])，返回一个整数值，表示指定字符表达式中首次出现的 pattern 的起始位置；如果未找到 pattern，则返回 0。默认情况下，patindex 返回的偏移以字符表示；若要返回以字节表示的多字节字符串的偏移，

请指定 using bytes。pattern 的前后必须添加通配符 "％"（搜索第一个或最后一个字符的情况除外）。这与操作字符串的 patindex 函数形式基本一样。

2. textptr (text_columname)，返回文本指针值（16 字节的 varbinary 值）。
3. textvalid (ìtable_name..col_nameî, textpointer)，检查给定的 text 指针是否有效。text、unitext 或 image 列的标识符必须包括表名。如果该指针有效，则返回 1，否则返回 0。

以上是三个操作 text/image 字段的内置函数，在用户应用程序中并不常用。

4. datalength 系统函数一样可以用于操作 TEXT/IMAGE 列，用于返回实际的字节长度。
5. 限制或扩展 TEXT/IMAGE 列的长度限制，使用命令：Set textsize <目标大小>，在 ASE 中，默认的 textsize 限制是 32KB，即 32768。要想超出此限制，可以将此值设大，比如设成 2MB：

```
Set textsize 2097152
```

要想恢复默认值，使用以下命令：

```
Set textsize 0
```

当然，我们也可以使用全局变量：@@textcolid、@@textdbid、@@textobjid、@@textptr 和 @@textsize 来处理 text、unitext 和 image 数据。区别在于前者是基于当前会话的，而全局变量的更新是需要相应权限的。

例如：设置 textsize 为 2MB，并从全局变量@@textsize 中读取以验证其大小：

```
1> set textsize 2097152
2> go
1> select @@textsize
2> go

 -----------
 2097152
(1 row affected)
1> set textsize 0
2> go
1> select @@textsize
2> go

 -----------
 32768
(1 row affected)
```

6. readtext/writetext，读取 TEXT/IMAGE 字段的文本内容。
   语法格式：

   ```
   readtext [[database.]owner.]table_name.column_name text_pointer offset size [readpast]
   [using {bytes | chars | characters}]
   writetext [[database.]owner.]table_name.column_name text_pointer [readpast] [with log] data
   ```

   实例：写一个字符串的内容到 TEXT 字段：

```
1> declare @pageptr varbinary(16)
2> select @pageptr=textptr(col2) from testblob where id=1
3> writetext testblob.col2 @pageptr "d:/MyDocument/sqlnet.log"
4> go
(1 row affected)

Return parameters:

txts
------------------
0x00000000000023e9
1> select col2 from testblob
2> go
col2
--------------------------------------
D:/MyDocument/sqlnet.log
```

读取 testblob 的文本列：

```
1> declare @pageptr varbinary(16)
2> select @pageptr=textptr(col2) from testblob where id=1
3> readtext testblob.col2 @pageptr 1 5 using chars
4> go
(1 row affected)
 col2
------------------
 D:/My
```

如果想将一个文件写入 TEXT/IMAGE，或者将 TEXT/IMAGE 字段的内容存到一个文件，则要借助于程序来实现。因为这里不能直接用文件名作参数。除非预告将文件读进一个文本串，然后调用此函数写入。

6.6.4 集合函数

ASE 中基本的集合函数如表 6-6 所示。

表 6-6 ASE 基本集合函数

函数	参数	结果
Avg	(all|distinct) expression	所有值的平均值
Count	(all|distinct) expression	以 int 型返回非空值的数量或行数
Count_big	(all|distinct) expression	以 bigint 型返回非空值的数量或行数
Max	Expression	返回表达式中的最大值
Min	Expression	返回表达式中的最小值
Sum	(all|distinct)expression	返回值的总和

这些函数意义比较明确，同时也有需要注意的地方。有些函数的实现有局限性。
请看下边的示例：

```
1> create table tavg(id int)
2> go
1> insert into tavg values(2147483647)
2> go
(1 row affected)
1> insert into tavg values(2)
2> go
(1 row affected)
1> select avg(id) from tavg
2> go
Msg 3606, Level 16, State 0:
Server 'XIONGHE', Line 1:
算术溢出发生
1>
```

在这个例子里，只是插入了两行值，第一个值是 $2^{31}-1$，第二个值是 2，两者都没有超过最大整数范围，均值怎么会溢出呢？这种错误目前出现在 ASE 数据库当中，而在其他数据库当中都能顺利执行，所以要引起注意。

解决的办法是：在 avg 之前，使用强制转换成域值范围更广的数据类型，例如：

```
1> select avg(convert(bigint, id)) from tavg
2> go
 --------------------
  1073741824
(1 row affected)
```

6.6.5 数学函数

ASE 中的内置数学函数返回经过数学计算后得到的值。

其一般语法形式为：function_name(arguments)

参数形式有以下几种：

1. approx_numeric，近似数值类型（float、real、double）的变量、列名或表达式以及它们的组合。
2. integer，任意整数类型（tinyint、smallint、int、bigint、unsigned smallint、unsigned int 或 unsigned bigint）。
3. numeric，任意精确数值（numeric、dec、decimal、整数类型系列、近似数值系列以及 money 类型）。

各个函数的函义如表 6-7 所示。

表 6-7　数学函数

函数	参数	结果
Abs	Numeric	求绝对值
Acos	Approx_numeric	反余弦值
Asin	Approx_numeric	反正弦值
Atan	Approx_numeric	反正切值
Ceiling	Numeric	大于或等于指定值的最小整数
Cos	Approx_numeric	三角余弦
Cot	Approx_numeric	三角余切
Degree	Numeric	弧度转换为度
Exp	Approx_numeric	指定值的指数值，即 e 的 n 次幂
Floor	Numeric	小于或等于目标值的最大整数值
Log	Approx_numeric	指定值的自然对数
Log10	Approx_numeric	指定值的以 10 为底的对数
Pi	()	常数值：3.1415926535897931
Power	(numeric, power)	numeric 的 power 次幂
Radians	Approx_numeric	将度转为弧度，degree 的逆运算
Rand	([Integer])	用可选整数作为源值，返回 0 到 1 之间的随机数
Round	(numeric, integer)	舍入 numeric 以使其具有 integer 个有效位数。正 integer 确定小数点右边的有效位数；负 integer 确定小数点左边的有效位数
Sign	Numeric	返回数值的符号 1、0 或 -1
Sin	Approx_numeric	三角正弦
Square	Numeric	平方值
Sqrt	Approx_numeric	指定值的平方根
Tan	Approx_numeric	三角正切

6.6.6　时间日期函数

时间日期函数主要涉及到列类型：datetime、smalldatetime、date、time 值的相关运算。当然也包括新的时间日期类型 bigdatetime 和 bigtime（ASE 15.5 开始引入）的相关运算。

ASE 将 datetime 类型的数据在内存存储为两个 4 字节的整数，前 4 个字节为基准日期（1900 年 1 月 1 日）以前或以后的天数。基准日期是系统的参考日期。datetime 类型不能早于 1753 年 1 月 1 日，内部 datetime 表示中的另 4 个字节用于存储一天中的时间，能精确到 1/300 秒。

date 数据以 4 个字节来存储，基准日期范围从 0001 年 1 月 1 日到 9999 年 12 月 31 日。time 数据类型的值范围则是从 12:00:00AM 到 11:59:59:999PM。

smalldatetime 数据类型存储的日期和时间精度低于 datetime，存储为两个双字节整数。前两个字节存储的是 1900 年 1 月 1 日以后的天数，后两个字节存储的是午夜后的分钟数。日期范围从 1900 年 1 月 1 日到 2079 年 6 月 6 日，可以精确到分钟。

bigdatetime 类型，列保存 0001 年 1 月 1 日至 9999 年 12 月 31 日的日期，以及 12:00:00.000000 AM 至 11:59:59.999999 PM 的时间。存储大小为 8 个字节。bigdatetime 的内部表示一个 64 位整数，其中包含自 0000 年 1 月 1 日以来所逝去的微秒数。

bigtime 列保存 12:00:00.000000 AM 至 11:59:59.999999 PM 的时间。存储大小为 8 个字节。bigtime 的内部表示一个 64 位整数，其中包含自午夜以来所逝去的微秒数。

如表 6-8 所示是几个主要的日期函数列表。

表 6-8 日期函数

函数	参数	结果
Current_date	Date	返回当前日期
Current_time	Date	返回当前时间
Day	(date_expression)	返回 datepart 中的日子
Datename	(datepart, date)	datetime、smalldatetime、date 或 time 值中以 ASCII 字符串表示的部分
Datepart	(datepart, date)	datetime、smalldatetime、date 或 time 值中以整数表示的部分
Datediff	(datepart, date,date)	第一个日期与第二个日期的差量，单位以 datepart 计
Dateadd	(datepart, number, date)	向日期值加入日期分量产生新日期值
Getdate	()	返回系统当前时间和日期
Getutcdate	()	返回格林威治标准时间
Month	(date_expression)	返回指定日期的月份
Year	(date_expression)	返回指定日期的年份

有关日期分量，其有效值范围如表 6-9 所示。

表 6-9 日期值各部分分量及其有效范围

日期分量	缩写	值
Year	yy	1753～9999
Quarter	qq	1～4
Month	mm	1～12
Week	wk	1～54
Day	dd	1～31
Dayofyear	dy	1～366
Weekday	dw	1～7（星期日～星期六）
Hour	hh	0～23

续表

日期分量	缩写	值
Minute	mi	0～59
Second	ss	0～59
Millisecond	ms	0～999
calweekofyear	cwk	1～52
Calyearofweek	cyr	1753～9999
Caldayofweek	cdw	1～7

关于 datepart，可以看看下边的实例：

```
1> select getutcdate()
2> go
 ------------------------
 四月 25 2012  2:13PM
(1 row affected)
1> select datename(mm, "1997/07/16")    //返回中间的月份值
2> go
 ------------------------------
 七月

(1 row affected)
1>
1> select dateadd(dd, 55, "2012/02/21")    //2012 年 2 月 21 日再过 55 天
2> go

 -------------------------
 四月 16 2012 12:00AM
(1 row affected)
1>
判断今天是星期几？
1> select getdate(), datepart(cdw, getdate())
2> go

 -------------------------------- -----------
 四月 25 2012 10:23PM         3
(1 row affected)
1> select datediff(dy, '2012/09/08', '2012/02/03')    //两个日期相差的天数
2> go
 -----------
 -218
(1 row affected)
```

由于 datetime 类型的精度为 1/300 秒，这种精度有时候还是会出现重复值，甚至出现让人迷惑的结果。

请看下例：

```
1> create table tdatetime(id datetime)
2> go
1> insert into tdatetime values('2008-10-10 12:05:07.777')
2> go
(1 row affected)
1> insert into tdatetime values('2008-10-10 12:05:07.775')
2> go
(1 row affected)
1> select * from tdatetime where id='2008-10-10 12:05:07.776'
2> go
 id
 ------------------------------
 十月 10 2008 12:05PM
 十月 10 2008 12:05PM
(2 rows affected)
1>
1> select convert(char(19), id, 19) from tdatetime
2> go

 ----------------------
 12:05:07:776PM
 12:05:07:776PM
(2 rows affected)
```

虽然两次插入不同的微秒数，但最终结果得到的都是776微秒的分量。在实际的应用中，有时候确实需要考虑到这种情况。

有关日期格式转换也非常重要，请看下例：

```
1> select convert(char(20), getdate(), 23)
2> go

 ----------------------------
 2012-04-25T22:49:02
(1 row affected)
```

这只是日期格式输出的一个极普通的例子，就是想让日期输出成比较标准的格式，可是即便这样，上述结果还是多了一个"T"字符，还得加一个函数来去掉它：

```
1> select str_replace(convert(char(20), getdate(), 23), 'T', ' ')
2> go
 ------------------------
 2012-04-25 22:50:25
(1 row affected)
```

6.6.7 数据类型转换函数

ASE 提供了 convert、inttohex 和 hextoint 三个类型转函数。

数据类型转换函数 convert 是 ASE 内置函数里使用频率很高的一个，它的语法形式为：

Convert(datatype, expression[, style])

在前一节的示例中，我们就用到了函数 select convert(char(20), getdate(), 23)，即将当前的日期转换为字符串类型。

本来，convert 也可以提供 int 到 hex 和 hex 到 int 类型的转换，但它不是跨平台的，即在各个平台上的执行结果不一样。因此额外提供了 inttohex 和 hextoint 两个转换函数。

请看下例：

```
1> select hextoint("0x00000100")
2> go

 -----------
       256

(1 row affected)
1> select convert(int, 0x00000100)
2> go

 -----------
     65536

(1 row affected)
1> select inttohex(256)
2> go

 --------
 00000100

(1 row affected)
1> select convert(binary(4), 256)
2> go

 ----------
 0x00010000

(1 row affected)
```

我们重点要提的是，关于日期时间和字符串之间转换的那些样式（style 值），如表 6-10 所示。

表 6-10　日期时间的字符串格式样式

两位年份	四位年份	标准	输出
-	0 或 100	默认值	Mon dd yyyy hh:mm AM（或 PM）
1	101	美国	mm/dd/yy
2	102	SQL 标准	yy.mm.dd
3	103	英语/法语	dd/mm/yy

续表

两位年份	四位年份	标准	输出
4	104	德语	dd.mm.yy
5	105		dd-mm-yy
6	106		dd mon yy
7	107		mon dd, yy
8	108		HH:mm:ss 如 07:48:57
-	9 或 109	默认值+毫秒	mon dd yyyy hh:mm:sss AM（或 PM） 如四月 26 2012　7:48:57
10	110	美国	mm-dd-yy
11	111	日本	yy/mm/dd
12	112	ISO	yymmdd
13	113		yy/mm/dd
14	114		mm/yy/dd
15	115		dd/yy/mm（如 26/2012/04）
16	116		mon dd yy HH:mm:ss（如四月 26 2012 07:48:57）
17	117		hh:mm AM(或 PM) 17，如 7:48AM 117，如 2012/04/26 07:48:57
18	118		HH:mm 18，如 7:48AM 118，如 2012/04/26　7:48AM
19	无 119		hh:mm:ss:zzzAM 19，如 7:48:57:536AM
20	无 120		hh:mm:ss:zzz，如 07:48:57:536
21	无 121		yy/mm/dd hh:mm:ss，如 12/04/26 07:48:57
22	无 122		yy/mm/dd hh:mmAM（或 PM） 如 12/04/26　7:48AM
23	无 123		yyyy-mm-ddThh:mm:ss 如 2012-04-26T07:48:57

以上函数样式都可以从下面的 SQL 语句检验得来：

select convert(char(23), getdate(), *style*)：style 值可以取前边两列中的值，小于 100 的或者大于 100 的。

有些不足的是，查遍全表，我们依然无法找到能够生成形如"2012-04-26 07:48:57"这样的标准时间格式。通常如何做呢？那就是依赖于样式值 23，然后用 str_replace 把 T 替换掉。例如：

```
1> select str_replace(convert(char(19), getdate(), 23), 'T', ' ')
2> go
```

```
--------------------
2012-04-26 08:01:51
```

111、108、117、23 这几个样式用得也比较多，因为它们生成的时间和日期格式比较符合我们的习惯。唯独没有生成 yyyy-mm-dd 这样格式的 style 值。

```
1> select convert(char(23), getdate(), 108)
2> go

--------------------
 08:02:55
1> select convert(char(23), getdate(), 111)
2> go

--------------------
 2012/04/26
1> select convert(char(23), getdate(), 117)
2> go

--------------------
 2012/04/26 08:05:19
1> select convert(char(23), getdate(), 23)
2> go

--------------------
 2012-04-26T08:06:17
```

至于日期时间数值的插入，我们可以始终以格式 yyyy-mm-dd HH:mm:ss 的字符串形式进行输入。例如：

```
1> insert into tdatetime values('2013-11-12')
2> go
(1 row affected)
1> insert into tdatetime values('99-11-12')
2> go
(1 row affected)
1> select * from tdatetime
2> go
 id
 --------------------------
    十月  10 2008 12:05PM
    十月  10 2008 12:05PM
    四月  26 2012 7:48AM
    十一月 12 2013 12:00AM
    十一月 12 1999 12:00AM
(5 rows affected)
```

在实际应用当中，还有在 Sybase 数据库中使用 int 型来存储具体时间的，来看一个实例：

```
1> select convert(char(20), dateadd(ss, 1265856544,'1970/01/01 08:00'), 23)
2> go
```

```
---------------------
2010-02-11T10:49:04
(1 row affected)
```

用户实际存储的是整数 1265856544,它使用的是从 1970 年 1 月 1 日 08:00 起的秒数来记数,应该是中国时区来计算的。

6.6.8 随机数据的生成

有了前边介绍的内置函数,我们就可以在这些函数的基础上为数据库表生成随机数据了。随机数据生成对于数据库管理员和开发人员来说,是非常有用的方法。

1. 数值类型值的随机生成。

    用 rand()函数可以随机生成小数,再与某一个固定的数相乘,就可以得到符合要求的数。

    比如,要生成 5 位整数以内的随机数:

    ```
    1> select convert(int, rand()*100000)
    2> go

    -----------
      80388
    ```

    生成 decimal(8,2)的随机数:

    ```
    1> select convert(decimal(8,2), round(rand()*1000000, 2))
    2> go

    -----------
      620749.85
    ```

2. 字符串值的随机生成。

    先看一个单字符的随机生成:

    ```
    1> select char(convert(int, round(rand() * 26, 0)+65) )
    2> select char(convert(int, round(rand() * 26, 0)+65) )
    3> go
     -
     N
    (1 row affected)
     -
     O
    ```

    生成一个长度为 10 的随机字符串(目前没有很好的方法,可以使用下述形式):

    ```
    1> select replicate(char(convert(int, round(rand() * 26, 0)+65) ), 10)
    2> select replicate(char(convert(int, round(rand() * 26, 0)+65) ), 10)
    3> go
    ------------------------
     OOOOOOOOOO
    ```

```
(1 row affected)
 ------------------------
 SSSSSSSSSS
```

3. 日期值的随机生成。

生成 2012 年的随机日期和时间值,先根据一年总共的秒数 31536000,然后用它生成随机数与 2012 年的第一天相加,即得到最终的随机日期:

```
1> select convert(char(20), dateadd(ss, rand()*31536000, '2012-01-01 00:00:00'), 23)
2> select convert(char(20), dateadd(ss, rand()*31536000, '2012-01-01 00:00:00'), 23)
3> select convert(char(20), dateadd(ss, rand()*31536000, '2012-01-01 00:00:00'), 23)
4> go
 --------------------
 2012-11-28T06:21:45
(1 row affected)
 --------------------
 2012-10-15T20:11:21
(1 row affected)
 --------------------
 2012-02-14T13:50:24
```

4. 二进制数据的随机生成。

binary(10)的随机生成,可以利用 256 以内的随机整数与 binary(1)间的转换,然后再调用 replicate 函数来实现。

```
1> select replicate(convert(binary(1), rand()*256), 10)
2> select replicate(convert(binary(1), rand()*256), 10)
3> select replicate(convert(binary(1), rand()*256), 10)
4> go

 ----------------------
 0x82828282828282828282

(1 row affected)

 ----------------------
 0xb0b0b0b0b0b0b0b0b0b0

(1 row affected)

 ----------------------
 0x52525252525252525252

(1 row affected)
```

以上是生成随机数据的基础。

那么为一张数据库表生成随机数据该如何进行呢？假设有一张表，其定义如下：

create table rand_t (id int not null primary key, col2 varchar(32), col3 decimal(8,2), col4 datetime)

1. 只要求快速生成 10000 条左右的记录，不要求随机，该如何进行？

    如果不要求随机，我们只需要插入几条记录，然后根据这几条记录，不断地生成新记录的主键值即可，其他各个字段的值都可以保留原来的值，其过程如下：

    ```
    1> create table rand_t (id int not null primary key, col2 varchar(32), col3 decimal(8,2), col4 datetime)
    2> go
    1> insert into rand_t values(1, 'wang', 134124.23, '2012-06-03 08:50:50')
    2> go
    (1 row affected)
    1> insert into rand_t values(2, 'hao', 244124.23, '2012-03-03 05:50:50')
    2> go
    (1 row affected)
    1> insert into rand_t select id+max(id), col2, col3, col4 from rand_t
    2> go 14
    (16384 rows affected)
    1> select count(*) from rand_t
    2> go

     -----------
          32768
    ```

    这里先插入两条记录，然后每次都以上一次的记录为基础，将 max(id) 值与 id 值相加得到新 id 值，插入一条记录，共执行 14 轮，相当于 2^{15}。总共插入 32768 条记录。

2. 如果要为它生成 10000 条左右的随机记录，该如何生成呢？

    依然利用前边的随机值生成方法，可以直接生成 16384 条记录，其过程如下：

    ```
    1> truncate table rand_t
    2> go
    1> insert into rand_t values(1, 'wang', 134124.23, '2012-06-03 08:50:50')
    2> insert into rand_t values(2, 'hao', 244124.23, '2012-03-03 05:50:50')
    3> go
    1> insert into rand_t select id+max(id), replicate(char(convert(int, round(rand() * 26, 0)+65) ), 32), convert(decimal(8,2), round(rand()*1000000, 2)), dateadd(ss, rand()*31536000, '2012-01-01 00:00:00') from rand_t
    2> go 13
    (8192 rows affected)
    13 xacts:
    1> select count(*) from rand_t
    2> go

     -----------
          16384
    ```

```
(1 row affected)
1> select top 3 * from rand_t order by id desc
2> go
 id          col2                                                              col3
 ----------- ----------------------------------------------------------------- -----------
 col4
 ---------------------------
 16384       WWWWWWWWWWWWWWWWWWWWWWWWWWWWWW                                    363938.87
 五月 26 2012   7:34PM
 16383       VVVVVVVVVVVVVVVVVVVVVVVVVVVVVV                                    909229.11
 一月 30 2012   4:39AM
 16382       BBBBBBBBBBBBBBBBBBBBBBBBBBBBBB                                    790105.39
 四月 21 2012   2:47AM

(3 rows affected)
1>
```

无论是哪种方法，都充分利用了 go <N>这个方法，它可以重复执行 N 轮，以指数形式进行迭代，能很快为表填充数据。

6.7 ASE 中的存储过程

ASE 数据库里，存储过程使用频度比较高，因为很多时候要想提高性能，都需要去创建存储过程。存储过程是一系列 SQL 语句的有序处理集合。

运行存储过程时，Adaptive Server 会准备一个计划，使该过程能非常快地执行。存储过程可以：

- 带参数；
- 调用其他过程；
- 把状态值返回给调用过程或批处理，以指明成功或失败，以及失败的原因；
- 把参数的值返回给调用过程或批处理；
- 在远程 Adaptive Server 上执行编写存储过程的能力大大提高了 SQL 的功能、效率和灵活性。编译后的过程显著提高了 SQL 语句的性能与批处理能力。另外，如果您的服务器和远程服务器都设置为允许远程登录，则可以执行其他 ASE 服务器上的存储过程。可以在本地 ASE 上编写触发器，当本地发生删除、更新或插入事件时，这些触发器执行远程服务器上的过程。

存储过程与普通的 SQL 语句和批 SQL 语句不同，它是预编译的。第一次运行过程时，Adaptive Server 查询处理器对它进行分析，准备最终保存到系统表中的执行计划。随后，过程将按照存储在系统表中的计划执行。由于大多数查询处理工作已被执行，存储过程的执行几乎是瞬时的。

ASE 提供多种存储过程，作为方便用户使用的工具。存储在 sybsystemprocs 数据库中的名称以"sp_"开头的过程通称为系统过程，因为它们插入、更新、删除并报告系统表中的数据。

6.7.1　创建并执行存储过程

创建简单的存储过程的语法是：

create procedure procedure_name as SQL_statement

比较完整的创建语法是：

```
create proc {proc name}
        ( @{param_name} {param_type}[=value],
          @{param_name} {param_type}[=value],
          {...}
        )
    as
    {statement}
```

其中参数列表的存储过程名，最好不要超过 30 个字符，这是因为老版本的 ASE（12.5.x 及以前）只允许最长存储过程名为 30 个字符。15.x 版本以后允许 128 个字符。

举一个最简单的示例，用一个存储过程来获取所有学生列表：

```
1> create procedure liststudent as select * from student
2> go
1> exec liststudent
2> go
```

执行存储过程时，使用命令 exec procedure_name 就行了，这里不带参数。

调用存储过程的完整语法是：

execute server_name.[database_name].[owner].procedure_name param1, param2, ...

execute 可以简写成 exec。上例的完整调用为：

```
1> execute XIONGHE.iihero.spring.liststudent
2> go
```

一个存储过程可以包含多条 SQL 语句。使用 sp_helptext procedure_name 可以显示过程的源文本内容。例如：

```
1> create procedure demo as
2> select 1
3> select getdate()
4> select * from student
5> go
```

```
1> sp_helptext demo
2> go
 # Lines of Text
 ---------------
 1
(1 row affected)
 Text
create procedure demo as
select 1
select getdate()
/* Adaptive Server 已扩展了以下语句中的所有 '*' 元素 */ select student.sno, stud
          ent.sname, student.sgender, student.sbirth, student.sage, student.sdept
           from student
(1 row affected)
```

6.7.2 存储过程的参数

ASE 中存储过程可以带一个或多个参数，每个参数都可以指定默认值，参数的语法形式是 @param *data_type* [=*value*]。参数类型 data_type 必须指定，默认值是可选的。

参数名的作用范围只在它所在的存储过程以内，最长不超过 255 个字符。

下面的存储过程将以用户名为参数，输出当前数据库中该用户所拥有的普通表（不含系统表）：

```
create proc list_tables @dbuser varchar(30)='dbo'
as
      select name from sysobjects where type='U' and uid=user_id(@dbuser)
go
```

观察执行结果，设当前数据库是 iihero，取用户名为 spring，如果不指定此值，则会使用默认值 dbo：

```
execute list_tables 'spring'
execute list_tables spring
exec list_tables   spring
list_tables spring
XIONGHE.iihero.dbo.list_tables spring
```

该存储过程的调用形式在 iihero 中执行结果为：

```
+-------------+
| name        |
+-------------+
| ULCustomer  |
| ULEmployee  |
| ULOrder     |
```

```
| ULProduct     |
| bonus         |
| course        |
| dept          |
| emp           |
| movie         |
| movieexec     |
| moviestar     |
| multitype_t   |
| mydummy       |
| salgrade      |
| st_course     |
| starsin       |
| student       |
| studio        |
| t0            |
| t1            |
| tavg          |
| tdatetime     |
| testblob      |
+---------------+
total count: 23
```

在调用存储过程时,如果以@param=value 的形式来提供,则可以以任意顺序来提供参数值。如果不是这样的显式形式提供,则必须按照定义存储过程时的参数顺序来提供。只要有一个值是以@param=value 的形式提供,则后边所有的参数值都必须这样提供。

当参数带有默认值时,如果用户提供错误的参数名,系统并不会提示错误,会直接使用默认值来调用存储过程。再如上例,如果调用 list_tables @dbusername='spring',实际上传递的还是@dbuser='dbo',因为并没有参数@dbusername。

存储过程还可以用 NULL 作为参数的默认值,此时如果用户没有提供参数,ASE 不会提示错误。

默认参数值还可以使用通配符的形式。如下边的存储过程,@tname 参数的默认值是开头为 sys 的字符串,表示以 sys 开头的表名。

```
create proc list_tables2 @tname varchar(33)='sys%',   @dbuser varchar(30)='dbo'
as
      select name from sysobjects where name like @tname and uid=user_id(@dbuser)
go
```

该存储过程在 iihero 数据库上的执行结果是:

```
+------------------+
| sysalternates    |
| sysattributes    |
```

```
| syscolumns        |
| syscomments       |
| sysconstraints    |
| sysdepends        |
| sysencryptkeys    |
| sysgams           |
| sysindexes        |
| sysjars           |
| syskeys           |
| syslogs           |
| sysobjects        |
| syspartitionkeys  |
| syspartitions     |
| sysprocedures     |
| sysprotects       |
| sysquerymetrics   |
| sysqueryplans     |
| sysreferences     |
| sysroles          |
| syssegments       |
| sysslices         |
| sysstatistics     |
| systabstats       |
| systhresholds     |
| systypes          |
| sysusermessages   |
| sysusers          |
| sysxtypes         |
+-------------------+
```

6.7.3 存储过程选项

with recompile 选项用于 create proc 和 exec proc。当它用于创建阶段，将指示 ASE 不为该过程保存执行计划。每次执行过程时都要创建一个新计划。如果没有 with recompile，则 ASE 存储其创建的执行计划。通常情况下，此执行计划会满足要求。

然而，如果提供给后续执行的数据或参数值发生变化，可能导致 ASE 创建一个与过程初次执行时创建的计划不同的执行计划。这种情况下，ASE 需要一个新的执行计划。当用户认为需要一个新计划时，在 create procedure 语句中可以使用 with recompile。

在 exec 中使用 with recompile，它指示 ASE 编译一个用于后续执行的新计划。每次使用 with recompile 时都生成新的计划。

存储过程内可以嵌套调用存储过程或触发器，最多嵌套级别不能超过 15 层。

存储过程内可以使用临时表，在 ASE 的存储过程里，临时表会经常用到。带"#"号前缀的临时表生命周期仅在存储过程的调用期间使用，脱离调用期间以后，该表将不存在。只有使用 create table tempdb..tablename 创建的临时表才可以超出存储过程的调用期而存在。

存储过程变量值的最大长度是 16KB，它可以是字符，也可以是二进制数据。

创建完存储过程以后，存储过程的源文本会保存到 syscomments 系统表的 text 列当中。因此不要删除该表的内容。

6.7.4　执行存储过程的方式

1. 使用 waitfor 命令延时执行。

例如：

```
Begin
    Waitfor delay "0:05:00"
    Exec myproc
End
```

延时 5 分钟执行存储过程 myproc。

2. 远程执行存储过程。

只要把服务器名作为标识符的一部分，就可以执行远程 Adaptive Server 上的任何过程。例如，执行服务器 Linux2 上的过程 remoteproc：

```
exec linux2.db1.dbo.remoteproc
```

3. 返回状态。

在执行完存储过程以后，它会返回一个状态值，用于表示该存储过程是否成功完成，如果没有，将报告失败原因。当调用过程时，这个值会保存在变量当中，并在以后的 SQL 语句中使用。ASE 定义的有效状态值范围为-99～0 之间。0 表示成功，负值表示失败。用户可以定义该范围外的状态值。

下边是一个简单的存储过程示例，返回字符串到 bigint 类型的转换。

```
1> create proc proc_0 @a varchar(32)='abc' as
2> select convert(bigint, @a)
3> go
1> declare @status int
2> exec @status = proc_0 '12345'
3> go

 -------------------
 12345

(1 row affected)
(return status = 0)
```

这次成功执行，因为输入参数是'12345'
```
1> declare @status int
2> exec @status = proc_0
3> go
Msg 249, Level 16, State 1:
Server 'XIONGHE', Procedure 'proc_0', Line 2:
在 varchar 的值'abc'到一个 bigint 域的显式转换期间出现语法错误

-------------------
(return status = -6)
```

这里因为输入参数是默认值 abc，不能转换为 bigint，所以出现错误，返回状态值-6。

ASE 中保留的那些状态值的基本含义如表 6-11 所示。

表 6-11　ASE 中的状态值

状态值	含义
0	执行成功
-1	缺少对象
-2	数据类型错误
-3	进程处于死锁状态
-4	权限错误
-5	语法错误
-6	杂类用户错误
-7	资源错误，如空间不足
-8	非致命内部问题
-9	达到系统限制
-10	内部严重不一致
-11	内部不一致
-12	表或索引损坏
-13	数据库损坏
-14	硬件错误

当然，我们也可以定义自己的返回值，例如：

```
1> create proc proc_2 (@a int = 0)
2> as
3> if (@a > 0) return 1
4> else return 2
5> go
1> proc_2
2> go
(return status = 2)
1> proc_2 1
```

```
2> go
(return status = 1)
```

> **注意**
> return 语句返回的是状态值，也只能是状态值。而存储过程中 select 返回的是中间的结果集。

如果存储过程涉及到系统管理或安全有关的任务，可能需要调用 proc_role 函数来判断该存储过程是否具有指定的 role（角色），角色名为 sa_role、sso_role、oper_role。

一个简单的示例如下：

```
1> drop proc test_role
2> go
1> create proc test_role
2> as
3> if (proc_role("sa_role") = 0)
4> begin
5>      print "You have not the right role to execute me."
6>      return -1
7> end
8> else
9>      print "You have SA role"
10>     return 0
11> go
1> test_role
2> go
You have not the right role to execute me.
(return status = -1)
```

函数 proc_role 返回 0，表明不具备某角色的权限；返回 1，表示拥有某角色的权限。

6.7.5 以参数形式作为返回值

状态值是一种返回值，表示存储过程的执行状态。而存储过程的参数也可以作为返回值。当创建存储过程语句和 execute 语句都包含带有参数名的 output 选项时，存储过程会将值返回给调用方。调用方可以是 SQL 批处理，也可以是其他存储过程，返回值可以用于批处理或调用过程中的其他语句。output 关键字可以缩写为 out。

output 参数必须是作为变量而不是常量来传递给存储过程。

如下示例表示两数相加的存储过程：

```
1> create proc test_add (@a int, @b int, @c int output)
2> as select @c = @a + @b
3> go
1> declare @result int
```

```
2> exec test_add 1, 2, @result output
3> go
(return status = 0)
Return parameters:
-----------
 3

(1 row affected)
(注意，在执行此类存储过程时，必须显示地加入 exec 指令，否会出现语法错误)
如果我们在第三个参数上使用常量
1> exec test_add 1,2,3
2> go
(return status = 0)
```

结果将只返回状态值。

传递参数时，@parameter = @variable，不能传递常量，要正确地接收返回值，则必须有变量名。参数可以是除了 text/image 以外的任意 ASE 数据类型。

一个存储过程可以有多个参数形式的返回值。执行时的语法形式是：exec proc_name @a=@val_a out, @b=@val_b out …，至于这些参数是不是真正的 output 参数，则必须要跟存储过程中的定义完全一致。

6.7.6 存储过程的限制

创建存储过程在使用上有以下限制：

1. 不能使用 use 语句，不能使用下列 create 语句：

   > Create view, create default, create rule, create trigger, create procedure

2. 可以在过程内创建其他数据库对象，可以引用本过程创建的对象，只要它是在引用之前创建的即可。
3. 在某对象创建并删除之后，不能在存储过程内定义相同的名字进行创建。
4. 对象的创建是在执行期创建的，而不是在编译期创建的，这一点很重要。
5. 被调用存储过程可以访问调用它的存储过程创建的对象。
6. 可以在存储过程中引用临时表，如果在过程中创建带"#"前缀的临时表，则该临时表只用于该过程，退出过程后它将消失。用 create table tempdb..tablename 创建的临时表不会消失，除非已将它们显式删除。
7. 存储过程最多有 255 个参数。
8. 过程中局部变量和全局变量的个数仅受可用内存的限制。

6.7.7 删除、重命名存储过程

删除存储过程的语法比较简单，一次可以删除多个存储过程：

Drop proc [owner].proc_name, [owner].proc_name

后边可以带多个存储过程名。

重命名存储过程的语法为：sp_rename proc_name, newproc_name。

要查看存储过程引用的对象，可以调用系统过程 sp_depends，请看下例：

```
1> create proc proc_depends
2> as select * from t where id>10
3> go
1> exec sp_depends proc_depends
2> go
```

在当前数据库中，对象引用情况如下：

```
 object                   type              updated           selected
 ------------------------ ----------------- ----------------- -----------------
 dbo.t                    用户表            否                否
(return status = 0)
```

表明存储过程 proc_depends 引用了数据库中的表 t。

6.7.8 游标的使用

游标（Cursor）一次访问 SQL 查询结果中的一行或者多行，它的概念类似于面向对象语言中的迭代器，通过游标不断地获取数据，直至游标结束，从而完成整个结果集的获取。

游标始终与 select 语句关联在一起，是描述 select 语句结果集的一个符号，它主要由以下几个部分组成：

（1）游标结果集：执行与游标相关联的查询后返回的限定行的集合。

（2）游标位置：指向游标结果集内某一行的指针。

游标的位置用于指示游标的当前行，对于可更新游标，可以将 update 或 delete 语句与命名游标的子句结合使用来显式修改或删除该行。

定义游标行为有两个重要属性：敏感性（semi_sensitive、insensitive）和可滚动性（scroll、no scroll）。

如果一个游标声明为 insensitive（非敏感），则在打开游标时，只显示结果集原来的数据；如果基础表中的数据这时发生了变化，游标里并不体现最新的变化。如果游标声明为半敏感（semi_sensitive），则打开游标后，基础表中的数据变化可能显示在结果集当中，能否一定看到数据变化是不确定的。默认属性是 semi_sensitive。

对于是否可滚动，如果游标声明为 scroll，意味着游标可以顺序读取，也可以非顺序读取，并且结果集可以被反复遍历；如果声明为 no scroll，则游标不可滚动，且结果集仅向前移动，每次一行。可滚动性的属性默认值为 no scroll。如果可滚动性和是否敏感都未指定，则该游标为半敏感的不可滚动游标。

对于 no scroll 游标，可用的取结果调用只能是 fetch next。而对于可滚动游标，可以使用如下的调用形式：fetch first、fetch last、fetch absolute、fetch next、fetch prior、fetch relative，从而将游标设置到结果集中的任意位置。要取结果集的最后一行：

Fetch last [from] <cursor_name>

取结果集的第 15 行：

Fetch absolute 15 [from] <cursor_name>

所有可滚动的游标都是只读的，所有可更新的游标都是不可滚动的。
定义一个游标的语法格式如下：

declare cursor_name [insensitive | semi_sensitive] [scroll | no scroll] cursor
for select_statement for {read only | update [of column_name_list]}

用 semi_sensitive 可滚动游标代替 insensitive 可滚动游标的主要好处是：结果集的第一行能迅速返回给用户，因为表锁是逐行应用的。如果读取一行并更新它，则该行将通过 fetch 进入工作表，且对基表进行更新。不需要等待结果集工作表被完全填充。

read_only 选项指定游标结果集不能更新，与此相反，for update 选项则指定游标结果集是可更新的。可用在 select_statement 中定义为可更新的列表指定 for update 后的 of column_name_list。一般而言，滚动游标只能是只读游标，不能用作可更新游标。

下面来看一组示例：

（1）可更新游标。

```
create proc first_cursor
as
declare @id int, @col2 varchar(32)
declare mycursor cursor for select id, col2 from t for update
open mycursor
fetch mycursor into @id, @col2
select @id, @col2
update t set col2='ttt' where current of mycursor
close mycursor
deallocate cursor mycursor
go
```

表 t 原来行的 col2 值为 wang，调用此存储过程以后，会被更新为 ttt。而 **where current of mycursor** 则表示当前游标所在的记录行。

（2）滚动游标。

```
create proc second_cursor
as
```

```
declare @sno int, @sname varchar(32), @sbirth datetime
declare mycursor scroll cursor for select sno, sname, sbirth from spring.student for read only
open mycursor
fetch absolute 3 from mycursor into @sno, @sname, @sbirth
select @sno, @sname, @sbirth
close mycursor
deallocate mycursor
```
这是一个可滚动游标，直接取结果集中的第三行
执行结果：
```
1> exec dbo.second_cursor
2> go

 ----------- --------------------------------
 2007003     王敏           二月    3 1990 12:00AM

(1 row affected)
(return status = 0)
```

使用完游标时，要及时关闭游标和释放游标。

关闭游标主要用于关闭游标中的结果集、删除所有剩余的临时表以及被游标占用的服务器资源。但是它为游标查询计划以便能够再次打开游标。

例如 close mycursor，当关闭该游标后重新打开它时，ASE 将重新创建游标结果集，并将游标放到第一个有效行之前，这样可以根据需要多次处理游标结果集。可以随时关闭游标，而不必处理整个结果集。

释放游标的工作主要是从内存中删除查询计划并消除对游标结构的所有跟踪。

例如 deallocate [cursor] mycursor，cursor 关键字在 15.0 或更高版本当中是可选的。在释放游标以后，游标必须重新声明才能使用。

游标在执行过程中，经常要检查游标状态，每次读取游标后，可以通过全局变量@@sqlstatus、@@fetch_status 或者@@cursor_rows 来访问该值。后两者只在 15.0 及以后的版本才支持。

我们来看看@@sqlstatus 的含义：

1. 值为 0，表示 fetch 操作执行成功。
2. 值为 1，表示 fetch 语句产生错误。
3. 值为 2，表示结果集中不再有数据。如果当前游标位于结果集的最末行，而客户端程序对该游标提交了 fetch 语句，就会出现这一警告。

因而在游标的实际使用中，经常用 while @@sqlstatus = 0 作为取记录行的条件。当该值不为 0 时，即结束取记录行。

再看看@@fetch_status 的含义：

1. 值为 0，表明 fetch 操作成功。
2. 值为-1，表明 fetch 操作失败。

3. 值为-2，保留供以后使用。

> **注意**
>
> 只有 fetch 语句能够设@@sqlstatus 和@@fetch_status 的值，其他语句对它们没有影响。@@cursor_rows 用于指示游标结果集中上次打开并读取的行数。

@@cursor_rows 值的含义：

1. 值为-1，表明：①游标是动态的，符合条件的行数是不断变化的，不能明确检索符合条件的行；②游标可能是可滚动的，但滚动工作表尚未填充，符合条件的行数未知。
2. 值为 0，没有打开任何游标，没有符合上次打开的游标条件的行，或者上次打开的游标已经关闭或释放了。
3. 值为 N，上次打开或读取的游标结果集被完全填充，返回的值（n）是结果集中的总行数。

通过语句 set cursor rows <number> for <cursor name>，可以设置每次提取记录集的行数。

下面是一个运用了 set cursor rows 的存储过程实例：

```
create proc third_cursor
as
declare @sno int, @sname varchar(32), @sbirth datetime
declare mycursor cursor for select sno, sname, sbirth from spring.student
open mycursor
set cursor rows 3    for mycursor
fetch from mycursor into @sno, @sname, @sbirth
while @@sqlstatus = 0
begin
    select @sno, @sname, @sbirth
    select @@cursor_rows
    fetch from mycursor into @sno, @sname, @sbirth
end
close mycursor
deallocate mycursor
```

将 set cursor rows 运用到 SQL 批处理，其结果如下：

```
1> declare mycursor cursor for select sname from spring.student
2> go
1> open mycursor
2> go
1> fetch mycursor
2> go
 sname
   --------------
```

李勇

(1 row affected)
1> set cursor rows 2 for mycursor
2> go
1> fetch mycursor
2> go
 sname

 刘晨
 王敏

(2 rows affected)
1> close mycursor
2> go
1> deallocate mycursor
2> go

6.8 ASE 中的触发器

触发器（trigger）在 ASE 中主要用于强制实施参照完整性，是在表中插入、删除或者更新数据时起作用的特殊存储过程。

6.8.1 触发器的工作原理

触发器可自动工作，只要数据发生改变，触发器就会工作，对 update、delete、insert 都会起作用，并且对每个 SQL 语句执行一次。

这里举一个简单的示例，如果想阻止用户删除 spring.emp 表中的数据，可以为 delete 操作创建一个触发器：

```
create trigger tg_emp_del on spring.emp for delete
as
begin
        rollback transaction
        print "Delete any rows on emp is not permitted."
end
go

1> delete from spring.emp where ename='SMITH'
2> go
Delete any rows on emp is not permitted.
```

这样，后续的删除名为 SMITH 的操作将会被回退，并输出一条提示信息，触发器的动作是在修改语句时完成，并且 ASE 对所有数据类型、规则或者完整性约束检查完毕之后才会"触发"。触发器和引发它的语句将被当作单个事务，在触发器里可以回退。如果 ASE 检测到严重错误，则回退整个事务。

ASE 中，触发器在下述情形中特别有用：

1. 级联更新或删除（目前，ASE 的表结构定义不能直接支持级联更新或删除），只能借助于触发器。
2. 实施比参照完整性更复杂的限制，因为它能直接引用到数据库对象，可以实现更复杂的逻辑。
3. 能比较修改前后的状态，根据比较结果采取相应的操作。

6.8.2 创建触发器

创建触发器的基本语法格式如下：

```
Create trigger [owner.]trigger_name on [owner.]table_name for {insert, update, delete}
As SQL_statements
```

如果使用 if update 子句，后边的 SQL 语句形式为：

```
As
    [if update (column_name) [{and|or} update (column_name)] ...] SQL_statements
    [if update (column_name) [{and|or} update (column_name)] ...] SQL_statements
```

这里 on 子句指定了触发器工作时的表名，也称为触发器表，触发器是针对具体的物理表而起作用的。触发器的用户名 owner 必须与触发器表是同一个用户。

由于触发器是作为事务的一部分执行的，因此，在触发器中不允许执行非事务语句，这些语句包括：所有的 create 语句、所有 drop 语句、alter table、alter database、truncate table、grant、revoke、update statistics、reconfigure、load database/transaction、select into 等。

ASE 触发器使用了两种特殊的表：deleted 表和 inserted 表，它们用作触发器测试时使用的临时表。测试表中的数据不能直接更改，但可以使用 Select 语句来检测 insert、update 或 delete 操作的影响。

deleted 表用于存储执行 update 或 delete 操作时受到影响的行的副本，在执行 update 或 delete 期间，相关的行会从触发器表中删除并被传送到 deleted 表，对 update 操作，传递的应该是 update 之间的旧行，deleted 表与触发器表没有公共行。

inserted 表用于存储执行 insert 和 update 操作时受到影响的行的副本，存储的是 insert 到触发器表中的新行以及 update 操作以后的新行。inserted 表与触发器表都有新行的副本。

update 操作是先删除再插入，因此是先删除触发器表中的旧行并把旧行复制到 deleted 表，再把新行复制到触发器表和 inserted 表。

我们再看看几种典型的触发器的应用：

1. 插入、更新触发器

这类触发器多用于对插入或者更新操作以后的检验，如果有效，则允许插入，否则回滚。如对学生成绩表 st_course 的 grade（成绩），限制其值为 0 到 100 之间，不在此范围则为无效值。这个可以通过插入触发器来实现：

```
1> create trigger spring.tg_stcourse on spring.st_course for insert, update
2> as
3> if (select count(*) from inserted where grade<0 or grade>100) > 0
4> begin
5>     rollback transaction
6>     print "grade value should be between 0 and 100"
7> end
8> go
1>
2> update spring.st_course set grade = 110 where sno=2007001
3> go
grade value should be between 0 and 100
```

最后更新一条 grade 新值为 110 的记录被触发器阻止。

2. 删除触发器

ASE 不支持 on commit delete rows 的级联删除语法，所以需要借助于触发器来实现。即当主表里删除一条记录时，从表自动删除对应的记录。

下边的示例中，当从部门表 dept 里删除一行记录（即删除一个部门）时，要求对应该部门的 emp 记录也自动删除。

```
create trigger spring.tg_dept_del on spring.dept for delete
as
begin
    delete from spring.emp where spring.emp.deptno in (select deptno from deleted)
end
go
```

这里也要说明一下，如果表 emp 里提前定义了外键依赖，那么上述触发器将不起作用，因为 ASE 中的触发器都是 after 触发，就是相应的 delete 动作发生以后才触发的，而触发器的动作是发生在所有完整性约束检查完成以后，有了外键依赖，ASE 是不允许先删除父表中的记录的。

6.8.3 ASE 中触发器的限制

ASE 中的触发器有以下限制：

1. 一个表最多有三个触发器，分别对应于更新、插入和删除，在编写触发器时一定要考虑这个限制。
2. 每个触发器只能应用于用户的一个表上。
3. 触发器可以引用视图和临时表，但是不能在视图或临时表上创建触发器。
4. Truncate table 语句可以删除所有行，但是它并不包含事务日志，也不引发触发器。
5. 系统表不允许创建触发器。

对于 if update 语句，如果是显示更新，则自动激活触发器；如果是隐式触发，则不会激活触发器。例如：

```
1> create table tact(id int, col2 varchar(32) null)
2> go

1> create trigger tg_tact on tact for insert
2> as
3> if update(id) and update(col2)
4>     print "tg_tact activated...."
5> go

1> insert into tact values(1, null)
2> go
tg_tact activated....
(1 row affected)

1> insert into tact(id) values(2)
2> go
(1 row affected)
```

6.8.4　触发器的禁用及删除

insert、update 和 delete 命令通常会引发它们所遇到的任何触发器，这会增加执行操作所需的时间。要在执行批量 insert、update 或 delete 操作期间禁用触发器，可以使用 alter table 命令的 disable trigger 选项。使用 disable trigger 选项可禁用与表关联的所有触发器，也可以指定要禁用的特定触发器。

- 禁用一个触发器，使用语句：alter table table_name disable trigger [trigger_name]即可。如果不指定触发器名，则会禁用该表上的所有触发器。
- 重新启用触发器，将 disable 关键字设为 enable 即可。
- 删除一个触发器，使用 drop trigger trigger_name, trigger_name, …..。一次可以删除多个触发器，这种语法与 drop table 和 index 等类似。

6.8.5 获取触发器的相关元信息

查找当前数据库中所有触发器的列表：

```
1> select name from sysobjects where type="TR"
2> go
 name
 ------------------------------------------------------------------------------- tg_emp_del
 tg_stcourse
 tg_dept_del
 tg_tact
(4 rows affected)
```

获取某个指定触发器的元信息，可以使用 sp_help trigger_name 语句，例如：

```
1> sp_help tg_tact
2> go
 Name            Owner            Object_type            Create_date
 ----------------  ----------------  ------------------------  ---------------------------------
 tg_tact          dbo              触发器                 五月  5 2012   6:38PM

(1 row affected)
触发器已被启用
(return status = 0)
```

显示一个触发器的源码实现，调用 sp_helptext trigger_name 命令，例如：

```
1> sp_helptext tg_tact
2> go
 # Lines of Text
 ---------------
  1

(1 row affected)
 text
 --------------------------------------------------
 create trigger tg_tact on tact for insert
as
if update(id) and update(col2)
    print "tg_tact activated...."

(1 row affected)
(return status = 0)
1>
```

同时，sp_depends 可以列出触发器所有引用的对象的列表，下例将显示 tg_tact 触发器引用的所有对象的列表：

```
1> sp_depends tg_tact
2> go
在当前数据库中对象引用情况。
 object              type              updated           selected
 ------------------  ----------------  ----------------  ----------------
 dbo.tact            用户表            否                否
(return status = 0)
```

7

Sybase ASE 的字符集

Sybase 国际化和本地化支持允许我们使用本地语言和格式来存储和检索数据。国际化的应用程序使用外部文件，在运行时提供特定语言的信息。因为不包含特定的语言代码，所以国际化的应用程序可在任何本地语言环境中配置而无须更改代码。单一版本的软件产品可适应不同语言和地区，无需工程方面的更改即可符合当地要求和习惯。通过国际化和本地化支持，Sybase ASE 可以支持多种语言及字符集，展示出数据库的强大处理能力。

在安装 ASE 数据库的过程当中，在如图 7-1 所示的界面中单击 Language 按钮，进入图 7-2，用于配置字符集及系统支持的语言。

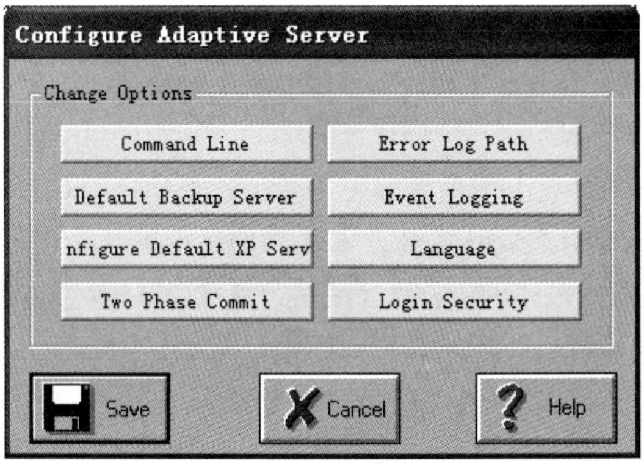

图 7-1 配置服务器

由于不同语言及字符集在设置上具有相当的复杂性，字符集问题一直存在于数据库系统创建和运行的整个生命周期。在 ASE 的应用开发中，字符集问题也是频频出现，成为大家普遍关心的一个问题。本章就字符集问题进行一次系统性的介绍和总结。

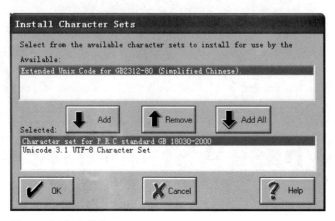

图 7-2　添加字符集

7.1　字符集的基本知识

什么是字符集？字符集是字符（包括字母和数字字符、符号和非打印控制字符）及为其指定的数值（二进制码）的特定集合。平台专用并支持一组语言（如西欧语言）子集的字符集称为本地（或国家）字符集。除了 Unicode UTF-8 之外，ASE 使用的所有字符集都是本地字符集。

字符集最早的编码方案来源于 ASCII（American Standard Code for Information Interchange，美国信息互换标准代码），这也是最常见的编码方式，最开始由美国使用（1967 年发表，最后一次更新发生在 1986 年）。计算机最初的编码方案采用的就是 ASCII 编码方案，一个字节 8 位，它采用了后 7 位（0～127），用于表示英文中一般的字符、数字、字母。在这 128 个字符当中，有 33 个字符无法显示，它们多数是控制字符，主要用于操作和控制已经处理过的文字，另外 95 个是可显示的字符，包括由空格键所产生的空白字符。回想一下，几乎所有编程语言默认都采用 ASCII 编码，即使有的编程语言确实支持更广泛的字符集，大家仍都采用 ASCII 范围内的字符集表示来编写程序，这是因为计算机对 ASCII 字符集的处理速度是最快的，没有任何字符集转换发生，刚好用一个字节（8 个二进制位）可以表示 ASCII 中任意字符。

然而，世界上的文字表示形式非常复杂，英文可以由 ASCII 字符集来表示，是因为所有的英文单词都可以由若干 ASCII 字符排列组合而成。而其他非英文语言则没有这么方便。ASCII 即便进行扩充，进行 8 位编码，也只能代表 256 个字符，这远远不能满足计算机发展的需要，对于亚洲国家复杂的字符存储，则需要更多的码位，这样，各种编码方案应运而生。

为了可以表示全世界各种语言的所有字符和符号，解决不同编码之间的兼容和转换问题，1991 年 1 月，Unicode 协会成立，随后产生了 Unicode 编码。Unicode 的口号是：给每个字符一个唯一的数值编码，不论是什么平台、什么程序和什么语言。

最初 Unicode 编码使用两个字节（16 个二进制位）来进行编码，最多可以容纳 65536 个字符，结果仍然不够使用。后来又进行了扩充，增加了额外的补充字符。详情可以参看 Unicode 官方网站（http://www.unicode.org）。

Unicode 在开始制定的时候，计算机的存储容量已得到极大的发展，存储空间似乎不是问题。于是 ISO 组织直接规定使用 16-bit 来统一所有的字符，即 UCS-2 方案，对于 ASCII 字符，Unicode 保持其编码值不变，只是将其长度由原来的 8 位变为 16 位，前边 8 位都是 0。对于其他语言的字符和符号，则一律重新进行编码，没有考虑到与任何一种非 ASCII 编码兼容，从而导致种种编码方案要想转到 Unicode 编码，需要通过查找表来进行转换。以中文字符编码 GB2312（信息交换用汉字编码字符集，基本集）为例，它由国家标准总局发布，早在 1981 年 5 月 1 日就开始实施，通告于大陆。而 Unicode 在 1991 年才开始陆续制定进制，整整晚了 10 年，两者编码竟然完全不同（Unicode 需要全盘考虑，而不仅仅是一种语言的编码）。后来发现 16-bit 仍不能完全表示全世界所有的字符和符号，又出现了 UCS-4 方案，即用 32-bit（四个字节）来对所有字符和符号进行编码。

由于 Unicode 客观上对 ASCII 采用首字节补 0 进行编码的形式，实际上非常不利于网络传输，因为它浪费了一倍的存储空间，而且传统 C 程序中以 0 作字符串结束符的处理显得非常不可靠；另一方面，对于 UCS-2 和 UCS-4 方案，由于是多字节数值表示，自然存在大端（Big-Endian[①]，高位在前）、小端（Little-Endian，低位在前）问题。为了节省存储空间，同时兼顾以前对字符串处理的通行办法，又出现了 UTF-8、UTF-16 编码方案。

UTF-8 编码是目前最常用的 Unicode 表示法，它可以与 ASCII 编码保持最大程度的兼容，所有的 ASCII 字符编码在 UTF-8 中完全不变。UCS-2 和 UTF-8 之间有固定的转换算法，其转换原理如表 7-1 所示。

表 7-1　UCS-2 与 UTF-8 间的转换

UCS-2 编码(16 进制)	UTF-8 编码(二进制)
0000 - 007F	0xxxxxxx
0080 - 07FF	110xxxxx 10xxxxxx
0800 - FFFF	1110xxxx 10xxxxxx 10xxxxxx

从上表可以看出，UCS-2 编码表示成 UTF-8，一个字符最短可以用一个字节来描述，那就是 ASCII 码，最长需要三个字节来描述。由于计算机对 UTF-8 编码的处理比 UCS-2 编码要方便快捷，因而 UTF-8 编码方案得到国际化支持，使用得更为普遍。

[①] Endian 这个词出自《格列佛游记》。小人国的内战就源于吃鸡蛋时是究竟从大头（Big-Endian）敲开还是从小头（Little-Endian）敲开，由此曾发生过六次叛乱，一个皇帝送了命，另一个丢了王位。

7.2 中文字符集

上一节说到，中国最早的字符集 GB2312（又称 GB2312-80）于 1981 年制定，通告于中国大陆。GB2312 收录简化汉字及符号、字母、日文假名等共 7445 个图形字符，其中汉字占 6763 个。它采用的是双字节编码，是迄今为止最常用的中文字符集。其编码范围是高位 0xa1～0xfe，低位也是 0xa1～0xfe；汉字从 0XB0A1 开始，结束于 0XF7FE，占用的码位是 72×94=6768。其中有 5 个空位是 D7FA-D7FE。

随着时间推移及汉字文化的不断延伸推广，有些原来很少用的字，现在变成了常用字，导致 GB2312 编码的 6763 个汉字不够用。

为解决这些问题并配合 Unicode 的实施,全国信息技术化技术委员会于 1995 年12月1日《汉字内码扩展规范》。GBK 向下与 GB2312 完全兼容，向上支持 ISO-10646 国际标准，在前者向后者过渡过程中起到承上启下的作用。GBK 是 GB2312 的扩展，包含了 20902 个汉字，编码范围是 0x8140～0xfefe,其所有字符都可以映射到 UCS-2 编码。包括 GB2312 中的全部汉字、非汉字符号、BIG5 中的全部汉字、与 ISO-10646 相应的国家标准 GB13000 中的其他 CJK（中日韩）汉字；还包括其他汉字、部首、符号，共计 984 个。

另外，为了支持繁体中文，中国台湾于 1984 年制定了专门用于表示繁体中文的 BIG5 编码，其编码范围为：0xA140～0xF97E、0xA1A1～0xF9FE。在 BIG5 编码中，每个字由两个字节组成，其第一字节编码范围为 0xA1～0xF9，第二字节编码范围为 0x40～0x7E 与 0xA1～0xFE，总计收入 13868 个字（包括 5401 个常用字、7652 个次常用字、7 个扩充字及 808 个各式符号）。

GB18030 编码（GB18030-2000）又称 GBK2K，是在 GBK 编码的基础上进一步扩展了汉字，增加了若干少数民族的字形，GBK2K 从根本上解决了字位不够、字形不足的问题。它有几个特点：它并没有确定所有的字形，只是规定了编码范围，留待以后扩充。编码是变长的，其二字节部分与 GBK 兼容；四字节部分是扩充的字形、字位，其编码范围是首字节 0x81-0xFE、二字节 0x30-0x39、三字节 0x81-0xFE、四字节 0x30-0x39。同时，它又是强制的国家标准。

如果我们想查一个简体汉字的 GB2312 编码值和对应的 UCS-2 Big Endian 编码值，可以运行 charmap 命令，进入如图 7-3 所示的界面，在字符集里选择 "Windows：中文简体"，这里我们查询 "何" 字的编码值，它的 Unicode 编码（UCS-2 Big Endian）值为 0x4F55，它的 GB2312 码值为 0xBACE。

在随书的代码（charset\GB2312UnicodeSimple.java）里，还给出了一个 GBK 编码到 Unicode 编码的对应表的 Java 生成程序，利用此程序生成的对应关系表，再依据上一节介绍的 Unicode 与 UTF-8 相互转换的算法，很容易实现 GB2312 到 UTF-8 之间的相互转换。关于 Unicode 与 UTF-8 互转的算法，在 charset\unicode2utf8.cpp 里有一个参考实现，稍加改动即可直接使用。其实现的伪码如下：

Sybase ASE 的字符集　第 7 章

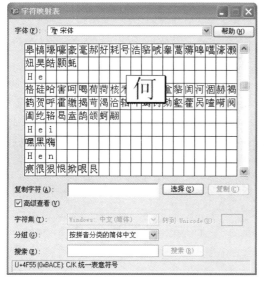

图 7-3　查询汉字对应的 GBK 编码和 Unicode 编码

```
#001  void Unic::Unicode16ToUTF8v2(const utf16string& src, tstring& utf8res)
#002  {
#003      int len = src.size();
#004      utf8res.resize(len*4);
#005      memset(&utf8res[0], 0, len*4);
#006      int pos = 0;
#007      for (int i=0; i<len; ++i)
#008      {
#009          utf16 a = src[i];
#010          if (a<(utf32)0x80)
#011          {
#012              utf8res[pos] = (tbyte)a;
#013              ++pos;
#014          }
#015          else if (a<(utf32)0x800)    // 0000 0111 1111 1111 (max)
#016          {
#017              utf8res[pos] = 0xD0 | (a>>6);
#018              utf8res[pos+1] = 0x80 | (a & 0x003F);
#019              pos+=2;
#020          }
#021          else                         // 0000 1000 0000 0000 (min)
#022          {
#023              utf8res[pos] = 0xE0 | (a>>12);
#024              utf8res[pos+1] = 0x80 | ((a&0x0FFF)>> 6);
#025              utf8res[pos+2] = 0x80 | (a & 0x003F);
#026              pos+=3;
#027          }
```

```
#028        }
#029    }
#030
#031    void Unic::UTF8ToUnicode16v2(const std::string& src, utf16string& utf16res)
#032    {
#033        utf16res.clear();
#034        int len = src.size();
#035        int pos = 0;
#036        utf16 t = 0;
#037        for (int i=0; i<len, pos<len; ++i)
#038        {
#039            //高位为 0 的单字节字符
#040            if ((tbyte)src[i]<0x80)
#041            {
#042                t = src[i];
#043                ++pos;
#044                utf16res.push_back(t);
#045            }
#046            else if ( (((tbyte)src[i]>>5) == 6 )    // "110", 2 bytes
#047            {
#048                if (pos+2 > len) throw DBException(__FILE__, __LINE__, "Not valid utf8 string");
#049                tbyte b1 = (src[i] & 0x1F) >> 2;
#050                tbyte b2 = (src[i+1] & 0x3F) | (src[i] << 6) ;
#051                t = (b1 << 8) | b2;
#052                pos+=2;
#053                utf16res.push_back(t);
#054            }
#055            else if ( (((tbyte)src[i]>>4) == 14 )    // "1110", 3 bytes
#056            {
#057                if (pos+3 > len) throw DBException(__FILE__, __LINE__, "Not valid utf8 string");
#058                tbyte b1 = (src[i]<<4) | ((src[i+1]>>2) & 0x0F);
#059                tbyte b2 = (src[i+1]<<6) | (src[i+2] & 0x3F);
#060                t = (b1 << 8) | b2;
#061                pos += 3;
#062                utf16res.push_back(t);
#063            }
#064
#065        }
#066    }
```

从 GB2312 编码的字符串转到 UTF-8，可以先依次查找 GB2312-Unicode 对照表（可以二分查找），找到对应的 Unicode Big Endian 值数组，然后调用 Unicode16ToUTF8v2 函数，即可得到对应的 UTF-8 字符串。

关于 GB2312 到 Unicode 的对照表，一种比较高效的方式是对汉字的 GB2312 码值进行无冲突 hash 处理，设码值为 a[0]a[1]，则 hash(a) = (a[0]-176)*94 + (a[1]-161)，它的值域刚好在 1 到 7000 以内，并且没有两个汉字的 GB2312 码值的 hash 值是相同的。这样，使用一个长度为 7000 的一维数组，即可存储对应的 Unicode 值。给定一个简体汉字的 GB2312 码值，先求出它的 hash 值，然后直接从数组里可以得到对应的 Unicode 值。从效率上看，它比二分查找要快得多。

在 Sybase ASE 中，表示中文的字符集有以下几种：

1. eucgb：它的全称是 Extended UNIX Code for GB2312-80（Simplified Chinese），完全是基于 GB2312-80 编码规范的。支持汉字 6763 个，编码范围：第一字节 0xA1～0xFE（实际只用到 0xF7），第二字节 0xA1～0xFE。
2. cp936：该字符集是基于 GBK 编码规范（实际上的国家标准是 GB13000-90），是对 GB2312 进行的扩展，第一字节为 0x81～0xFE；第二字节分两部分，一是 0x40～0x7E，二是 0x80～0xFE。其中和 GB2312 相同的区域表示的字符完全相同。
3. gb18030：字符集（国家标准号是 GB18030-2000）是 2000 年 3 月 17 日发布的新的中文编码标准。它是 GB2312 的扩充，采用单/双/四字节编码体系结构，收录了 27000 多个汉字以及藏文、蒙文、维吾尔文等主要的少数民族文字。Sybase 从 ASE 12.5.0.3 之后开始支持 GB18030 字符集。

当然，还有更为通用的字符集表示方式，那就是使用 UTF-8 字符集，Sybase ASE 里将其略称为 UTF8 字符集编码。因为是一种 Unicode 字符集方案，它可以用来存储所有类型的字符，而不仅仅是中文。

一般来说，由于 eucgb 不支持国标一、二级字库以外的汉字，所以我们推荐用户在服务器端和客户端都使用 CP936 字符集，或者在 ASE 12.5.0.3 之后还可以使用 GB18030 字符集，它可以支持一些比较生僻的汉字。CP936 字符集的不足是只有一种排序方式，即区分大小写的二进制排序方式。所以，如果需要使用支持中文字符集且不区分大小写的数据库，就只能使用 UTF-8 作为服务器端字符集，而客户端使用 CP936 或 GB18030 字符集。

7.3　Sybase ASE 中的字符集文件

在 Sybase ASE 中，有两个目录下边的内容与字符集密切相关，它们分别是%SYBASE%/charsets 和%SYBASE%/locales。

charsets 子目录用于存储服务器端各种字符集的相关信息，其目录结构形式是：$SYBASE/charsets/<character set type>。如 ASE 支持 eucgb（GB2312）字符集，charsets 下边会出现 eucgb 子目录。每个字符集下边会对应有以下文件：

1. *.srt：该文件主要用来描述字符集的排序方式，比较该字符集字符串时，采用何种排序方式，基本上每种字符集都有一个 binary.srt，以支持二进制排序方式，有的字符集还支持其他多种排序方式，如上一节所述，EUCGB、CP936 和 GB18030 只支持二进制排序方式，而 UTF-8 字符集则有两种排序文件的支持（binary.srt 和 nocase.srt）。
2. charset.loc：描述字符集的主文件。

3. **<charset>.cfg**：该文件用于描述本字符集到 UTF-8 字符集的转换方式，内容通常只有一行：

   [conversion]
   convertto = utf8, table, MATCH, 3F

4. **utf8.ctb**：这个文件非常关键，里边存储了本字符集到 UTF-8 的转换对照表，ASE 通过对照表，内部实现了本字符集与 UTF-8 字符集之间的相互转换。

locales 目录下边的文件主要用来定义客户端连接时所用的语言及支持的字符集表示。在 locales 目录下边有一个文件 locales.dat，里边记录着 Sybase ASE 当前版本在各个平台下所能支持的语言和字符集信息，其支持的平台如表 7-2 所示。

表 7-2　locales 中的平台名称

[aix]	IBM AIX 操作系统
[axposf]	Alpha 操作系统
[hpia]	HP Italium（安腾）操作系统（基于 Intel Italium）
[hp ux]	HP UNIX 操作系统
[linux]	Linux
[linuxamd64]	Linux for AMD64 位操作系统
[linux_s390]	Linux for Z Series S/390
[sun_svr4]	Sun Solaris for Sparc 操作系统
[sunx86]	Sun Solaris for x86 操作系统
[NT]	Windows NT/2000/XP/2003 系列
[macosx]	Mac OS

locales.dat 中，各个平台下边会逐行记录所能支持的语言及字符集，其格式如下：

locale = vendor_locale, syb_language, syb_charset[, sort_order]

其 locale 标识符主要供 Sybase 内部解析文件使用，按照 posix 规范，vendor_locale 的格式是：语言[_国家][.代码集]。它的值在外部通常通过环境变量 LANG 可以指定，Sybase ASE 根据具体平台类型及 LANG 的值，很容易找到对应的字符集。

（1）syb_language：用于指定 Sybase 认可的国家语言，外部环境变量 SYBLANG 也用于描述此值。

（2）syb_charset：指定 Sybase 认可的字符集名称，用以确定字符集的转换方法和确定本地化文件的具体位置。

另外，Sybase ASE 在 locales 目录下边，按照语言不同，将本地化的消息文件按照各种语言所支持的字符集的不同，置于不同的字符集子目录当中。以中文为例，其目录结构如图 7-4 所示。

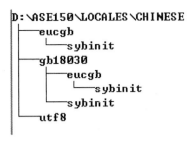

图 7-4　字符集相关目录结构

在 CHINESE 子目录下边，分别有 EUCGB、GB18030 以及 UTF-8 子目录，在这些子目录下，存储着所有的本地化消息文件。

7.4　Sybase ASE 的字符集设置

首先，如前边所述，所有操作系统下有一个重要的环境变量 LANG，它指明系统的"语言[_国家][.代码集]"（符合 posix 标准的 locale 值）。Sybase 通过这个变量并查找 locales.dat 文件，从而决定 Sybase ASE 客户端最终所采用的字符集。

在 UNIX 下运行命令 locale，会得到具体的 locale 值，如以 Linux 为例：

```
LANG=en_US
LC_COLLATE="en_US"
LC_CTYPE="en_US"
LC_MONETARY="en_US"
LC_NUMERIC="en_US"
LC_TIME="en_US"
LC_MESSAGES="en_US"
LC_ALL=
```

LANG 对应的设置值为 en_US，再查找 locales.dat 文件，发现[linux]配置项下边有一行：

locale = en_US, us_english, iso_1

由此可以断定，Linux 下的系统客户端将使用 ISO_1 字符集与服务器端交互。

要查看 Sybase ASE 服务器端采用的字符集，执行如下 SQL 命令：

```
1> sp_helpsort
2> go
```

Collation Name	Collation ID
altdict	45
altnoacc	39
altnocsp	46
binary	25
cyrnocs	64

defaultml	20
dict	51
elldict	65
espdict	55
espnoac	57
espnocs	56
hundict	69
hunnoac	70
hunnocs	71
iso14651	22
noaccent	54
nocase	52
nocasep	53
rusnocs	59
scandict	47
scannocp	48
thaidict	21
turknoac	73
turknocs	74
utf8bin	24

Loadable Sort Table Name	Collation ID
cp932bin	129
cyrdict	140
dynix	130
eucjisbn	192
euckscbn	161
gb2312bn	137
gbpinyin	163
rusdict	165
sjisbn	179
turdict	155
big5bin	194

Sort Order Description

Character Set = 171, cp936
 CP936 (Simplified Chinese).
 Class 2 Character Set
Sort Order = 50, bin_cp936
 Binary sort order for simplified Chinese using **cp936**.
(return status = 0)

这里可以看出，服务器端采用的是 CP936 字符集。

查看客户端采用的字符集，执行如下 SQL 命令：

```
1> select @@client_csname
2> go
```

```
------------------
 cp850

(1 row affected)
1>
```

当服务器端与客户端字符集不一致时，便会产生"Error converting client characters into server's character set."的错误，也就是我们通常所说的乱码现象。如在这里，我们查询 iihero 数据库中的 spring.student 表，因为 sname 字段值全是中文，因而最终会出现乱码，结果如下：

```
1> select sname from student
2> go
 sname
 ------------------
 ??
 ??
 ??
 ???
WARNING!   Some character(s) could not be converted into client's character set.

Unconverted bytes were changed to question marks ('?').

(4 rows affected)
```

要将客户端的字符集改成与服务器端的一致，接上边的例子，只需将客户端调成与 CP936 一致即可（可以是 EUCGB、CP936、GB18030 或者 UTF-8）。

如果是 Windows 平台，我们找到 locales.dat 文件当中的[NT]项，内容如下：

```
[NT]
    locale = enu, us_english, iso_1
    locale = fra, french, iso_1
    locale = deu, german, iso_1
    locale = rus, russian, cp1251
    locale = hun, us_english, cp1250
    locale = ell, us_english, cp1253
    locale = heb, us_english, cp1255
    locale = ara, us_english, cp1256
    locale = trk, us_english, cp1254
    locale = esp, spanish, iso_1
    locale = jpn, japanese, sjis
    locale = japanese, japanese, sjis
    locale = chs, chinese, cp936
    locale = cht, tchinese, big5
    locale = kor, korean, cp949
    locale = us_english.utf8, us_english, utf8
    locale = default, chinese, eucgb
```

我们发现有两行符合要求，它们是：

```
locale = chs, chinese, cp936
locale = default, chinese, eucgb
```

首先，要看 ASE 服务器端环境变量 $LANG（或者 NT 下的%LANG%）值，假设这里它的值为 zh_CN，语言集为 chinese，那么我们在 locales.dat 中的[nt]下，添加一行"locale = zh_CN, Chinese, cp936"即可。

否则，只需要设置环境变量 LANG 为 chs 或者 default 即可。但有一点需要提及的是，如果值为 default，那么最终采用的字符集会与客户端操作系统里正使用的字符集相同，除非我们同时显示地设置了 SYBLANG 的值。比如，当前操作系统是英文操作系统，代码页是 CP850，要想使用 EUCGB，同时需要设置 SYBLANG 值为 chinese。如果只是在当前会话里改变字符集，可以运行 chcp [代码页]，例如恢复代码 CP936，只需要运行 chcp 936 即可。

要想 Sybase ASE 客户端不出现乱码，唯一有效的办法是确保客户端操作系统、客户端程序（locales.dat 中代表的语言及字符集设置）及 ASE 服务器三者字符集保持兼容。

以上说的都是如何调整客户端的字符集，那么，该当如何设置服务器端的字符集呢？这里具体要分两种情况：①服务器刚装完，没有任何用户数据库，或者目前用户数据库中的数据对实际的应用影响微乎其微；②服务器已经装了有一段时间，用户数据库里也有"关键"的数据，不能忽略其实际影响。下面将分别介绍这两种情况下，如何去调整设置服务器端的字符集。

作为对中文字符集 CP936 或 GBK、EUCGB 一种附加的验证信息，我们可以用"中国"一词对应的编码值作为验证，它的 CP936 编码值为 0xd6d0b9fa，可以用 charmap 中的码表查到（如图 7-5 所示）。

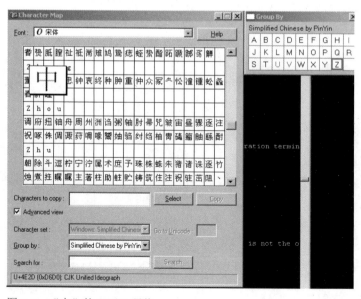

图 7-5 "中"的 CP936 码值

在一个以 CP936 为默认字符集的 ASE 服务器里，我们执行以下 SQL：

```
1> select convert(binary(4), '中国')
2> go

 -----------
 0xd6d0b9fa
(1 row affected)
```

可以看到，结果与码表里查到的编码是一致的。但是如果 Server 端的字符集是 ISO_1，那么得到的就不是这个编码值。

```
D:\SybaseASE\OCS-15_0\bin>isql -Usa -SSEANLINUX -zus_english -Jcp936
Password:
1> use tempdb
2> go
1> select convert(binary(4), '中国')
2> go

 -----------
 0x2d4efd56
```

7.4.1 直接设置字符集

没有用户数据库的情况下，我们只需要安装目标字符集，并对服务器进行设置即可。基本步骤如下（以目标字符集为 CP936 为例）：

Step 1 检查 $SYBASE/charsets 目录下是否有目标字符集目录 CP936，如果没有，则直接从安装盘中 CP936 相关文件拷贝到此目录下。

然后运行：

```
$SYBASE/bin/charset -Usa -Ppassword -Sserver_name binary.srt cp936
```

Step 2 以 sa 用户进入 isql，查询 CP936 对应的字符集 ID 值。

```
1> select name, id from syscharsets where name like '%936%'
2> go
 name                        id
 -------------------------- -------------
 cp936                       171
 bin_cp936                   50
(2 rows affected)
```

从而得知 CP936 对应的字符集 ID 值为 171，其实，如果仔细研究 $SYBASE/charset/cp936 下边的文件 charset.loc，会发现有以下描述信息：

```
[charset]
    class = 0x02
```

```
        id = 0xab
        status = 0x0001
        name = cp936
        description = "CP936 (Simplified Chinese)."
```

其中，id = 0xab，刚好为 171。

Step 3 设置字符集。

```
sp_configure "default character set id",171
go
```

Step 4 重启 Sybase ASE 服务器，一般都要启动两次。第一次，服务器进行一次完整的字符集转换工作，然后自动 shutdown；第二次，再启动，恢复正常。

Step 5 使用 sp_helpsort 验证新字符集是否设置成功。

如本例，将服务器字符集设为 CP936 以后，会看到 sp_helpsort 结果的最后一部分为：

```
排序顺序描述
 ---------------------------------------------------------------
 字符集  = 171, cp936
         CP936 (Simplified Chinese).
         Class 2 字符集
 排序顺序 = 50, bin_cp936
         Binary sort order for simplified Chinese using cp936.
(return status = 0)
```

Step 6 可能遇到的问题是，运行 isql 时，系统提示 cache 过小，转换字符失败，可以增大系统参数 size of unilib cache 的值。做法如下（这里将 unilib cache 调成 1MB）：

```
sp_configure "size of unilib cache", 1024000
go
```

当然，也有图形化的方法来设置字符集。

Windows 下，直接运行$SYBASE\ASE-15_0\bin\syconfig.exe，即可进入如图 7-1 所示的界面，选择语言，然后直接选择语言集、字符集、排序顺序，便很容易地就完成了配置工作。配置完需要重启服务器。

在 UNIX/Linux 下，$SYBASE\ASE-15_0\bin\sqlloc 命令非常实用，它会打开相应的图形界面，也是选择语言集、字符集、排序顺序，最终完成字符集配置工作。配置完需要重启服务器。

7.4.2 有重要用户数据的情况下如何调整

如果 Sybase ASE 服务器中有重要的用户数据，这时要调整字符集，必须要将这些重要的用户数据先行按照原有的服务器端字符集导出,并清空表里的数据。

```
F:\>bcp iihero..student out f:\student.out -Sseanlaptop -Uspring -Pspring1 -Jcp936 -c

启动拷贝...

已拷贝 4 行.
时钟时间(ms.): 共计= 16 平均 = 4 (250.00 行每秒.)
```

然后,按照上一节介绍的方法设置新的字符集,并重启服务器进行字符集验证。最后,重新导入表中的数据。

在这种情形下,只要字符集之间不能正常转换,则很容易出问题。

所以,字符集问题是在数据库建库之前,就应该考虑清楚的问题。否则,越往后调整,付出的代价则越高。

7.5 乱码的产生

第 3 节提到,当客户端操作系统、客户端应用程序(用到的 locales)及 ASE 服务器中的字符集不兼容时,则会产生乱码现象。

UTF-8 字符集可以与任意字符集兼容,因此,当服务器端字符集是 UTF-8 时,则可以支持任意语言的字符。

如果服务器端字符集不兼容于客户端操作系统字符集,这时如果不修改 locales.dat,则会出现乱码。

例如,ASE 服务器字符集为 CP850,而客户端操作系统字符集为 CP936(中文 GBK),当不指定应用程序字符集时:

```
1> select trim(col2) from t
2> go
col2
----------------------
 ????

1> select @@client_csname
2> go

 ------------------------------------
 cp936

(1 row affected)
```

我们会发现,系统客户端用的是 CP936 字符集。

当我们显示指定客户端应用的字符集时,则不会出现乱码。isql 命令行后带"-J<字符集>"名称，则可以指定 isql 要用到的字符集。如果不指定，则会读取 locales.dat 中相应的字符集作为实际要用的参数。

```
C:\Documents and Settings\hex>isql -Uspring -Pspring1 -Diihero -Jcp850
1> select col2 from t
2> go
 col2
 ------------------
 中国
1> select @@client_csname
2> go

 ------------------
 cp850

(1 row affected)
```

这里，我们显示地指定了客户端的字符集为 CP850，从而与服务器端完全一致，避免了乱码的产生。

还有一种不常见的问题，就是关于 EUCGB 仅支持 GB2312-80 字符集的简体汉字，如第 2 节所述，共有 6763 个汉字，对于一些不太常用的汉字存储来说，则明显不够。而 CP936 和 GB18030 字符集则能提供更好的支持，并且与 EUCGB 完全兼容。为说明问题，我们看看下边的实验示例："岠"字属于 CP936 字符集，而不属于 EUCGB。

首先，将数据库服务器的字符集调整为 CP936。建一个示例表 t2，并将"岠"字插入到表 t2 中。

```
F:\>isql -Uspring -Pspring1 -Diihero
1> select @@client_csname
2> go

 ------------------
 cp936

(1 row affected)
1> create table t2(col varchar(4))
2> go
1> insert into t2 values('岠')
2> go
(1 row affected)
1> select * from t2
2> go
 col
 ----
 岠

(1 row affected)
```

启动另一个窗口，以 EUCGB 字符集来连接服务器。

```
C:\Documents and Settings\hex>isql -Uspring -Pspring1 -Diihero -Jeucgb
1> select @@client_csname
2> go

 ----------------
 eucgb

(1 row affected)
1> select * from t2
2> go
 col
 ----
 ??
警告！一些字符不能被转换到客户端的字符集。不可转换的字符被转换为问号标记（'?'）

(1 row affected)
1>
```

我们会看到，"岖"并不能从 CP936 字符集转换到 EUCGB 字符集，因而出现"?"标记。

8

ASE 中的空间管理

8.1 安装完 ASE 后的物理空间调整

在安装完 ASE 之后，系统创建新数据库时默认都用 master 设备文件。这对于实际的生产环境是不可接受的。另外，已经安装的 tempdb 数据库容量非常小，通常都是 4MB，这同样也是不可接受的。

因此，安装完 ASE 之后，实际上并不算真正的结束，还有一些工作要做。本来这些工作应该是 ASE 的安装程序可以默认做完的，这里却要交给安装用户自己去完成。

Step 1 将 tempdb 从 master 设备中剥离。

基本思想就是，把 system、default、log 段全部设置到新的设备文件上。

临时库的基本信息如下：

```
1> sp_helpdb tempdb
2> go
Warning: Row size (2619 bytes) could exceed row size limit, which is 1962
bytes.
 name              db_size           owner           dbid            created
 status
 ----------------- ----------------- --------------- --------------- --------------------
 -----------------------------------------------------------------------------------------
 tempdb            4.0 MB            sa              2               Apr 14, 2012
 select into/bulkcopy/pllsort, trunc log on chkpt, mixed log and data
 (1 row affected)
 device_fragments          size            usage           created                         free kbytes
 ------------------------- --------------- --------------- ------------------------------- ---------------
 master                    4.0 MB          data and log    Apr  8 2009  1:02PM             2200
```

(return status = 0)

这就是默认安装时创建的临时库的基本信息。4MB 大小的临时库对于稍大一些的数据库事务，就连中间数据都无法存放。

先创建一个设备文件给 tempdb 专用：

1> disk init name="tempdevice0", physname="d:/SybaseASE/tempdevice0.dat", size="4M"
2> go

这里创建了一个新的设备文件 4MB 大小（注意：带引号，设备文件名路径在 Windows 下使用"/"或"\"分隔符均可）。

接着将 tempdb 扩充到这个新的设备上，使得 default、system、logsegment 段有切换的空间。否则无法删除已有的这几个段。

1> use master
2> go
1> alter database tempdb on tempdevice0="4MB"
2> go
Extending database by 2048 pages (4.0 megabytes) on disk tempdevice0

移除过程如下：

1> use tempdb
2> go
1> sp_dropsegment "default", tempdb, master
2> go
DBCC execution completed. If DBCC printed error messages, contact a user with System Administrator (SA) role.
Segment reference to device dropped.
(return status = 0)
1> sp_dropsegment "system", tempdb, master
2> go
DBCC execution completed. If DBCC printed error messages, contact a user with System Administrator (SA) role.
Segment reference to device dropped.
(return status = 0)
1> sp_dropsegment "logsegment", tempdb, master
2> go
DBCC execution completed. If DBCC printed error messages, contact a user with System Administrator (SA) role.
DBCC execution completed. If DBCC printed error messages, contact a user with System Administrator (SA) role.
The last-chance threshold for database tempdb is now 16 pages.
Segment reference to device dropped.
WARNING: There are no longer any segments referencing device 'master'. This device will no longer be used for space allocation.
(return status = 0)

这里已经提示这三个段全部从 master 设备里删除了，以后只能在设备 tempdevice0 上使用这三个段：

```
1> sp_helpsegment "default"
2> go
 segment         name            status      device              size           free_pages
 --------------- --------------- ----------- ------------------- -------------- --------------
   1             default           1         tempdevice0         4.0MB           1993
1> sp_helpdb tempdb
2> go
 name       db_size     owner       dbid        created
 status
 ---------- ----------- ----------- ----------- -----------------------
 ---------------------------------------------------------------------
 tempdb     8.0 MB      sa          2           Apr 14, 2012
 select into/bulkcopy/pllsort, trunc log on chkpt, mixed log and data

(1 row affected)
 device_fragments       size                   usage
 created                        free kbytes
 ---------------------- ---------------------- -------------------
 -------------------------------- -------------------
 master                 4.0 MB                 data only
 Apr  8 2009  1:02PM            2360
 tempdevice0            4.0 MB                 data and log
 Apr 14 2012  3:35PM            3956
 device
         segment
 ---------------------------------
 master
              -- unused by any segments --
 tempdevice0
              default
 tempdevice0
              logsegment
 tempdevice0
              system
(return status = 0)
```

这里提示到 master 未被任何段所使用，当然这里为了演示的目的，将新的设备文件仍然设为 4MB，我们看看会引发什么样的问题？

示例如下：

```
1> create table t (id int, name varchar(32))
2> insert into t values(1, 'sql9')
```

```
3> insert into t values(2, 'sql6')
4> go
(1 row affected)
(1 row affected)
//接着会插入 2^(18+1) 行数据到表 t 当中
1> insert into t select id + (select max(id) from t), name from t
2> go 18
(262144 rows affected)
18 xacts:
1> select count(*) from t
2> go

 -------------
  524288

(1 row affected)
```

使用临时表将这 524288 行数据插入，以便排序分页：

```
1> select syb=identity(11), * into ##tempT from t
2> select * from ##tempT where syb>=10 and syb<20
3> go
Space available in the log segment has fallen critically low in database
'tempdb'.   All future modifications to this database will be suspended until the
log is successfully dumped and space becomes available.
The transaction log in database tempdb is almost full.    Your transaction is
being suspended until space is made available in the log.
```

这时提示 tempdb 的事务日志差不多已经满了，我们必须为 tempdb 的设备文件扩容，再扩展 tempdb 的大小才能正常使用。

使用另外一个会话，用 sa 用户登录：

```
1> use master
2> go
1> disk resize name="tempdevice0", size="50M"
2> go
```

先扩容 50MB，加上原来的 4MB，设备文件变为 54MB 大小。

```
1> alter database tempdb on tempdevice0 = "54MB"
2> go
Extending database by 25600 pages (50.0 megabytes) on disk tempdevice0
```

将 tempdb 的大小从 4MB 变为 54MB。

这时我们再看看前边执行的 SQL 语句的情况：

```
being suspended until space is made available in the log.
(524288 rows affected)
```

```
syb             id              name
--------------  ------------    ----------------
10              10              sql6
11              11              sql9
12              12              sql6
13              13              sql9
14              14              sql6
15              15              sql9
16              16              sql6
17              17              sql9
18              18              sql6
19              19              sql9

(10 rows affected)
1> select count(*) from ##tempT
2> go

 -------------
      524288

(1 row affected)
```

这样就可以顺利执行了。

临时表、排序等操作都会用到临时数据库。因此临时数据库 tempdb 必须足够大，以适应实际的运行环境。

Step 2 创建新的设备文件，用作新建数据库的默认设备文件。

对于即将要新建的用户数据库，最好为它们集中创建一个默认的设备文件，甚至是多个设备文件。

```
1> disk init name="defaultdevice" ,physname="D:/sybasease/data/defaultdevice.dat", size="100MB"
2> go
1> sp_diskdefault "master", defaultoff
2> go
(return status = 0)
```

到这里，我们已经创建了一个新的设备，名为 **defaultdevice**，并将 **master** 设备禁用作默认设备。接着看：

```
1> create database tt
2> go
Msg 1808, Level 17, State 1:
Server 'XIONGHE', Line 1:
Crdb_disk: Getnext SCAN_NOINDEX on sysdevices.status=DEFAULT failed to find
default rows
```

这时，因为找不到默认设备，创建数据库失败。

```
1> sp_diskdefault "defaultdevice", defaulton
2> go
(return status = 0)
1> create database tt
2> go
CREATE DATABASE: allocating 1536 logical pages (3.0 megabytes) on disk
'defaultdevice'.
Database 'tt' is now online.
```

将 defaultdevice 设为默认设备以后,创建数据库成功,默认大小为 3MB。也可以这样:

```
1> drop database tt
2> go
1> create database tt on default="10MB"
2> go
CREATE DATABASE: allocating 5120 logical pages (10.0 megabytes) on disk
'defaultdevice'.
Database 'tt' is now online.
```

通过指定默认的设备和在此设备上占用的空间大小来创建数据库。

实际上,Sybase ASE 数据库支持同时拥有多个默认设备,在本章的后边将有介绍。

8.2 用户数据库的容量管理

ASE 在安装完后,一般会有两到三个设备文件,如 master.dat、sybproc.dat,有的甚至还有 sybsysdb.dat 文件,这几个设备文件是系统数据库 master、model、sybsystemdb、sybsystemprocs 和 tempdb 的容器。

在 isql 命令行里,通过如下查询可以看到各数据库的创建状态:

```
1> sp_helpdb
2> go
 name              db_size         owner          dbid          created
status
--------------------  ---------------------  ---------------  ---------------  ---------------------
-------------------------------------------------------------------------------
 iihero            240.0 MB        sa             4             Feb 16, 2012
select into/bulkcopy/pllsort, trunc log on chkpt, ddl in tran, abort tr an on log full

 master            220.0 MB        sa             1             Apr 08, 2009
mixed log and data

 model             3.0 MB          sa             3             Apr 08, 2009
mixed log and data
 pubs2             4.0 MB          sa             5             Dec 09, 2010
trunc log on chkpt, mixed log and data
```

sybsystemdb	8.0 MB	sa	31513	Apr 08, 2009
trunc log on chkpt, mixed log and data				
sybsystemprocs	132.0 MB	sa	31514	Apr 08, 2009
trunc log on chkpt, mixed log and data				
tempdb	58.0 MB	sa	2	Apr 15, 2012
select into/bulkcopy/pllsort, trunc log on chkpt, mixed log and data				

这个查询结果分别列出了已有的数据库名、数据库大小、数据库属主、数据库 ID 号、数据库的创建时间、数据库的一些选项属性。

这些信息通过 Sybase Central 客户端工具，可以一目了然地看出来，如图 8-1 所示。

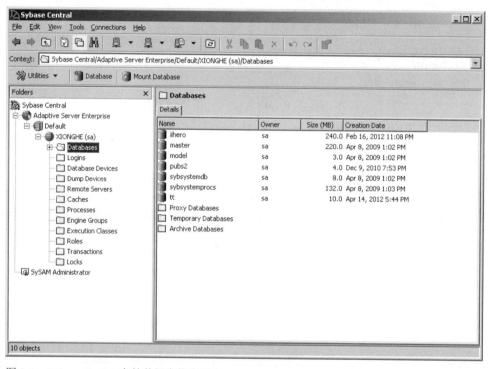

图 8-1 Sybase Central 中的数据库信息列表

我们可以看到系统数据库大多默认采用的是混合日志和数据。

具体到用户数据库，其创建及容量管理基本涉及到以下内容：

1. 为用户数据库规划好一个或者多个设备文件。如果是一个设备文件，通常其日志和数据都放到这个设备文件中。通常，设备文件最好是数据库专属的，最好不要多个用户数据库共用一个设备文件，为了避免 I/O 争用。
2. 为系统创建默认设备文件，指定默认数据库创建属性。

3．为用户数据库指定属主用户。

4．当容量或者空间不够时，对设备文件容量和数据库容量同时进行有效管理。

5．用户有时候可能需要将数据库从一个设备文件迁移到另一个设备文件。

6．如果想对数据库空间进行精细化管理，可能还要借助于段来进行处理。

下面针对上述内容分别进行介绍：

（1）用户数据库设备文件的创建。

设备文件和数据库是多对多的关系，在实际应用中，为了不致于引起混乱，最好弄成多对一的关系，即一个数据库对应一个或多个设备文件，数据和日志分别对应不同的设备文件，可以避免I/O 争用，以达到优化的目的。

假设这里要创建一个用户数据库 foo，数据空间为 500MB，对应设备文件为 D:\SybaseASE\foo.dat，日志空间大小为 500MB，对应的设备文件为 D:\SybaseASE\foo_log.dat。可以使用如下脚本进行创建：

```
D:\SybaseASE>isql -Usa -s "|" -w500
```

保密字：

```
1> disk init name="foo",physname="D:/SybaseASE/foo.dat",size="500MB"
2> go
1> disk init name="foo_log",physname="D:/SybaseASE/foo_log.dat",size="500MB"
2> go
1> create database foo on foo=500 log on foo_log=500
2> go
CREATE DATABASE: allocating 256000 logical pages (500.0 megabytes) on disk 'foo'.
CREATE DATABASE: allocating 256000 logical pages (500.0 megabytes) on disk 'foo_log'.
Database 'foo' is now online.
```

这里分两步：第一步创建两个设备文件；第二步创建数据库。指定这两个设备分别用于 data 和 log，各 500MB，也就是说，这两个设备文件被数据库 foo 独占了。

这里提示一下，语法确实有点怪，在 disk init 创建设备时，设备名可以加引号，而在 create database 指定设备名时，却不能加引号，希望读者注意。

（2）指定默认数据库创建属性，创建默认的设备文件给后边的用户数据库使用。

默认数据库创建时，它的大小是怎么确定的呢？上一节我们看到，创建的默认大小是 3MB。它的值由以下选项决定：

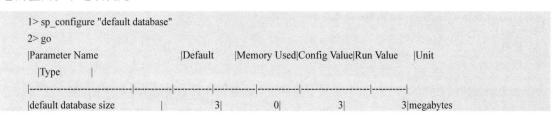

```
            |dynamic    |

(1 row affected)
(return status = 0)
```

在上一节已经创建了一个默认的设备，名为 defaultdevice，并且将 master 的默认设备设置给取消了。也就是说，后边新建的数据库如果没有指定设备名，指向的都是这个设备。

其实，ASE 是支持使用多个默认设备的。当第一个默认设备用完之后，就会自动使用第二个默认设备。那么怎么决定谁是第一个，谁是第二个呢？ASE 采用设备名的字母序。为了说明问题，我们接着上一节的 defaultdevice 来展开，在 defaultdevice 设备上已经创建了数据库 tt，占用了 10MB 大小，defaultdevice 总共为 100MB，还余下 90MB。我们再创建一个默认设备 alphadefault，大小为 50MB，创建一个数据库 test2，大小为 50MB，我们看看这个数据库到底创建到哪个设备上。

```
1> disk init name="alphadefault", physname="D:/SybaseASE/alphadefault.dat", size="50MB"
2> go
1> sp_diskdefault   "alphadefault", defaulton
2> go
(return status = 0)
1> create database test2 on default=50
2> go
CREATE DATABASE: allocating 25600 logical pages (50.0 megabytes) on disk 'alphadefault'.
Database 'test2' is now online.
```

我们看到，创建的数据库 test2 直接使用了 alphadefault 作为它的存储容器。

```
1> drop database test2
2> go
1> create database test2 on default=60
2> go
CREATE DATABASE: allocating 25600 logical pages (50.0 megabytes) on disk 'alphadefault'.
CREATE DATABASE: allocating 5120 logical pages (10.0 megabytes) on disk 'defaultdevice'.
Database 'test2' is now online.
1> drop database test2
2> go
1> sp_dropdevice "alphadefault"
2> go
Device dropped.
(return status = 0)
1> xp_cmdshell "del d:\SybaseASE\alphadefault.dat"
2> go
(0 rows affected)
(return status = 0)
```

如果新创建的数据库容量超过了第一个默认设备的容量大小，超出的部分将采用第二个默认设备。在此例中，数据库 test2 就从第一个默认设备 alphadefault 中使用了 50MB 空间，又从第二个默认设备 defaultdevice 中使用了 10MB 空间。

如果说新建的数据库要求的空间超过了默认设备可用空间的总和,会出现什么结果呢?答案是实际创建的数据库空间大小是默认设备可用空间的总和,并且不会提示出错。

请看示例:

```
1> disk init name="alphadefault", physname="D:/SybaseASE/alphadefault.dat", size="50MB"
2> go
1> sp_diskdefault  "alphadefault", defaulton
2> go
(return status = 0)
1>   create database test2 on default=180
2> go
CREATE DATABASE: allocating 25600 logical pages (50.0 megabytes) on disk 'alphadefault'.
CREATE DATABASE: allocating 46080 logical pages (90.0 megabytes) on disk 'defaultdevice'.
Database 'test2' is now online.
```

这里原本要求 test2 创建为 180MB 大小,可实际上创建的大小为 140MB。因为实际上默认的设备可用总空间只有 50MB + 90MB = 140MB。

对于默认数据库大小,我们可以指定其大小为 50MB,并创建一个新的数据库 test3 以作测试:

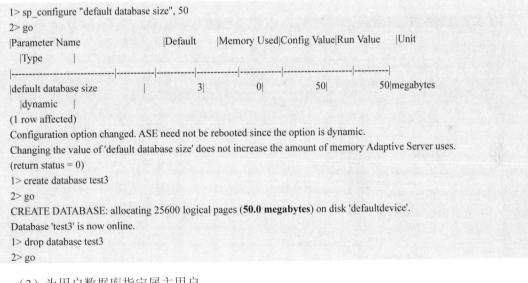

(3)为用户数据库指定属主用户。

假定我们在默认的设备 defaultdevice 上创建了一个新库 foo,并创建了对应的登录用户(login)foo_user。

示例如下:

```
1> use master
2> go
```

```
1> create database foo
2> go
CREATE DATABASE: allocating 25600 logical pages (50.0 megabytes) on disk 'defaultdevice'.
Database 'foo' is now online.
1> sp_addlogin foo_user, foo123, foo
2> go
Password correctly set.
Account unlocked.
New login created.
(return status = 0)
1>use foo
2> go
1> sp_changedbowner "foo_user"
2> go
Warning: The stored procedure 'sp_thresholdaction' may not execute; check database owner's threshold authorization.
DBCC execution completed. If DBCC printed error messages, contact a user with System Administrator (SA) role.
Database owner changed.
(return status = 0)
```

如果用 Sybase Central 来观察，也能看出数据库 foo 的属主已经从 sa 变为 foo_user 了，如图 8-2 所示。

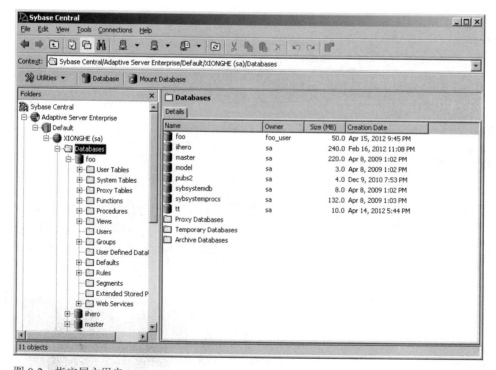

图 8-2 指定属主用户

这里能将 foo_user 设为数据库 foo 的属主用户的前提是，foo_user 首先必须是 ASE 服务器的登录用户，另外，它不能是数据库 foo 的普通用户。如果已经是普通用户，则不能将其设为属主。同时，系统数据库是不能改变属主的，它们的属主都是 master。

（4）当数据库空间不够时，可以对设备空间和数据库大小都进行调整。

如果只是数据库空间不够，而对应的设备文件还有足够的可用空间，那么只需要调整数据库的空间大小即可。示例如下：

```
1> create database test2
2> go
CREATE DATABASE: allocating 25600 logical pages (50.0 megabytes) on disk 'alphadefault'.
Database 'test2' is now online.
1> alter database test2 on default=20
2> go
Extending database by 10240 pages (20.0 megabytes) on disk defaultdevice
```

这里将数据库 test2 的空间大小扩充了 20MB。当然，这些操作都可以在 Sybase Central 中完成，如图 8-3 所示。

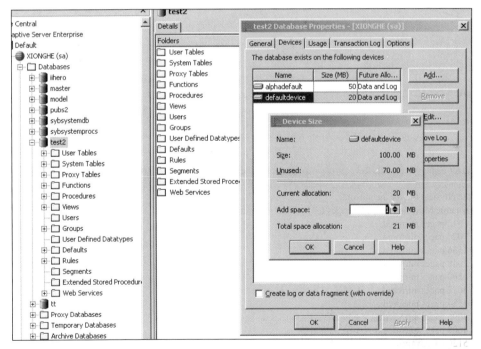

图 8-3　扩充设备容量

只需要单击指定的数据库→属性→设备（devices），然后编辑指定的设备名，单击 Add space 按钮，输入想要增加的大小即可。

（5）将数据库从一个设备文件转移到另一个设备文件。

有时数据库管理人员确实会遇到这样一种情况：最开始设计的设备位置不妥，需要改变一下位置。重新建库当然是一种方法，但是太费时间，而且中间也容易出错。这里推荐一种比较稳妥的方法，利用设备的镜像功能进行设备文件转移。

比如在上一节，我们已经创建了一个数据库 tt，使用的是设备 defaultdevice。现在，想将其转移到一个新建的设备文件上。我们可以查到 defaultdevice 的大小为 100MB。

首先，我们要启用镜像功能，再重启 ASE 使其生效。

```
1> sp_configure 'disable disk mirroring',0
2> go
|Parameter Name                  |Default    |Memory Used|Config Value|Run Value  |Unit
   |Type         |
|--------------------------------|-----------|-----------|------------|-----------|---------|
|disable disk mirroring          |          1|          0|           0|          1|switch
   |static       |

(1 row affected)
Configuration option changed. Since the option is static, Adaptive Server must be rebooted in order
for the change to take effect.
Changing the value of 'disable disk mirroring' does not increase the amount of memory Adaptive Server uses.
(return status = 0)
```

重启 ASE 之后，为 defaultdevice 设备创建一个镜像。

```
D:\SybaseASE>isql -Usa -s "|" -w500
```

保密字：

```
1> use master
2> go
1> disk mirror name="defaultdevice", mirror="d:\SybaseASE\beta.dat"
2> go
Creating the physical file for the mirror...
Starting Dynamic Mirroring of 51200 pages for logical device 'defaultdevice'.
        512 pages mirrored...
       1024 pages mirrored...
       1536 pages mirrored...
       2048 pages mirrored...
       2560 pages mirrored...
       3072 pages mirrored...
       3584 pages mirrored...
       4096 pages mirrored...
       4608 pages mirrored...
       5120 pages mirrored...
       5632 pages mirrored...
       6144 pages mirrored...
       6656 pages mirrored...
       7168 pages mirrored...
       7680 pages mirrored...
```

```
        8192 pages mirrored...
        8704 pages mirrored...
        9216 pages mirrored...
        9728 pages mirrored...
       10240 pages mirrored...
       10752 pages mirrored...
       11264 pages mirrored...
       11776 pages mirrored...
       12288 pages mirrored...
       12800 pages mirrored...
       13312 pages mirrored...
       13824 pages mirrored...
       14336 pages mirrored...
       14848 pages mirrored...
       15360 pages mirrored...
The remaining 35840 pages are currently unallocated and will be mirrored as they are allocated.
```

之后，beta.dat 文件则相当于是 defaultdevice 的另一个备份的设备文件，使用 unmirror 命令可以把 defaultdevice 的主设备文件去掉。

```
1> disk unmirror name="defaultdevice",side="primary",mode="remove"
2> go
```

之后，我们再看看 defaultdevice 的构成，如图 8-4 所示。

图 8-4　defaultdevice 的设备属性

defaultdevice 不再由 D:\SybaseASE\data\defaultdevice.dat 来作设备文件了。使用下述命令即可删除这个冗余文件：

```
1> xp_cmdshell "del D:\SybaseASE\data\defaultdevice.dat"
2> go
(0 rows affected)
(return status = 0)
```

（6）使用段对数据库空间进行精细化管理，在下一节我们会进行详细介绍。

8.3　使用段管理数据库空间

在 ASE 数据库中，要想有效管理用户数据库和系统数据库，使用段来管理是其中一个重要手段。其实，段是一个逻辑上的概念，它是指向一个或多个数据库设备（文件）的标签。通过段的指定，可以将数据库的表数据、索引、日志等的存储位置进行有效的分离，从而使 I/O 达到最好的并行，提高性能，这是最终目的，同时可以让 DBA 对这些对象的位置、大小和空间使用情况进行有效的控制。目前，ASE 数据库可以包含多达 32 个段。

8.3.1　段与其他数据库对象的关系

虽然我们知道段是一个指向一个或多个设备的标签，但是它与其他数据库对象有什么样的关系呢？图 8-5 充分解释了它们之间的关系。

ASE 数据库中最基本地可以使用多个物理页来存储的对象有日志、表和索引，同时，ASE 为 TEXT/IMAGE（其他 DBMS 系统也称 BLOB/CLOB，即大的二进制或文本对象）使用单独的页存储，表分区也可以指定不同的位置进行存储。这样，这 5 种类型数据都可以为它们定义不同的段来进行存储。每个段可以指定不同的设备甚至多个设备。

每当一个新的数据库创建完毕时，系统默认为它创建了三个段，这三个段分别是：

1. logsegment（日志段）：存储数据库的事务日志，实际上它在 ASE 中也是以表的形式存在，可以称之为日志表 syslogs。
2. system 段：存储该数据库的系统表，这些表大多以 sys 为前缀。
3. default 段：用于存储普通的用户表数据，如果某些表含有 TEXT/IMAGE 字段，这些大对象也存储于该段当中。

我们再来看看创建数据库的命令：

```
create [temporary] database database_name
    [on {default | database_device} [=size]
        [, database_device[=size]]...]
    [log on database_device [=size]
        [, database_device[=size]]...]
```

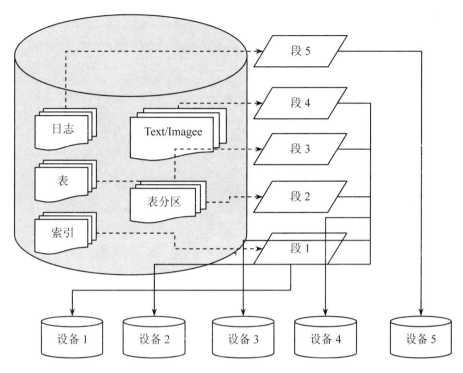

图 8-5 段与其他数据库对象间的关系

这里如果创建数据库时只指定了 on 参数，后边跟一个或多个设备，那么，三个段都指向这些设备。如果同时指定了 log on 参数，那么 system 段和 default 段指向 on 指定的数据设备，而 logsegment 指向 log on 参数设定的日志设备。也就是说，默认情况下，system 段和 default 段都指向的是数据设备，而 logsegment 段则指向日志设备，如果没有指定日志设备，它也指向数据设备。

例如，创建两个设备 mydata 和 mydata_log，然后创建数据库 db1，这两个设备分别用作它的数据设备和日志设备：

```
1> use master
2> go
1> disk init name="mydata", physname="d:\SybaseASE\mydata.dat", size="50M"
2> go
1> disk init name="mydata_log", physname="d:\SybaseASE\mydata_log.dat", size="50M"
2> go
1> create database db1 on mydata="4M" log on mydata_log="3M"
2> go
CREATE DATABASE: allocating 2048 logical pages (4.0 megabytes) on disk
'mydata'.
CREATE DATABASE: allocating 1536 logical pages (3.0 megabytes) on disk
```

```
'mydata_log'.
Database 'db1' is now online.
1>
```

这时我们看看 db1 的状态：

```
1> sp_helpdb db1
2> go
 name         db_size       owner         dbid        created                status
 ------------ ------------- ------------- ----------- --------------------- ---------------------
 db1          7.0 MB        sa            8           Apr 18, 2012          no options set

 device_fragments         size          usage         created                free kbytes
 ------------------------ ------------- ------------- --------------------- ---------------------
 mydata                   4.0 MB        data only     Apr 18 2012  7:02AM   2376
 mydata_log               3.0 MB        log only      Apr 18 2012  7:02AM   not applicable

 ----------------------------------------
 log only free kbytes = 3058
(return status = 0)
```

我们能看到 mydata 仅作数据用（系统表数据、用户表数据对应于 system 段和 default 段），而 mydata_log 作为日志存储用，对应于 logsegment 段。

8.3.2 创建数据库段

用户可以通过创建自定义段来控制对指定设备的使用，例如将用户创建的表位置指定到特定的段上。定义新段的语法如下：

```
sp_addsegment   name, dbname, devicename
```

该存储过程后边的参数依次是段名、数据库名和设备名。需要说明的是，对段的操作必须先调用 use <目标数据库>。

继续上例，我们将数据库 db1 的数据设备扩充到三个设备上：

```
1> use master
2> go
1> disk init name="mydata2", physname="D:\SybaseASE\mydata2.dat", size="50MB"
2> disk init name="mydata3", physname="D:\SybaseASE\mydata3.dat", size="50MB"
3> go
1> alter database db1 on mydata2=4, mydata3=4
2> go
Extending database by 2048 pages (4.0 megabytes) on disk mydata2
Extending database by 2048 pages (4.0 megabytes) on disk mydata3
```

扩充以后，db1 的空间存储情况如下：

```
1> sp_helpdb db1
2> go
 name        db_size      owner        dbid    created         status
 ----------  -----------  -----------  ------  --------------  --------------------
 db1         15.0 MB      sa           8       Apr 18, 2012    no options set

(1 row affected)
 device_fragments    size       usage       created              free kbytes
 ------------------  ---------  ----------  -------------------  --------------
 mydata              4.0 MB     data only   Apr 18 2012  7:02AM  2376
 mydata_log          3.0 MB     log only    Apr 18 2012  7:02AM  not applicable
 mydata2             4.0 MB     data only   Apr 18 2012  7:24AM  4080
 mydata3             4.0 MB     data only   Apr 18 2012  7:24AM  4080

 --------------------------------
 log only free kbytes = 3058
(return status = 0)
```

现在，db1 在扩充以后，段的分布情况如下：

1．mydata 设备：仅数据，system 段和 default 段，存储系统表和用户表数据。
2．mydata_log 设备：仅日志，logsegment 段。
3．mydata2 设备：仅数据，system 段和 default 段，存储系统表和用户表数据。
4．mydata3 设备：仅数据，system 段和 default 段，存储系统表和用户表数据。

这时，我们在数据库 db1 中创建一个新段 db1_seg1，让它指向 mydata2，可执行以下命令：

```
1> use db1
2> go
1> sp_addsegment "db1_seg1", "db1", "mydata2"
2> go
DBCC execution completed. If DBCC printed error messages, contact a user with System Administrator (SA) role.
Segment created.
(return status = 0)
```

现在的状况是，mydata2 被 system 段和 default 段占用外，同时它还用作自定义的段 db1_seg1。从这里我们也可以看出，一个设备可以被多个段"征用"，同时一个段也可以跨多个设备。作为 DBA 来讲，应该尽量避免一个设备被一个数据库的多个段征用。

8.3.3 改变数据库段的指定

我们继续前一小节的内容，创建一个新的设备 mydata4，为扩充段 db1_seg1 使用：

```
1> use master
2> go
1> disk init name="mydata4", physname="d:\SybaseASE\mydata4.dat", size="50MB"
2> go
```

扩充段如下：

```
1> use db1
2> go
1> sp_extendsegment "db1_seg1", "db1", "mydata4"
2> go
Msg 17281, Level 16, State 1:
Server 'XIONGHE', Procedure 'sp_extendsegment', Line 173:
The specified device is not used by the database.
(return status = 1)
1>
```

这里出现错误，设备 mydata4 在被扩充到 db1 的段 db1_seg1 之前，并未被 db1 使用，因而出错。正确的顺序是先将 db1 扩充，以使用设备 mydata4，再改变段 db1_seg1。我们不能将一个未经使用的设备指定为某个数据库的段。

```
1> use master
2> go
1> alter database db1 on mydata4=4
2> go
Extending database by 2048 pages (4.0 megabytes) on disk mydata4
1> use db1
2> go
1> sp_extendsegment "db1_seg1", "db1", "mydata4"
2> go
DBCC execution completed. If DBCC printed error messages, contact a user with System Administrator (SA) role.
Segment extended.
(return status = 0)
```

扩充成功。这时候，段 db1_seg1 应该包含了对设备 mydata2 和 mydata4 的指定。同时这两个设备也作为 system 段和 default 段存在。这样混合使用显然不利于管理。ASE 提供了删除段指定的命令 sp_dropsegment，可以将某段从某个设备的指定中删除。

比如，这里我们可以从 db1_seg1 段中删除对 mydata2 的指定：

```
1> use db1
2> go
1> sp_dropsegment "db1_seg1", "db1", "mydata2"
2> go
DBCC execution completed. If DBCC printed error messages, contact a user with System Administrator (SA) role.
Segment reference to device dropped.
(return status = 0)
```

这时 db1_seg1 只剩下对 mydata4 的指定了。我们可以通过调用 sp_dropsegment，将 mydata4 中对应的 system 段和 default 段删除，让它只用于自定义段 db1_seg1。

```
1> sp_dropsegment "system", "db1", mydata4
2> go
DBCC execution completed. If DBCC printed error messages, contact a user with System Administrator (SA) role.
Segment reference to device dropped.
(return status = 0)
1> sp_dropsegment "default", "db1", mydata4
2> go
DBCC execution completed. If DBCC printed error messages, contact a user with System Administrator (SA) role.
Segment reference to device dropped.
(return status = 0)
```

这样，mydata4 就专属于段 db1_seg1 了。

做了这样的改变以后，db1 中的段分布情况如下：

```
------------------------ ------------- ----------------- ------------------------------ --------------------
mydata           4.0 MB    data only    Apr 18 2012  7:02AM   2376
mydata_log       3.0 MB    log only     Apr 18 2012  7:02AM   not applicable
mydata2          4.0 MB    data only    Apr 18 2012  7:24AM   4080
mydata3          4.0 MB    data only    Apr 18 2012  7:24AM   4080
mydata4          4.0 MB    data only    Apr 18 2012  7:53AM   4080


------------------------------------------------------------------------------------------
log only free kbytes = 3058
 device                  segment
------------------------  ------------------------
 mydata                  default
 mydata                  system
 mydata2                 default
 mydata2                 system
 mydata3                 default
 mydata3                 system
 mydata4                 db1_seg1
 mydata_log              logsegment
(return status = 0)
```

我们可以看到，mydata4 只属于 db1_seg1 段。

```
4> use db1
5> go
1> sp_helpsegment
2> go
 segment             name                  status
------------------  --------------------  -------------
 0                  system                0
 1                  default               1
```

```
2           logsegment       0
3           db1_seg1         0
(return status = 0)
```

上述信息描述了 db1 中段的基本情况，依次是段号、段名和状态。状态值为 1，意指此为数据库的默认段。

要获得某个具体段的具体信息，使用 sp_helpsegmemt <段名>，示例如下：

```
1> use db1
2> go
1> sp_helpsegment db1_seg1
2> go
 segment               name                  status
 --------------------  --------------------  ------------------
 3                     db1_seg1              0
 device                size                  free_pages
 --------------------  --------------------  ------------------
 mydata4               4.0MB                 2040
 total_size   total_pages    free_pages    used_pages    reserved_pages
 ----------   -----------    ----------    ----------    --------------
 4.0MB        2048           2040          8             0
(return status = 0)
```

可以看到，db1_seg1 段指向设备 mydata4，大小为 4MB，2048 页，空闲页为 2040 页，已经使用了 8 页。

要查看数据库中设备与段的对应关系，直接使用命令 sp_helpdb <数据库名>：

```
1> sp_helpdb db1
2> go
……
 device                        segment
 ----------------------------  ------------------
 mydata                        default
 mydata                        system
 mydata2                       default
 mydata2                       system
 mydata3                       default
 mydata3                       system
 mydata4                       db1_seg1
 mydata_log                    logsegment
(return status = 0)
```

8.3.4　在段中存放数据库对象

除了 logsegment、system、default 段存放对象比较明确以外，如果没有为创建的数据库对象指定段，则它们会被创建到默认的段上。我们可以为索引、普通数据及大对象字段指定特定的段来存放。这是分离 I/O 并优化性能的一个重要举措。

1. 表数据及其索引的存放

通常，我们可以在建表和建索引时，为它们指定不同的段，以达到并行 I/O 的目的。

例如，我们为数据库 db1 创建表 test，将其放置到段 db1_seg1 上，执行如下命令：

```
1> create table test(id int primary key, address varchar(128) null) on db1_seg1
2> go
1> create nonclustered index ind_test_1 on test(address) on db1_seg1
2> go
1> sp_helpsegment db1_seg1
2> go
Objects on segment 'db1_seg1':

 table_name      index_name                    indid     partition_name
 --------------- ----------------------------- --------- -----------------------------------
 test            test_id_6080021661            1         test_id_6080021661_608002166
 test            ind_test_1                    2         ind_test_1_608002166
Objects currently bound to segment 'db1_seg1':

 table_name      index_name                    indid
 --------------- ----------------------------- -------------
 test            test_id_6080021661            1
 test            ind_test_1                    2

 total_size   total_pages    free_pages     used_pages     reserved_pages
 ------------ -------------- -------------- -------------- ----------------
 4.0MB        2048           2017           31             0
(return status = 0)
```

从上述结果可以看出，段 db1_seg1 上存放了表 test 及其索引 ind_test_1。

我们还可以把表和索引分到不同的段进行存放，例如：

```
1> drop index test.ind_test_1
2> go
1> sp_addsegment db1_seg2, db1, mydata2
2> go
DBCC execution completed. If DBCC printed error messages, contact a user with
System Administrator (SA) role.
Segment created.
(return status = 0)
1> create nonclustered index ind_test_1 on test(address) on db1_seg2
2> go
```

这会将非聚簇索引 ind_test_1 存放到新段 db1_seg2 上。因为非聚簇索引和表数据存放到不同的物理页上。

如果创建的是聚簇索引，那么数据和索引必须存放在相同的物理页上连续存储，这时它们必须使用相同的段，也就是说，一旦创建了聚簇索引，数据也必须位于创建索引时所使用的段上。利用这一特性，可以实现段的迁移。

接上例，表 test 原来在 db1_seg1 上，普通非聚簇索引 ind_test_1 位于 db1_seg2 上，而默认的主键 id 对应有一个聚簇索引 test_id_6080021661。

```
1>alter table test drop constraint test_id_6080021661
2>go
1> create clustered index ind_test_2 on test(address) on db1_seg2
2> go
Non-clustered index (index id = 2) is being rebuilt.
1> sp_helpsegment db1_seg2
2> go
 segment              name                   status
 --------------       --------------------   ----------------
       4              db1_seg2               0
 device               size                   free_pages
 --------------       --------------------   ----------------
 mydata2              4.0MB                  2016

 table_name           index_name             indid
 --------------       --------------------   -------
 test                 ind_test_2             1
 test                 ind_test_1             2

 total_size   total_pages   free_pages   used_pages   reserved_pages
 ----------   -----------   ----------   ----------   --------------
 4.0MB        2048          2016         32           0
(return status = 0)
1> sp_helpsegment db1_seg1
2> go
 segment              name                   status
 --------------       --------------------   ----------------
       3              db1_seg1               0
 device               ize                    free_pages
 --------------       --------------------   ----------------
 mydata4              4.0MB                  2040

(return status = 0)
1>
```

我们看到 db1_seg1 中已经没有存放东西了。表 test 的数据及其索引 ind_test_2 都由聚簇索引 ind_test_2 迁移到段 db1_seg2 上了。这也是移动表的一个过程。

即要完全移动表，可以删除其聚簇索引（如果有），然后在所需的段上创建或重新创建聚簇索引。要完全移动非聚簇索引，可先删除该索引，然后在新段上重新创建。

2. 大对象的存放

我们也可以将表中大对象字段指定存放到不同的段上。这要用到命令 sp_placeobject。创建带有 text 或 image 列的表时，数据存储在单独的文本页链上。具有 text 或 image 列的表在 sysindexes 中有一个额外行用于文本链，名称列设置为前面带有字母 t 的表名称，且 indid 为 255。可以使用 sp_placeobject 在单独的设备上存储文本链（需给出表名和 sysindexes 中文本链的名称）：

示例如下：

```
1> create table tblob(id int primary key, col2 image null, col3 text null) on db1_seg1
2> go
1> sp_placeobject db1_seg2, "tblob.ttblob"
2> go
DBCC execution completed. If DBCC printed error messages, contact a user with
System Administrator (SA) role.
'tblob.ttblob' is now on segment 'db1_seg2'.
(return status = 0)
1>
```

这个单独的文本链名称就是表名附上前缀 t：ttblob。

3. 将表分区置于不同的段上

当表数据规模非常庞大而需要分区时，可以为不同的分区指定不同的段。

示例如下：

```
1> use db1
2> go
1> create table books(name varchar(64) not null, btype varchar(12) not null, pubdate datetime not null, price money null, pubs varchar(64) null) partition by hash(btype) (p1 on db1_seg1, p2 on db1_seg2)
2> go
```

这里将 books 表按 btype 进行 hash 分区，分到两个段上分别进行存储。

使用分区表特性时，需要使用 sa 用户先激活这个 option：

```
D:\SybaseASE>isql -Usa
```

保密字：

```
1> sp_configure 'enable
2> semantic partitioning
1> sp_configure 'enable semantic partitioning', 1
```

```
2> go
Parameter Name                    Default    Memory Used Config Value    Run Value    Unit      Type
--------------------------------  ---------  --------------------------  -----------  --------  --------
enable semantic partitioning      0          0                           1            1 switch  dynamic

(1 row affected)
Configuration option changed. ASE need not be rebooted since the option is
dynamic.
Changing the value of 'enable semantic partitioning' does not increase the
amount of memory Adaptive Server uses.
(return status = 0)
```

8.3.5 使用 Sybase Central 客户端工具管理段

使用图形化工具 Sybase Central 来管理段非常方便。假设有数据库 db1，对应的设备文件为 dev1.dat，大小为 50MB，另有一个设备文件为 dev2.dat，大小为 50MB。现在想为 db1 创建和扩展段。

使用 sa 用户连接数据库以后，首先选中数据库设备（Database Devices），创建数据库设备：dev1.dat 和 dev2.dat。为设备指定设备名和路径如图 8-6 所示。

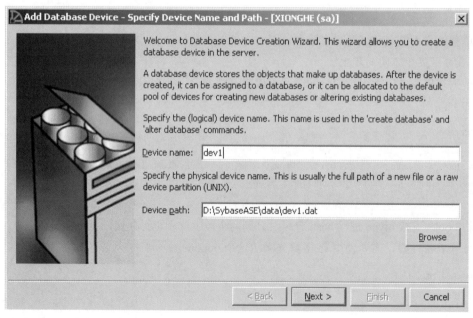

图 8-6　指定设备名和路径

单击 Next 按钮，如图 8-7 所示。

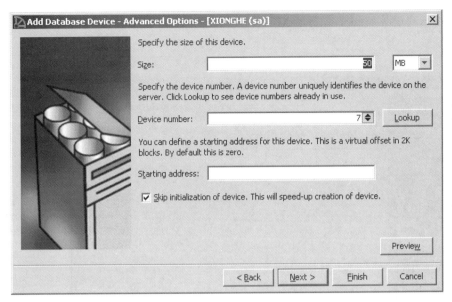

图 8-7　指定设备大小

单击 Next 按钮，如图 8-8 所示。

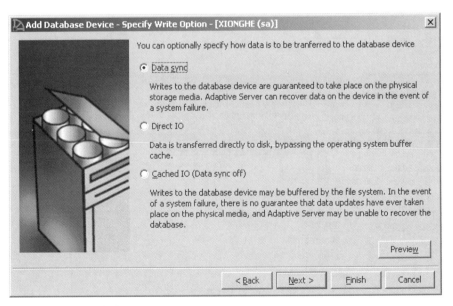

图 8-8　指定同步模式

单击 Next 按钮，直至完成。
用同样的步骤创建设备文件 dev2.dat。

然后创建一个示例数据库 db1，服务器 XIONGHE→Databases，然后选择，创建新数据库，如图 8-9 所示。

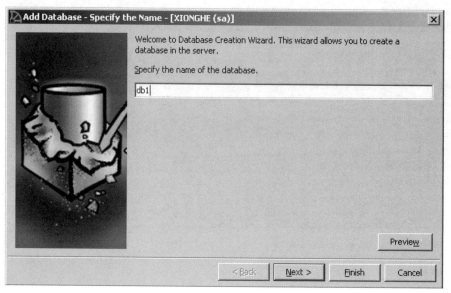

图 8-9　指定数据库名

单击 Next 按钮，如图 8-10 所示。

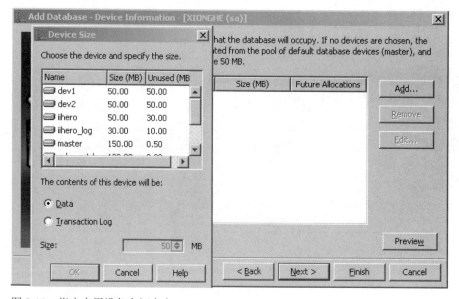

图 8-10　指定占用设备空间大小

单击 Add 按钮，选择 dev1，数据 30MB，日志 10MB；选择 dev2，数据 20MB，如图 8-11 所示。

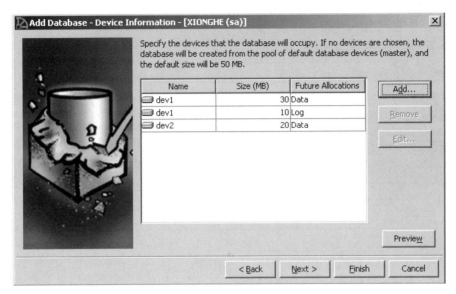

图 8-11　为数据库指定设备

在 Add 弹出的对话框里选择 with override，然后结束。

再看看数据库 db1 中段的管理：

打开数据库 db1，选择 Segments，段的基本情况如图 8-12 所示。

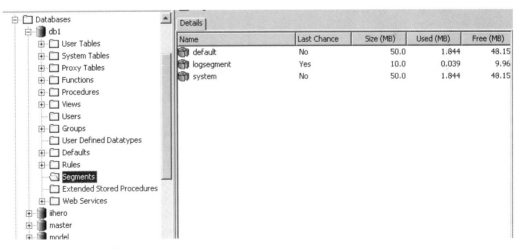

图 8-12　数据库的段信息

现在为 db1 新建一个段 db1_dev2_seg，如图 8-13 所示。

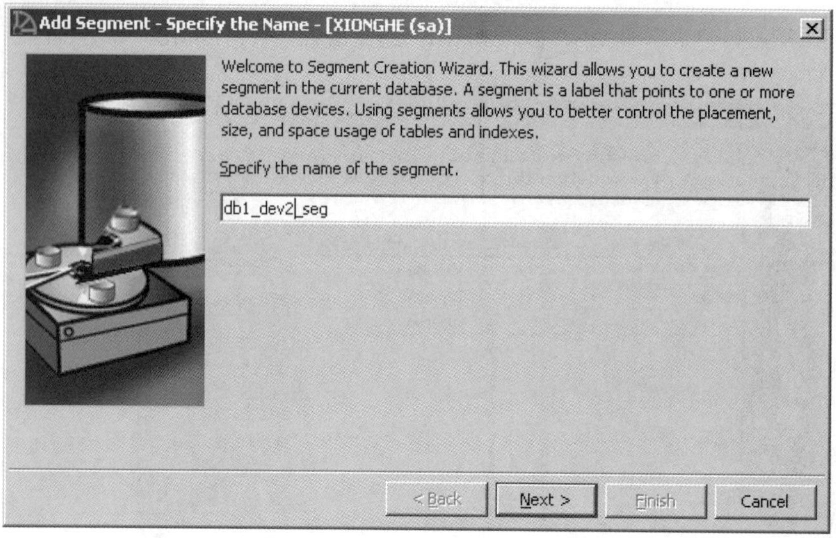

图 8-13　指定段名

单击 Next 按钮，选择设备名 dev2，如图 8-14 所示。

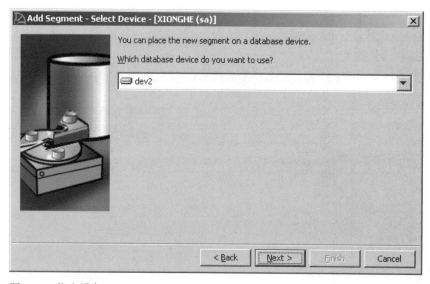

图 8-14　指定设备

单击 Next 按钮，结束。这样，dev2 就专属于段 db1_dev2_seg 了，它只存数据。

如何将 dev2 从 system 和 default 段中剥离呢？

在 Segments（段）的右边面板中，选择 default→properties→设备（devices）命令，得到如图 8-15 所示的界面。

ASE 中的空间管理　第 8 章

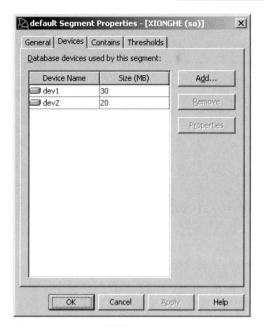

图 8-15　default 段属性

选中 dev2，然后单击 Remove 按钮即可除去 dev2 在 default 段中的指定，对 system 需要执行同样的过程。在执行 Remove 的过程中，我们可以选择 Preview（预览），看看它到底执行了什么 SQL 语句，如图 8-16 所示。

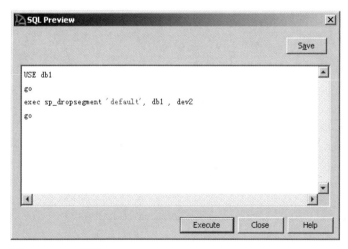

图 8-16　SQL 预览

我们会发现，它与我们单独执行 SQL 语句的效果是一样的。

9

ASE 的用户及安全管理

所有的数据库，安全管理是非常重要的一环，ASE 也不例外。数据的保护与安全管理是密不可分的，数据损坏的因素中，除了物理损坏以外，人为损坏或破坏主要就是被外界突破了安全防线导致的。

ASE 提供了各种粒度的安全管理来保护数据库里的数据，免受外界非法访问。

如果仔细划分，它可以分成下边 5 个粒度的安全级别：操作系统级的安全、ASE 数据库服务器级的安全、ASE 数据库级的安全、数据库对象级的安全和更细粒度对象的安全，如图 9-1 所示。

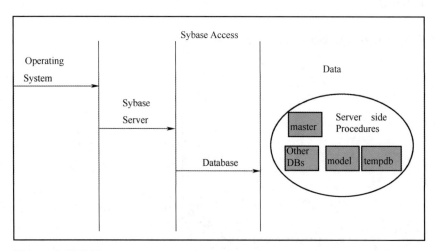

图 9-1　权限分级

首先，如果你得不到操作系统的权限，你就无法访问 ASE 的数据文件。ASE 通过监听固定的端口，可以让你通过远程的客户端来查询访问 ASE 中的数据。

一旦你拥有操作系统的权限，你需要能够登录 ASE 服务器的用户密码信息才能正常访问 ASE

服务器。这些用户信息是通过 sp_addlogin 存储过程添加进去的,每个用户信息都登记在 master 数据库的 syslogins 系统表中。

能够访问 ASE 服务器,并不意味着就能访问具体的某个用户数据库,要访问具体的某个数据库,你还得拥有能够访问这个数据库的用户信息,在它通过 sp_adduser 存储过程中,最终用户信息会保存在该数据库的 sysusers 表中。

即便能够访问某个数据库,也不参保证你能访问该数据库中所有的对象。ASE 还可以控制具体的数据库对象对某个用户的访问权限。

上述几个级别是按照对象的粒度和层级进行划分,来确定各个层级的安全,它们都是在常规的网络通信的情况下,即客户端和服务器端都是在不加密的通信协议的基础上进行交互的。要达到更高级别的安全,Sybase ASE 还支持 Socket 传输层加密(SSL)的高级特性,以达到最大程度的安全性,即使用户在中途截取了通信的报文,也达不到破坏系统的目的。关于 SSL 加密的管理,将作为独立的一节进行介绍。

下面,我们就按照这些安全级别分别进行介绍。

9.1 操作系统级别的安全

操作系统级别的安全,是操作系统管理员的首要职责,至少他应该确保 root 用户信息不能被攻破,同时服务器只能被限制性的访问,如只能通过内网的一台或几台机器远程访问。如果 DBA 同时兼作操作系统的系统管理员(SA),则更应该考虑内外网隔离等因素。

除此以外,ASE 的管理员在服务器中,必须有相应的访问文件的权限,他应该有一个单独的操作系统用户,切不可用操作系统的超级用户(尤其是 UNIX/Linux 环境下),通常使用的是 Sybase 用户,专用于管理 Sybase ASE 数据库服务。

该用户应该拥有访问 ASE 数据库所用的设备文件的权限,以及访问 ASE 数据库安装路径所在目录的所有权限。

9.2 ASE 服务器级别的安全

ASE 数据库服务器级别的安全,主要是通过数据库登录用户(login)的 ID 和密码来控制的。使用的命令是:

```
sp_addlogin    loginame, passwd [, defdb]
    [, deflanguage] [, fullname] [, passwdexp]
    [, minpwdlen] [, maxfailedlogins] [, auth_mech]
```

下面的示例是添加一个登录用户 iihero,密码为 iihero1:

sp_addlogin iihero, iihero1, iihero, null, 'iihero', 90, 6, 3

默认情况下，该用户使用的数据库是 iihero，语言这里设成 NULL，它将采用系统默认的语言，密码 iihero1 的有效期为 90 天，至少为 6 个字符长，并且最多只能尝试 3 次错误登录（有效地保证了有人恶意多次尝试登录）。

ASE 中，登录 ID 和密码的最大长度为 30 个字符，在创建密码的时候，如果是以数字开头，或者密码中的字符超出基本字符范围（a-z、A-Z、0-9），则需要将密码用双引号引起来。例如：

sp_addlogin   hana, "&abc_34"

我们来看看 sp_addlogin 的参数及其含义，如表 9-1 所示。

表 9-1   sp_addlogin 参数

参数	含义
loginname	用户名（登录名）
passwd	用户密码
defdb	访问的默认数据库
deflanguage	默认的语言
fullname	用户名全名
passwdexp	用户密码过期的天数，默认为 0（表示永不过期）
minpwdlen	最短密码长度，默认值为 6，有效值是 0~30
maxfailedlogins	在被锁定之前，连续输入错误密码（登录失败）的最大次数，默认值为 0，即永远不会被锁定，有效值是 0~32767
auth_mech	登录验证机制，默认值为 ANY，该参数有如下有效值： ASE：使用 ASE 的验证机制 LDAP：外部的 LDAP 服务器验证 PAM：通过插件式的验证模块来验证 ANY：上述任意验证方法，先使用外部的验证方法，如果没有，则使用 ASE 来验证

考虑到可选参数，我们不必每次都把这 9 个参数完全输入，可以显示地指定参数名及其对应值，这样就不用考虑参数的顺序，请看下边的示例：

```
1> sp_addlogin 'demo', 'demo_pwd', @passwdexp=90, @defdb='db0', @deflanguage='chinese', @fullname='demo1'
2> go
口令设置正确
账户被解锁
新的登录被创建
(return status = 0)
```

可以看到，使用这种方法，通过指定参数名（如@passwdexp=90），不用考虑参数的顺序，可以方便解决多个参数输入的顺序问题。

在 ASE 中，登录用户信息将集中存储到 master 数据库的 syslogins 系统表中，每个登录用户对应一条记录，记录此用户的各种配置信息，其中，suid 列是 ASE 为登录用户指定的标识号，它是在整个 Server 级唯一的标识。登录用户 sa 对应的 suid 值是 1，其他用户的 suid 值依次加 1。

下面的查询列出了当前 ASE 服务器中所有登录用户的元信息，其中：

```
1> select suid, name, status, dbname, password from syslogins
2> go
 suid         name              status               dbname
password
 -----------  ----------------  -------------------  ----------------
 -----------------------------------------------------------------------
   1          sa                0                    master
0xc007a554664cfbfe51c034fe3709830c8b0707db23da7737ddfdbaa1a388a8046150045b27ab0ef87450
   2          probe             0                    sybsystemdb
0xc0076dc16d5f3783d8153f65c8144764fd89f42ab2065cff207bf7bc0a7c73ba6f33e6bfe386da19cd3e
   3          jstask            0                    master
0xc00781ad8a7527cb1b82acbab18befe9b6d44da9260f0569243b6cb11fb1709d05dd1e124506cf7d2748
   4          spring            0                    iihero
0xc00788ef93c16f0fbc87f2156add4369f6ba6bdfb4bc4e8f584db5661557cd15ab8f9ed7f9db9e31d264
   5          test              0                    db0
0xc007645bab6b7733bc45a1888f952aa78ba91ff16e3b1a372db0a39e63911f16a038690d0f2cf4f5a38c
   6          demo              0                    db0
0xc0078af824e9241a52a5b64ebbe96f893d5f25787de1b49c1562a017b30a0d4386c53074f8ad08418b28
(6 rows affected)
```

其中，status 列表示该用户是否被锁定，0 表示未锁定。password 列存储的是该用户的密码信息，以十六进制的形式加密存储，采取的是一种单向加密算法，只通过密文，基本上没法得到原始明文。probe 和 jstack 两个用户都是安装 ASE 时系统内部添加的，其他几个用户都是 DBA 添加的。

9.2.1　调整修改登录用户

登录用户创建以后，有时候可以修改和调整该用户。通过调用存储过程 sp_modifylogin 可以完成此任务。该系统存储过程的语法格式如下：

sp_modifylogin {loginame | "all overrides"}, option, value

其中，各参数的含义和用法如下：

loginame：要修改的登录用户的名字。

all overrides：删除已使用 passwd expiration、min passwd length 或 max failed_logins 参数设置的系统替换值。若要删除所有特定于登录名的值，请指定 sp_modifylogin、all overrides、option、-1。

option：指定要更改的选项的名称。可使用的选项如下：

- authenticate with，验证登录方式，有 ASE、LDAP、PAM、ANY。
- defdb，用户连接到 ASE 服务器时默认使用的数据库。
- deflanguage，默认使用的语言。

- fullname，用户全名。
- add default role，用户登录时被激活的角色。
- drop default role，用户登录时，从激活的角色列表中删除的角色。
- passwd expiration，密码有效期，有效值 0~32767。
- min passwd length，密码最小长度，有效值 0~30，默认值是 6。
- max failed_logins，指定登录名的允许失败登录的尝试次数，有效值 0~32767。
- login script，用户登录时自动运行的存储过程的名称。

这些参数修改完成以后，新的参数会在下一次用户登录时生效。从上述参数可以看出，除了登录用户名和密码不能修改以外，其他信息都能修改。修改密码需要调用另外的存储过程 sp_password。下面我们通过示例来看看如何修改登录用户的相关信息。

【例 1】修改登录用户 demo 的默认数据库为 iihero，原来的默认数据库为 db0。

```
1> sp_modifylogin demo, defdb, 'iihero'
2> go
```

默认数据库被改变。

(return status = 0)

【例 2】修改登录用户 demo 的最小密码长度为 3，最大失败登录次数为 5，有效期为 30 天，登录时运行一个自定义的存储过程 echo_demo。

```
1> use iihero
2> go
1> create proc echo_demo
2> as
3> print 'this is me: demo'
4> go
1> use master
2> go

1> sp_modifylogin demo, 'min passwd length', '3'
2>
3> go
选项已改变
(return status = 0)
1> sp_modifylogin demo, 'max failed_logins', '5'
2> go
选项已改变
(return status = 0)
1> sp_modifylogin demo, 'passwd expiration', '30'
2> go
选项已改变
(return status = 0)
```

```
1> sp_modifylogin demo, 'login script', 'echo_demo'
2> go
选项已改变
(return status = 0)
```

当然，如果要改变登录用户的默认数据库，我们还可以调用 sp_defaultdb loginname 和 defaultdb 来实现。

尽管改变了默认数据库，用户要想登录时能够连接到设定的新的默认数据库，必须事先已经是该数据库的用户。

接例 2，用户 demo 新的默认数据库是 iihero，登录以后：

```
C:\>isql -Udemo
```

保密字：

```
Msg 10351, Level 14, State 1:
Server 'XIONGHE':
Server user id 6 is not a valid user in database 'iihero'
Msg 4001, Level 11, State 1:
Server 'XIONGHE':
Cannot open default database 'iihero'.
1> select db_name()
2> go

 --------------------
 master

(1 row affected)
1>
```

我们发现 demo 用户并没有成功地登录到 iihero 数据库，这是因为 iihero 数据库并没有这个用户。要想成功登录，必须为其添加用户 demo 或 guest 用户，请看：

```
1> use iihero
2> go
1> select name from sysusers where suid>=-1
2> go
 name
 -----------------
 dbo
 spring
1> setuser 'demo'
2> go
Msg 4604, Level 16, State 2:
Server 'XIONGHE', Line 1:
```

没有这样的用户 demo
```
1> sp_adduser demo
2> go
新用户被加入
(return status = 0)

1> select name from sysusers where suid>=-1
2> go
 name
 ----------------
 dbo
 spring
 demo
(3 rows affected)
1> setuser 'demo'
2> go
```

这时使用另一个会话，用 demo 用户登录，也能成功连接到 iihero 数据库。

9.2.2　密码的强化管理

在 ASE 服务器级别的安全管理当中，如果是生产环境，对登录用户的密码安全强度进行设置是有必要的。实际的生产环境（客户的实际运行环境）对安全的要求要比测试开发环境严格得多。

我们可以设置全局性的密码设定要求，比如要求密码中必须至少有一位数字，可以使用过程"sp_configure "check password for digit", {0|1}"来控制。

如果此值设为 0，则不必要求密码中含有数字；如果为 1，则密码中必须要有一位数字。

请看下边的示例：

```
1> sp_configure 'check password for digit', 1
2> go
 Parameter Name            Default    Memory Used Config Value    Run Value    Unit        Type
 -----------------------   --------   -------------------------   ----------   ---------   ---------
 check password for digit  0          0                           1            1 switch    dynamic

(1 row affected)
配置选项已更改。由于该选项是动态的，无须重新启动 ASE
更改 check password for digit 的值不会增加 Adaptive Server 使用的内存量
(return status = 0)
1> sp_addlogin demo1, 'demopwd'
2> go
Msg 9551, Level 16, State 1:
Server 'XIONGHE', Procedure 'sp_password', Line 118:
指定的口令中没有数字字符。新口令中必须至少有一个数字字符
Msg 17720, Level 16, State 1:
Server 'XIONGHE', Procedure 'sp_password', Line 128:
```

```
错误: 不能设置口令
(return status = 1)
```

在设定口令必须有一个数字以后，添加用户 demo1，口令 demopwd 失败。新的规则只对新登录用户有效。对已经添加的数据库用户，则不加限制。

作为数据库的 DBA，有责任将登录用户的口令设置得足够安全，至少是大小写字母及数字的强组合，并且口令中不应该有字典式的英文单词。

9.2.3 ASE 中的特殊登录用户

在 ASE 数据库中，有几个特殊的登录用户，它们对于 DBA 的日常管理和维护非常重要。

1. sa 用户

首先，我们要区分一下 sa 用户和 SA（系统管理员），sa 只是一个登录用户，在安装的时候就已经创建好了，它拥有所有可能的数据库权限，因为系统给了它所有的角色（三个标准角色）。而 SA 是 ASE 的系统管理员，他拥有系统管理员的职责，控制了服务器所有的资源，包括 ASE 的环境及其所有功能。出于安全的考虑，在很多生产环境当中，sa 用户基本上都被锁定，由 SA 创建另一个合适的用户并赋予管理员的角色。

推荐禁止使用默认的 sa 用户有两个原因：一是默认的 sa 用户是没有密码的，直接可以登录。其实，安装程序可以在安装阶段就强行让用户设定 sa 口令的。这个任务也只能交给 DBA 在安装完成以后再行设定；另一个原因是，这个用户的用户名成了众所周知的用户，同时它具有三种角色权限：sa_role（管理员权限）、sso_role（安全管理权限）、oper_role（系统备份相关权限），具体的职责无法分清。因此，作为对 ASE 服务器级别的最低安全要求，首要的是在安装完 ASE 之后，立即设置 sa 用户的高强度密码。至于是否禁用 sa 用户，取决于具体的运行环境，如果有多名 DBA 同时维护一套数据库，那么可以创建三个登录用户，分别授予 sa_role、sso_role、oper_role，然后禁用 sa 用户。

在所有的数据库中，sa 用户（或者具有 sa_role 角色权限的用户）会被自动当作该数据库的 DBO（数据库的拥有者）。

通过查询系统函数 suser_name 和 user_name，可以随时得到登录用户及当前的数据库用户名。例如：

```
1> select suser_name()
2> go

 --------------
  sa
(1 row affected)
1> select user_name()
2> go
 --------------
```

dbo
(1 row affected)

这里给出一个示例,创建三个分属三种角色的管理员用户,用于数据库管理,在创建完这三种角色的管理员用户之后,再将 sa 用户锁定(好禁用)。

```
1> sp_addlogin sa1, 'sa1_pwd'
2> go
```

口令设置正确。

账户被解锁。

新的登录被创建。

(return status = 0)
```
1> sp_addlogin sa2, 'sa2_pwd'
2> go
```

口令设置正确。

账户被解锁。

新的登录被创建。

(return status = 0)
```
1> sp_addlogin sa3, 'sa3_pwd'
2> go
```

口令设置正确。

账户被解锁。

新的登录被创建。

(return status = 0)
```
1> grant role sa_role to sa1
2> go
1> grant role sso_role to sa2
2> go
1> grant role oper_role to sa3
2> go

1> sp_locklogin sa, 'lock'
2> go
```

警告:指定的一个或多个账户处于活动状态。

账户被加锁。

(return status = 0)

你会发现,再次以 sa 用户登录时会失败:

C:\>isql -Usa

保密字：

Msg 4002, Level 14, State 1:
Server 'XIONGHE':
Login failed.
CT-LIBRARY error:
   ct_connect(): 协议特定层: 外部错误: 试图连接服务器失败

要想重新启用 sa 用户，必须使用用户 sa2（具有 sso_role 安全管理权限），对 sa 用户进行解锁。如何解锁后边会有介绍。这种做法有点类似于"三权分立"，三种角色各司其职。

2. dbo 用户

数据库属主拥有完全掌控当前数据库的所有权限。dbo 不是登录用户，它是隶属于具体数据库的能完全控制该数据库的数据库用户。简言之，它是属于数据库级别的用户，不是属于服务器级别的 ASE 登录用户。

默认情况下，数据库的属主是创建该数据库的登录用户。默认情况下，只有具备 sa_role 角色的登录用户可以创建数据库。当然这种 dbo 用户之后是可以转移的。

dbo 用户控制数据库的方式与 SA 控制整个 ASE 服务是类似的，dbo 决定谁可以访问它拥有的数据库，以及每个用户在该数据库下拥有什么样的权限。在生产环境中，通常由 dbo 用户创建所有的数据库对象。

dbo 用户可以对数据进行装载（load）或备份（dump），这对应于角色 oper_role，还可以执行检查点操作，并使用 dbcc（数据库一致性检查）命令对数据库进行检查。

dbo 用户可以执行 setuser 命令，允许它以另外一个用户的身份来使用当前数据库，它在某些特殊的情况下非常有用，比如某些数据库对象只有某个用户才能访问，就连 dbo 都无法直接访问，dbo 要想访问该对象，只有先 setuser 到该用户，才能访问该对象。

3. 对象属主用户

对象的属主完全控制着它所拥有的对象，如谁能访问某一个表、存储过程等，就像前边提到的一样，就连 dbo 也无法直接访问该对象。

所谓对象的属主，就是创建该对象的数据库用户。在生产环境里，我们通常推荐使用 dbo 用户作为对象的属主，这样会把用户管理简化。尽管如此，这并不是强制性的要求。在我们的示例数据库中，iihero 数据库中的数据库表就不是由属主用户 dbo 创建的，而是由普通的数据库用户 spring 创建的。

由普通的数据库用户成为对象属主用户，造成的一个不好的后果就是，这种属主关系是无法转移的。dbo 可以转移（使用 sp_changedbowner 存储过程可以实现），而对象属主却不能直接转移，这也是生产环境里不推荐使用普通用户创建数据库对象的一个原因。

创建数据库对象的对象属主，同时具备了修改对象元信息的权限。具备 sa_role 权限的登录用户或者 dbo 用户，在 setuser 到这个对象属主以后，同样拥有相应的权限。

如 dbo 用户 sa 登录到 iihero 数据库下，直接访问 student 表时会出错，但是在 setuser 'spring' 以后，则可以直接访问：

```
c:\>isql -Usa
```

保密字：

```
1> use iihero
2> go
1> select sname from student
2> go
Msg 208, Level 16, State 1:
Server 'XIONGHE', Line 1:
```

未发现 student，指定 owner.objectname 或使用 sp_help 检查对象是否存在（sp_help 可产生很多输出）。

```
1> setuser 'spring'
2> go
1> select sname from student
2> go
 sname
 --------------------
 李勇
 刘晨
 王敏
 张铁林
(4 rows affected)
```

9.2.4 ASE 中的标准角色（role）

角色（role）就是一个或若干个具体权限的组合，它可以是系统安装完成以后内建的角色，也可以是用户自定义的角色（由管理员在安装数据库完成以后自行创建）。而一个登录用户可以同时拥有多种角色。

在 ASE 安装完成以后，会为系统创建几个标准的角色。sa 用户就同时拥有三种角色，如图 9-2 所示。

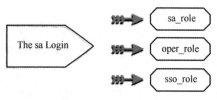

图 9-2 sa 的三种角色

这三种角色相互之间没有交集，分属于三个方面的权限集合。下面介绍这三个标准的系统角色分别有哪些权限。

1. sa_role

主要职责有：

- 管理硬盘资源（创建设备、创建数据库等）。
- 对 sa_role 进行授权或撤销授权。
- 执行 dbcc 命令。
- 配置服务器级的参数（包括服务器级别的内存分配）。
- 关闭 Server。
- 其他一些服务器管理任务。

2. oper_role

操作员（operator）角色，主要职责就是备份和装载数据库，当然，数据库的属主用户仍然可以备份和装载数据库。它的职责也仅限于这个，不能直接使用数据库中的对象。

请看示例，sa3 是一个拥有 oper_role 角色的用户，它可以对任意用户数据库进行备份和装载，但是不能使用特定的数据库。例如：

```
c:\>isql -Usa3
```

保密字：

```
1> use iihero
2> go
Msg 10351, Level 14, State 1:
Server 'XIONGHE', Line 1:
: 在数据库 'iihero' 中，服务器用户 ID 9 是无效用户
```

3. sso_role

sso_role 主要用于安全配置和管理，拥有此角色的人称为 SSO（系统安全管理员），主要职责如下：

- 创建 ASE 的登录用户，包括设置初始密码。
- 改变任意一个用户的密码。
- 对角色进行授权和撤销授权，但是不能对 sa_role 进行授权和撤销授权。
- 创建角色。
- 设置口令及其他安全有关的选项。
- 管理 ASE 的审计（audit）系统。

sso_role 不同于 sa_role，拥有 sso_role 角色的 SSO 拥有整个 ASE 的安全职责，这也包括设置服务器范围内的口令配置参数。如要求口令有效期为 90 天：

```
sp_configure 'password expiration', 90
```

这就强制要求用户每隔 90 天要修改其密码。

SSO 可以在不删除用户的前提下，阻止某用户对服务器访问，那就是通过锁定该用户来实现。在 ASE 中，只要数据库中拥有某用户创建的对象，你就不能直接删除此用户，但是你可以锁定该用户以禁止该用户对数据库的访问。锁定某一个用户的命令格式如下：

sp_locklogin   login_name, "lock" | "unlock"

如锁定登录用户 spring，之后 spring 用户将无法访问 ASE 数据库。

sp_locklogin   spring, "lock"

对 spring 用户进行解锁中，只需要调用：

sp_locklogin   spring, "unlock"

在锁定以后，虽然 spring 用户无法访问数据库，但是它所拥有的数据库仍然可以被 dbo 访问，通过 setuser 就可以达到目的，这个并不矛盾，例如：

C:\>isql -Usa

保密字：

```
1> sp_locklogin spring, 'lock'
2> go
账户被加锁
(return status = 0)
1> use iihero
2> go
1> setuser 'spring'
2> go
1> select sname from student
2> go
 sname
 --------------------
 李勇
 刘晨
 王敏
 张铁林
(4 rows affected)
```

SSO 另一个常见的任务是修改登录用户的密码，通过存储过程 sp_password 来实现：

sp_password   caller_pwd, new_pwd [, login name]

每个登录用户都可以修改自己的密码，不用指定登录名，而 SSO 可以通过指定登录名，修改任意一个用户的密码。

在第 1 章里提到的忘记了 sa 用户的密码，可以通过-psa 额外的启动选项让 ASE 启起来，最终生成一个新的 SSO 口令，这里的实际原理就是通过在-p 后边加上一个具有 sso_role 角色的用户名

220

（这里是 sa，也可以是其他）来实现的。

SSO 有职责删除一些不用的登录用户，语法是：

sp_droplogin login_name

当然，删除登录用户有如下限制条件：不能删除已经拥有数据库对象（表、存储过程等）的用户，想禁止此类用户，可以通过 sp_locklogins 达到目的。只有要这些数据库对象完全转移（通过重新创建并重新装载）之后，才能删除此类用户。

同时，最后一个未锁定的 SSO 或者 SA 用户是你不能删除的，这个限制是合理的，如果这两个权限的用户删除了，最后系统就无法创建一个具备 SSO 或者 SA 权限的用户了。还有一个限制，如果某个用户正在使用 ASE 期间，它就不能被删除。

要查看某个登录用户的详细信息，可以通过存储过程 sp_displaylogin 来实现，如下例，查看登录用户 sa3 的相关信息：

```
1> sp_displaylogin sa3
2> go
Suid: 9
登录名: sa3
全名:
默认数据库: master
默认语言:
自动登录脚本:
已配置的特权:
         oper_role (default ON)
被锁定: 否
最后口令改变日期: 五月   6 2012   1:48
口令截止间隔: 0
口令到期: NO
最小口令长度: 6
登录失败次数的最大值: 0
当前已失败的登录次数: 0
请用  AUTH_DEFAULT  进行鉴定
登录口令加密:SHA-256
上一次登录日期:
(return status = 0)
```

9.2.5 查看已连接用户

通过 sa 用户，我们始终可以使用 sp_who 来查看已经连到 ASE 的哪些用户，如下是一个示例输出：

```
exec sp_who
go
```

fid	spid	status	loginame	orig-host-spid	blk_xloid	dbname	cmd	block_	name
0	2	sleeping			0	NULL	NULL	NULL	0
master		DEADLOCK TUNE							
0	3	sleeping			0	NULL	NULL	NULL	0
master		SHUTDOWN HANDLER							
0	4	sleeping			0	NULL	NULL	NULL	0
master		ASTC HANDLER							
0	5	sleeping			0	NULL	NULL	NULL	0
master		CHECKPOINT SLEEP							
0	6	sleeping			0	NULL	NULL	NULL	0
master		HK WASH							
0	7	sleeping			0	NULL	NULL	NULL	0
master		HK GC							
0	8	sleeping			0	NULL	NULL	NULL	0
master		HK CHORES							
0	9	sleeping			0	NULL	NULL	NULL	0
master		DTC COMMIT SVC							
0	10	sleeping			0	NULL	NULL	NULL	0
master		PORT MANAGER							
0	11	sleeping			0	NULL	NULL	NULL	0
master		NETWORK HANDLER							
0	15	sleeping			0	NULL	NULL	NULL	0
master		LICENSE HEARTBEAT							
0	18	recv sleep				sa	sa	Motoko	0
testdb		AWAITING COMMAND			0				
0	24	recv sleep				sa	sa	Motoko	0
master		AWAITING COMMAND			0				
0	25	running				sa	sa	MOTOKO	0
master		INSERT			0				

(14 rows affected)

(return status = 0)

我们来看看 status 列各种值的具体含义：

- background, ASE 后台进程。
- infected, 服务器发现一个严重的错误。
- alarm sleep, 等待一个事件。
- lock sleep, 等待获取一个锁。

- latch sleep，等待获取一个自旋锁。
- running，当前进程。
- PLC sleep，等待访问用户日志缓存。
- recv sleep，等待一个网络读。
- remove io，执行一个远程 IO。
- runnable，等待一个引擎空闲。
- send sleep，等待一个网络写。
- sleeping，等待磁盘 IO 或者其他资源。
- stopped，进程终止。
- sync sleep，等待来自其他进程的同步消息。

管理员在查看了已连接用户的进程状态以后，可以调用 kill 命令杀死某些进程。这个命令一般情况下不推荐使用。kill 命令的语法如下：

```
kill spid [with statusonly]
```

当指定选项 with statusonly 时，并不真正杀死相应的进程，而是报告该进程的回退状态。

9.3 数据库级别的安全

当我们通过 sp_addlogin 添加一个登录用户以后，它只是在 syslogins 表中添加了一行记录，并授权该用户可以访问 master 和 tempdb 两个数据库，该用户并不能立即访问用户数据库。

9.3.1 新建数据库用户

用户数据库的访问主要是基于数据库用户 ID 来控制的，最直接的方式是在当前数据库下，使用存储过程 sp_adduser 来添加新的用户 ID 来实现，其语法格式如下：

```
sp_adduser login_name
    [, name_within_database
    [, group_name]]
```

例如：

```
1> use iihero
2> go
3> 1> sp_adduser demo
4> 2> go
5> 新用户被加入
6> (return status = 0)
```

这个存储过程将为当前数据库 iihero 添加登录用户 demo，相应的数据库用户名也叫 demo，当然，你可以为其换一个名字 demo_iihero。

```
1> sp_adduser demo, demo_iihero
2> go
```

新用户被加入。

(return status = 0)

删除一个数据库用户 ID，直接使用 sp_dropuser user_name 即可，如从 iihero 数据库中删除用户 demo_iihero。

```
1> sp_dropuser demo_iihero
2> go
```

用户已从当前数据库中被删除。

(return status = 0)

当用户添加到数据库中时，会在 sysusers 系统表中添加一条记录，用户 ID 会被映射到登录用户的 ID。用户 ID 和登录用户的 ID 不一定完全相等。用户 ID 在每个数据库下是唯一的，但是在 Server 范围内并不一定是唯一的。而登录用户的 ID 是在 Server 范围内唯一的。

为数据库添加用户是 dbo 的职责，dbo 控制着数据库级别的资源，与 SA 控制着 Server 级别的资源类似。具备 sa_role 的登录用户，在默认情况下是所有数据库的 dbo，因而可以为任何一个数据库添加用户，当然，同时也可以为任何一个数据库删除某用户。删除用户有一个前提条件，该用户在当前数据库下所拥有的所有数据库对象必须先被删除。

9.3.2 guest 用户

当一个登录用户试图去使用一个数据库时（如 use <dbname>），如果数据库里并没有对应该登录用户的数据库用户 ID，该用户仍可能有机会访问到该数据库，前提是数据库的 dbo 向该数据库添加了 guest 用户，如下所示：

```
sp_adduser guest
```

当用户数据库包含一个 guest 用户时试图去使用这个数据库，你将自动获得访问该数据库的权限，默认权限是 public 组权限。关于权限本章后边将有相关介绍。这也解释了为何登录用户能够自动访问 master 和 tempdb 数据库，因为这两个数据库中都有一个 guest 用户，并且不能删除。如果 model 数据库中有 guest 用户，那么所有新建的数据库里都会自动包含一个 gueste 用户。sa 用户的 ID 值为 1，dbo 用户的 ID 值为 1，而 guest 用户对应的 suid（登录 uid）值为-1，数据库用户 id 值为 2，这都是固定的。

9.3.3 别名

ASE 中引入了别名的概念，它可以让多个数据库用户看起来像是同一个用户，这对于开发环境来说特别有用。

语法如下：

sp_addalias login_name, user_within_database

示例：

```
1> use iihero
2> go
1> sp_addalias demo, dbo
2> go
```

别名用户被加入。

(return status = 0)

别名是数据库相关的，首先，你必须是在某一个数据库下执行此操作，确定使用的是哪个数据库，直接 select db_name() 可以得到。另外，我们要注意，这里的 demo 用户（登录用户）要被设成 dbo 的别名，任何一个 dbo 的别名用户将假定拥有 dbo 权限。这是一个非常有用的方法，用来确保开发人员之间可以共享访问一个专供开发使用的数据库。

设想一下，在一个多人组成的几个开发小组里，只有一个开发数据库供大家使用，但是出于某种原因，大家又不能共享一个 dbo 用户的用户密码信息，于是每个开发小组创建一个 dbo 的别名，使用不同的用户及密码信息，但是他们访问的是同一个数据库，拥有相同的 dbo 权限，从而达到以相同权限访问同一个数据库的目的。

还有一种情形，就是开发环境与生产环境不一致的情况，比如，生产环境中使用的某个特定用户，在目标数据库下创建了一系列数据库对象（表、存储过程等），这个特定用户与开发环境中的不一样。开发环境中使用的不是同一个用户，出于测试的目的，开发环境可以创建一个针对这个特定用户的别名，最终以特定用户的身份达到访问数据库资源的目的。甚至可以反过来，在生产环境中创建一个针对开发环境用户的别名。

使用 sp_addaias 存储过程时，login_name 不能是用户数据库里已经存在的用户，它必须是 master 数据库 syslogins 表中已经注册的登录用户。新的别名会添加到 sysalternates 表里。通过 sp_dropalias 可以删除一个别名。

我们看看表 sysalternates 表的内容：

```
1> select * from sysalternates
2> go
 suid       altsuid
 ---------- ----------
```

```
     6         4
(1 row affected)
1> select suid, name from master.dbo.syslogins where suid=6 or suid=4
2> go
 suid        name
 ----------- -----------
     4       spring
     6       demo
(2 rows affected)
```

可以看到 sysalternates 表存储的不过是两个 suid 的对应关系。suid 存储的是别名对应的 suid，altsuid 存储的是原始登录用户的 suid 值。

删除刚刚创建的别名 demo：

```
1> sp_dropalias demo
2> go
```

别名用户被删除。

```
(return status = 0)
```

获取用户以及别名的元信息，可以使用存储过程 sp_helpuser [user_name]来获得。

如获取用户 spring 的相关信息：

```
> sp_helpuser spring
2> go
 Users_name           ID_in_db           Group_name            Login_name
 -------------------- ------------------ --------------------- --------------------
 spring               3                  public                spring
```

别名到一个用户的多个用户。

```
 Login_name
 --------------------------
 demo
(return status = 0)
```

9.3.4 访问检查顺序

ASE 如何检查数据库访问的权限呢？首先，ASE 会检查用户是否拥有 sa_role，如果有这个角色，则直接授权，sa_role 是默认的 dbo 用户，拥有所有权限。

如果没有 sa_role，那么会检查 sysusers 里是否有此数据库用户，如果有此用户，也会直接授权。

如果 sysusers 没有此用户，ASE 会检查 sysalternates 里是否存储着该别名，如果有，也会授权。

如果 sysalternates 里也没有该别名，ASE 会检查该数据库是否有 guest 用户，如果有，因为所有用户都可以用 guest 身份访问数据库，所以也会授权。

如果目标数据库里也没有 guest 用户，ASE 会进行最后一道检测，看看该用户是否拥有 sso_role（安全管理员角色），如果有此角色权限，一样授权通过。如果该用户也没有 sso_role，最终会被拒绝访问。

检测的流程图如图 9-3 所示。

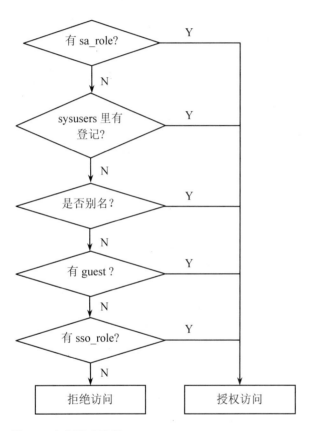

图 9-3　权限检查流程

9.3.5　数据库访问的设置途径

总结起来，大概有如下 6 种方式使某登录用户能够访问指定的目标数据库：

- sp_adduser login：直接将 login 添加到目标数据库中。
- sp_adduser guest：为目标数据库添加 guest 用户。
- sp_addalias login, <dbuser>：为目标数据库的 dbuser 添加一个别名，该别名就是当前 login。

- sp_changedbowner login：为目标数据库设定 dbowner。
- grant role sa_role to login：为当前 login 授予 sa_role 角色。
- set proxy。

前 4 种我们已经介绍过，其中 sp_changedbowner 必须在目标数据库下执行。

第 5 种方法显而易见，拥有 sa_role 角色就能够访问数据库里所有的资源。

最后一种方法 set proxy，我们将在后边进行介绍。

9.3.6 组 group

组基于指定的数据库，由若干用户组成，并授予一些公共的权限。它简化了数据库用户的创建。这意味着我们可以一次性地创建所有要用的权限组合给一个组，然后直接往这个组里边添加用户。

如果要创建跨多个数据库的组，只能用 Server 范围的角色来代替了，它比 group 来得更实用。

一个用户除了属于 public 组以外，最多只能属于一个组，但是一个用户可以有多个角色。

在 ASE 12.5 以前的版本里，sysusers 表中 uid>16383 的行都是为 group 创建的，12.5 版本及以后的版本中，新的 group 对应的 uid 值介于全局变量@@mingroupid 和@@maxgroupid 之间。

例如：

```
1> select "group name"=name from sysusers where uid between @@mingroupid and @@maxgroupid
2> go
 group name
 ------------------------------
 sa_role
 sso_role
 oper_role
 sybase_ts_role
 navigator_role
 replication_role
 dtm_tm_role
 ha_role
 mon_role
 js_admin_role
 messaging_role
 js_client_role
 js_user_role
 webservices_role
 keycustodian_role
(15 rows affected)
```

下边是操作 group 的一些示例：

```
1> sp_addgroup market
2> go
```

新组被加入。

```
(return status = 0)
1> sp_addgroup develope
2> go
```

新组被加入。

```
(return status = 0)
1> sp_dropgroup develope
2> go
```

组已经被删除。

```
(return status = 0)
1> sp_changegroup market, spring
2> go
```

组被改变。

```
(return status = 0)
1> sp_changegroup "public", spring
2> go
```

组被改变。

```
(return status = 0)
```

首先，创建了一个组 market，删除组 develop，将 spring 数据库用户添加到组 market 当中，接着又将 spring 用户放回到 public 组当中（public 是关键字，所以要加引号）。

通过 sp_helpgroup [group_name]可以获取一个组的元信息，如获取当前数据库的 public 组的相关信息：

```
> sp_helpgroup "public"
2> go
 Group_name                Group_id            Users_in_group                Userid
 ------------------------- -------------------- ----------------------------- ----------------
 public                    0                   dbo                           1
 public                    0                   guest                         2
 public                    0                   spring                        3

(3 rows affected)
(return status = 0)
```

在 sysusers 表中有一个特征，uid 值为 1 的是 dbo 用户，值为 2 的是 guest 用户，而所有的组对应的 id 值都大于 16383。

9.3.7　用户名对传输数据库的影响

由于 syslogins 和 sysusers 表之间是通过 suid 进行关联的，当数据库从一台服务器 dump 出来，再 load 到另一台服务器，很难保证两个表之间完全对应起来。管理员有时候需要做一些额外工作。

例如，master..syslogins 有下边的记录：

```
1> select name, suid from master..syslogins
2> go
 name                  suid
 -------------------- ------------
 sa                    1
 spring                4
 demo                  6
```

sysuers 表有如下记录：

```
1> select name, suid, uid from sysusers where uid<16383
2> go
 name                  suid         uid
 -------------------- ------------ ------------
 public                -2           0
 dbo                   1            1
 guest                 -1           2
 spring                4            3
```

别名相关信息：

```
1> select *from sysalternates
2> go
 suid         altsuid
 ------------ -----------
  6            1
```

demo 用户是 dbo 别名，当数据库 iihero 备份完，恢复到另一台机器上时，谁是 iihero 的属主呢？本应该是 ID 值为 6 的用户，可是这个时候，ID 值为 6 的用户很可能已经是其他用户了。所以，比较实用的办法是，始终使用用户 SA 作为数据库的属主，如果用户 demo 想拥有 dbo 权限，直接创建别名给它。

如果不是这样，可以事后添加一些用户给新机器上的数据库：

```
select "exec sp_adduser " + name + ", " + name + ", "
    + user_name(group)
  from sysusers where uid between 3 and 16383
```

这样即可生成必要的脚本。这些 user 对应的 login 也必须在 master 数据库里事先生成好。

9.4　数据库对象级别的安全

数据库对象级别的安全是通过 grant 和 revoke 语句来实现的。

对象属主可以授予以下权限：

- select
- insert
- update
- delete
- execute
- references
- update statistics
- delete statistics
- truncate table

语法形式如下：

```
grant all | permission_list on object [column_list]
    to user_list
    [with grant option]
```

示例：

```
grant exec on echo_demo to spring
```

这里的 user_list 可以是一个用户、角色、组，甚至是它们的组合。grant 操作还可以通过 with grant option 选项传递给下一个用户，使得下一个用户也有授予此权限的权力。column_list 参数可以指定对象的某些列的访问权限。

除了 execute 权限以外，其他所有权限都可以用于表的访问授权。

回收权限使用 revoke 语句，其语法形式如下：

```
revoke [grant option for] all | permission_list on
    object [column_list] from user_list
    [cascade]
```

例如：

```
revoke update on student (sname) from test
```

通过 grant 和 revoke 的组合，可以创建比较细致的权限集合：

```
grant select on employees to public
revoke select on employees (salary) from public
```

我们不必为每一列创建 grant select 权限，只去掉那些不让访问的列即可。

要查询某用户或者对象的权限信息，使用存储过程 sp_helpprotect 即可得到。

语法如下：

```
sp_helpprotect [user_name | object_name]
```

要想获取表 student 的权限信息，可以执行如下语句：

```
1> sp_helprotect student
2> go
 grantor                     grantee                      type
 action                      object
 column                      grantable
 --------------------------  --------------------------  ----------------

(1 row affected)
(return status = 0)
```

9.4.1 系统中的默认角色

ASE 中除了 sa_role, oper_role, sso_role 三个比较常用的默认角色以外，还有下述系统角色，我们从系统表 syssrvroles 表中可以得到所有结果：

```
1> select srid, name, status from syssrvroles
2> go
 srid        name                            status
 ----------- ------------------------------- ----------------
 0           sa_role                         NULL
 1           sso_role                        NULL
 2           oper_role                       NULL
 3           sybase_ts_role                  NULL
 4           navigator_role                  NULL
 5           replication_role                NULL
 7           dtm_tm_role                     NULL
 8           ha_role                         NULL
 10          mon_role                        NULL
 11          js_admin_role                   NULL
 12          messaging_role                  NULL
 13          js_client_role                  NULL
 14          js_user_role                    NULL
 15          webservices_role                NULL
 16          keycustodian_role               NULL

(15 rows affected)
```

这些角色的作用如表 9-2 所示。

表 9-2　角色说明

角色	作用
sa_role	系统管理员
sso_role	安全管理
oper_role	操作管理（备份、恢复等）

续表

角色	作用
replication_role	复制进程
sybase_ts_role	执行一些内部维护任务
dtm_tm_role	分布式事务角色
ha_role	高可用服务器角色
mon_role	提供对监视表的访问
js_admin_role	管理任务调度器
js_client_role	执行任务调度器中的任务
js_user_role	创建和运行调度器中的任务
messaging_role	管理和执行实时消息服务
navigator_role	使用于导航服务器
web_services	管理 Web Services

有三个系统表与角色有关，它们之间的关系如图 9-4 所示。

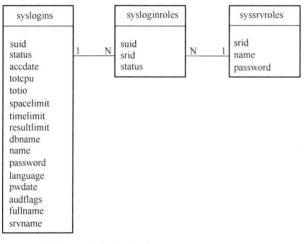

图 9-4　角色相关表间的联系

主要是通过 syslogins 表中的 suid，sysloginroles 表中的 suid、srid 以及 syssrvroles 表中的 srid 进行关联的。通过上述关联，我们可以查询到所有 suid 所拥有的角色名。

例如：

```
1> select a.name, c.name, b.status from syslogins a, sysloginroles b, syssrvroles c where a.suid=b.suid and b.srid=c.srid
2> go
 name                name                          status
 -------------       ----------------------------  --------
 sa                  sa_role                       1
 sa1                 sa_role                       1
```

sa	sso_role	1
sa2	sso_role	1
sa	oper_role	1
sa3	oper_role	1
sa	sybase_ts_role	1
sa	js_admin_role	1
jstask	js_admin_role	1
jstask	js_user_role	1

(10 rows affected)

9.4.2 角色方式授权

通过自定义角色进行授权，当系统需要对表或其他数据库对象进行细致的权限划分，多个用户共用某一套权限时，完全可以将这套权限定义成一个或多个角色，然后将这些角色授予给那些用户。

角色是服务器范围内的对象，所以要在 master 数据库中执行。也只有 SA 或者 SSO 才有权限创建角色。

创建角色的命令语法如下：

```
craete role role_name [ with passwd "password" [, {passwd expiration | min passwd length | max failed_logins} option_value]]
```

其中，各参数含义说明如下：

- role_name，新建角色名称。
- password，激活该角色时要用的密码。
- passwd expiration，密码有效天数，有效值范围为 0 到 32767 之间。
- min passwd lengh，密码最短长度。
- max failed_logins，允许最大登录失败次数。
- option_value，就是 passwd expiration、min passwd length、max failed_logins 这几个选项对应的值。

下边的示例创建了两个自定义角色，角色 role_nopwd 不带任何密码，角色 role_2 带有密码 pse_role2：

```
1> use master
2> go
1> create role role_nopwd
2> go
1> create role role_2 with passwd "pse_role2"
2> go
```

删除角色使用如下命令：

```
drop role role_name [with override]
```

其参数含义说明如下：
- role_name，要删除的角色名。
- with override，不检查该角色在所有数据库中的访问权限是否已被先行删除。

下面的两个命令分别删除角色 role_nopwd 和 role_2：

```
1> drop role role_nopwd
2> go
1> drop role role_2 with override
2> go
```

删除角色 role_2 时，会从所有数据库中收回与该角色相关的权限授予。删除角色有如下限制：
- drop role 权限不会包含在 grant all 命令中。
- 只能在 master 数据库中删除自定义角色。
- 不要试图删除任何系统角色，如 sa_role、sso_role、oper_role 等。

ASE 中的自定义角色的工作状态有两种：激活和未激活。授予用户的角色只有在处于激活状态时，其拥有的权限才在这个用户上有效。

激活和关闭新创建的角色，其语法说明如下：

```
set role role_name [with passwd "password"] [on | off]
```

参数意义如下：
- role_name，激活或关闭的角色名。
- with passwd "password"，激活指定角色时需要的密码。
- on | off，激活（on）或关闭（off）的开关。

下面我们来看一个综合示例，我们先创建两个自定义角色：role1 和 role_pwd：

```
1> create role role1
2> go
1> create role role_pwd with passwd 'role2_pwd'
2> go
```

创建两个登录用户：

```
1> sp_addlogin user1, 'user1_pwd'
2> go
```

口令设置正确。
账户被解锁。
新的登录被创建。

```
(return status = 0)
1> sp_addlogin user2, 'user2_pwd'
2> go
```

口令设置正确。
账户被解锁。
新的登录被创建。

(return status = 0)

对这两个角色分别进行授权：

```
1> grant role role1 to user1
2> go
1> grant role role_pwd to user2
2> go
```

再对这两个角色分别进行权限定义：

```
1> use iihero
2> go
1> grant update on iihero.spring.student to role1
2> go
1> grant update on iihero.spring.emp to role_pwd
2> go
```

这样在角色被激活之后，权限定义就可以生效了：

```
1> use iihero
2> go
1> set role role1 on
2> go
1> set role role_pwd with passwd 'role2_pwd' on
2> go
```

set role on 或 off 一定是被授予角色权限的用户自己去执行。为避免每次都要执行 set role on 的操作，可以通过修改 login 的 default role 来自动置 role 为 on 的状态。

例如：

```
1> sp_modifylogin demo, 'add default role', 'role1'
2> go
Option changed.
(return status = 0)
```

这样，在用户 demo 登录以后，不用再显式地设置 set role on 了。

从实际应用的角度来讲，如果一个数据库涉及到很多物理用户（即数据库用户，不是基于业务逻辑的用户），并且每个用户的权限各不相同，则可以使用角色化管理进行权限细分。否则，完全可以通过 sp_changedbowner 直接将 DBO 赋给某个登录用户或使用 sp_addalias login_name（dbo）进行别名设置。这样做的好处是维护非常方便。当然，这时的登录用户拥有对该数据库里所有对象的所有权限。

9.5　对 SSL 协议的支持、配置管理及使用

Sybase ASE 对 SSL 协议的支持是一项非常重要的功能，官方文档上介绍也不是很详细，但是作为数据库安全的一个重要功能，它的用途却是非常重要，尤其是在涉及安全的关键性信息系统中，对 SSL 协议的支持就显得非常重要了。

由于本书不是专门介绍安全方面的知识，这里只对涉及 SSL 通信的知识进行简要的介绍。

一个涉及客户端和服务器的通信系统要想做到传输加密，并且不被外界破解其中的报文，通常的解决方案是，有一套公钥（公开）加密，私钥来解密（不公开）报文。而公钥要由已授权机构进行授权，表示公钥不是第三方伪造。通过一个已授权的根证书，可以创建服务器端使用的证书，里边有对应于根证书的签名，形成可信的证书链。这样，客户端通过根证书就可以与服务器端使用 SSL 加密通信了。

在 Sybase ASE 中，要支持 SSL 加密需要具备如下条件：

1. 要有 ASE_ASM 的许可（License），Developer 版默认情况下是有这个许可的。
2. 要在服务器级别启用 ssl 选项，即要在 sa 用户下执行：

sp_configure "enable ssl", 1

执行完后，检查一下服务器端日志，看看是否有错。
上述准备工作做完以后，就可以进行 SSL 相关的配置了。

9.5.1　服务器端 SSL 的配置

在准备了上述工作以后，在服务器端要支持 SSL，需要配置两套含有加密密钥的证书，一个是获得授权的根证书，另一个是服务器端的专用证书。

Sybase ASE 有两个命令可以用于这些证书的创建，其步骤是先用 certreq 来创建证书的请求文件及密钥文件，再用 certauth 命令来生成相应的证书。

以制作根证书 root.cer 为例，看看一个 certreq 是如何使用的，进到服务器的 $SYBASE/$SYBASE_ASE/certificates 目录，执行命令 certreq：

```
xionghe@seanlinux2:~/ase1503/ASE-15_0/certificates$ certreq
Choose certificate request type:
    S - Server certificate request.
    C - Client certificate request.
    Q - Quit.
Please enter your request [Q] : S
Choose key type:
    R - RSA key pair.
    D - DSA with ephemeral Diffie-Hellman key exchange.
```

```
    Q - Quit.
Please enter your request [Q] : R
Enter key length (512,768,1024 for DSA; 512-2048 for RSA) : 512
Country: CN
State: beijing
Locality: haidian
Organization: sql9
Organizational Unit: rd
Common Name: SEANLINUX2
Generating key pair (please wait)...
Enter password for private key (max 64 chars):
Enter file path to save request: root.req
Enter file path to save private key: root.key
```

这里最关键的就是 Common Name 一项，它的值必须设置成跟你安装时的服务名一样，即你的 interfaces 文件中的 ASE 服务 5000（默认端口）对应的服务名。

做完这一步之后，会在当前目录下生成 root.req（请求文件）及 root.key（密钥）文件。

利用这两个文件以及 certauth 命令可以生成根证书，命令如下：

```
xionghe@seanlinux2:~/ase1503/ASE-15_0/certificates$ certauth -r -C root.req -Q root.req -K root.key -P sybase -T 365 -O root.cer
-- Sybase SSL Certificate Authority Utility --
Certificate Validity:
 startDate = Thu Aug 30 22:55:21 2012
 endDate = Fri Aug 30 22:55:21 2013
setting serial number 0x632bffffa5336298fff
Could not sign certificate using signature type 20, error 'No error string returned.' (3000).
Could not sign certificate using signature type 22, error 'No error string returned.' (3000).
CA sign certificate SUCCEED using signature type 2, return 'SSLNoErr' (0).
```

得到根证书文件 root.cer，我们将其重命名为"<服务名>.txt"，放入 certificates 目录当中：xionghe@seanlinux2:~/ase1503/ASE-15_0/certificates$ mv root.cer SEANLINUX2.txt。

接着，要创建服务器端要用的专用证书 server.cer，也分为生成请求文件和密钥文件，然后再生成证书文件两大步。

Step 1 server.req 及 server.key 文件生成。

```
    xionghe@seanlinux2:~/ase1503/ASE-15_0/certificates$ certreq
    Choose certificate request type:
      S - Server certificate request.
      C - Client certificate request.
      Q - Quit.
    Please enter your request [Q] : S
    Choose key type:
      R - RSA key pair.
```

```
        D - DSA with ephemeral Diffie-Hellman key exchange.
        Q - Quit.
Please enter your request [Q] : R
Enter key length (512,768,1024 for DSA; 512-2048 for RSA) : 512
Country: CN
State: beijing
Locality: haidian
Organization: sql9
Organizational Unit:
Common Name: SEANLINUX2
Generating key pair (please wait)...
Enter password for private key (max 64 chars):
Enter file path to save request: server.req
Enter file path to save private key: server.key
```

这一步与 root.req 文件生成基本上一样。

Step 2 生成 server.cer 证书文件。

```
xionghe@seanlinux2:~/ase1503/ASE-15_0/certificates$ certauth -C SEANLINUX2.txt -Q server.req -K root.key
-P sybase -T 365 -O server.cer
-- Sybase SSL Certificate Authority Utility --
Certificate Validity:
        startDate = Thu Aug 30 23:05:06 2012
        endDate = Fri Aug 30 23:05:06 2013
setting serial number 0xffffd0e0ffff8ae3fff
Could not sign certificate using signature type 20, error 'No error string returned.' (3000).
Could not sign certificate using signature type 22, error 'No error string returned.' (3000).
CA sign certificate SUCCEED using signature type 2, return 'SSLNoErr' (0).
```

我们来看看这个命令中几个参数的含义：

- -C: 后边指定的是根证书文件名（本来是 root.cer，已经重命名为 SEANLINUX2.txt）。
- -Q: 证书请求文件。
- -K: 根证书的密钥文件。
- -P: 新证书的密码。
- -T: 期限，以天为单位。
- -O: 指定最终生成的证书文件名，这里为 server.cer。

Step 3 经过上述两个步骤，server.cer 终于生成，但并不能立即用于服务器当中，在 server.cer 文件里，需要将它的密钥内容添加到其末尾，即要执行：

```
xionghe@seanlinux2:~/ase1503/ASE-15_0/certificates$ cat server.key >> server.cer
（Windows 下，可以执行命令 type server.key >> server.cer）
```

Step 4 接着进行证书的安装工作，使用 isql 命令使 sa 用户连接到服务器，使用 sp_ssladmin 进行相关管理，请看：

```
1> sp_ssladmin
2> go
sp_ssladmin Usage: sp_ssladmin command [, option1 [, option2]]
sp_ssladmin commands:
sp_ssladmin 'addcert', 'certificatepath', 'password'
sp_ssladmin 'dropcert', 'certificatepath'
sp_ssladmin 'lscert'
sp_ssladmin 'setcipher', { 'FIPS' | 'Strong' | 'Weak' | 'All' |
quoted_list_of_ciphersuites
sp_ssladmin 'lscipher'
sp_ssladmin 'help'
(return status = 0)
```

这里需要安装 server.cer 证书，执行：

```
1> sp_ssladmin 'addcert', '/home/xionghe/ase1503/ASE-15_0/certificates/server.cer', 'sybase'
2> go
(return status = 0)
```

这里第二个参数是 server.cer 证书全路径，第三个参数是这个证书的密码。

Step 5 安装成功以后，停止 ASE 服务，修改 interfaces 文件中的主服务项。

```
SEANLINUX2
        master tcp ether seanlinux2 5000 ssl
        query tcp ether seanlinux2 5000 ssl
```

在主服务项每行的末尾添加"ssl"（UNIX 平台）或", ssl"（Windows 平台）。

Step 6 再次启动 ASE 服务。

在服务启动以后，我们如果仔细查看服务的日志文件，可以看到下述内容：

```
00:00000:00001:2012/08/30 23:22:33.39 kernel   Certificate load from file `/home/xionghe/ase1503/ASE-15_0/certificates/server.cer`: succeeded.
00:00000:00001:2012/08/30 23:22:33.39 kernel   Trusted root certificates loaded from file '///home/xionghe/ase1503/ASE-15_0/certificates/SEANLINUX2.txt': succeeded.
00:00000:00008:2012/08/30 23:24:03.19 kernel   network name seanlinux2, interface IPv4, address 192.168.0.67, type ssltcp, port 5000, filter ssl
```

这意味着服务器以支持 SSL 的方式启动起来了。

这时如果我们用 isql 命令直接连接数据库，一般都会出现以下错误：

```
xionghe@seanlinux2:/etc$ isql -Usa -Jutf8
```

保密字：

```
CT-LIBRARY error:
    ct_connect(): 网络包层: 内部 net library 错误: 连接两个结束点的 Net-Library 协议驱动程序调用失败
```

第 9 章　ASE 的用户及安全管理

这是什么原因呢？这是因为客户端不能再直接和 Server 端以旧的方式连接 5000 端口了。它要依赖于下一节介绍的客户端 SSL 配置。

9.5.2　ASE 客户端 SSL 配置

这里主要介绍两种重要的客户端，只要知道了这两种客户端的 SSL 配置，再配置其他客户端就可以触类旁通了。

1. 命令行客户端 isql（或 Openclient 运行客户端）

首先介绍 Openclient 客户端，isql 命令行就是基于此客户端。

Windows 下，进入到%SYBASE%\ini 目录，将根证书（即前边提到的 SEANLINUX2.txt）里边的内容追加到已有文件 trusted.txt 末尾即可，这样会形成一个可信任的证书链，供客户端选择。

这里，将 Linux 下 SEANLINUX2.txt 根证书的内容复制到 Windows 下 ASE 的 ini 目录 trusted.txt 当中，然后修改 sql.ini 文件，添加注册项：

```
[SEANLINUX2]
master=NLWNSCK,192.168.0.67,5000,ssl
query=NLWNSCK,192.168.0.67,5000,ssl
```

使用命令"isql -Usa -SSEANLINUX2"很容易就能连接到目标数据库。

需要注意的是，Windows 下，要修改的 trusted.txt 文件位于目录 ini 当中，而 UNIX/Linux 下，该证书文件位于 $SYBASE/config 目录当中。

那么，如果连接出错，服务器端有没有什么可以诊断的开关呢？默认情况下，服务器端只给出很少的提示，要想得到更多的信息，可以在 startserver 的脚本中加入开关"-T7827 -T7829 -T3605"，这样会输出更多的日志供你诊断。

看一个示例 RUN_SEANLINUX2：

```
/home/xionghe/ase1503/ASE-15_0/bin/dataserver -T7827 -T7829 -T3605 \
-d/home/xionghe/ase1503/data/master.dat \
-e/home/xionghe/ase1503/ASE-15_0/install/SEANLINUX2.log \
-c/home/xionghe/ase1503/ASE-15_0/SEANLINUX2.cfg \
-M/home/xionghe/ase1503/ASE-15_0 \
-sSEANLINUX2 \
```

这样启动服务器以后，当用户连接出错时，会出现类似下边的诊断信息：

```
……
ssl_readcb() returns sslerr -6992, processed 0
ssl_writecb(0x1fd078c, 7, 0x24328dc, 0x36126c0) entry
ssl_writecb() returns sslerr 0, processed 7
ssl_handshake(14) return err -2, sslerr -6992
ssl_nclose(0x36126c0) entry
00:00000:00018:2012/08/30 23:40:47.11 kernel  ssl_close(14): SSLClose(0x1f9e3b0) failed (-6989).
ssl_nclose(0x36126c0) return 0
```

2. Java 运行时客户端

这里包括两类，一类是 Sybase Central 类的工具，另一类是用户自己编写的基于 JDBC 的应用程序。

如果使用 Sybase Central 工具，我们要进到 Sybase Central 的 JRE 相关目录 jre\lib\security 里，先删除原来的 cacerts 文件（可以将其重命名），然后将根证书导入，重新生成一个，命令如下：

```
D:\Sybase15\Shared\JRE-6_0_6_32BIT\lib\security>c:\shared\jdk1.6.0_02\bin\keytool  -import  -alias  SEANLINUX2
-keystore cacerts -file D:\Sybase15\ini\root.cer
Enter keystore password:
Re-enter new password:
Owner: CN=SEANLINUX2, OU="
", O="
", L="
", ST=CA, C=US
Issuer: CN=SEANLINUX2, OU="
", O="
", L="
", ST=CA, C=US
Serial number: ffff9eed65edffffc55
Valid from: Thu Aug 30 05:26:00 CST 2012 until: Fri Aug 30 05:26:00 CST 2013
Certificate fingerprints:
         MD5:   96:3B:37:0D:37:E4:45:3D:70:C4:FF:AB:22:9A:5B:F5
         SHA1: 1B:37:95:63:C9:2B:53:17:E6:6E:76:02:12:50:92:06:6F:34:06:35
         Signature algorithm name: SHA1withRSA
         Version: 1
Trust this certificate? [no]:   y
Certificate was added to keystore
```

这样导入以后，就可以使用 Sybase Central 来连接数据库了。

另一类是基于 ASE JDBC 的 Java 应用程序，它需要一个专门为 SSL 通信进行处理的注册类来创建数据用于连接，同 Sybase Central 客户端一样，它也需要将根证书导入一个 cacerts 文件，并且要在程序里指定这个文件的位置。

下边的内容本属于第 16 章 Java 编程接口，但它与安全性密切相关，所以将其放于本章。

这里通过一个实例，展示如何通过 Java 实现 SSL 连接 ASE 数据库。

主程序类 TestSSLConnection（见代码 code\java\src\ssl\TestSSLConnection.java）：

```
#001 package ssl;
#002
#003 import java.sql.Connection;
#004 import java.sql.DriverManager;
#005 import java.util.Properties;
#006
#007 public class TestSSLConnection
```

```
#008 {
#009
#010        public static void test() throws Exception
#011        {
#012
#013            DriverManager.setLogStream(System.out);
#014            Class.forName("com.sybase.jdbc3.jdbc.SybDriver");
#015
#016            Properties props = new Properties();
#017            props.put("SYBSOCKET_FACTORY", "ssl.MySSLSocketFactoryASE");
#018            props.put("user", "sa");
#019            props.put("password", "");
#020
#021            Connection conn = DriverManager.getConnection("jdbc:sybase:Tds:192.168.0.67:5000/master", props);
#022            conn.close();
#023        }
#024
#025        /**
#026         * @param args
#027         */
#028        public static void main(String[] args) throws Exception
#029        {
#030            test();
#031        }
#032
#033 }
```

这里最关键的就是第 17 行，要在连接属性里指定 SYBSOCKET_FACTORY 的实现类。再看实现类 ssl. MySSLSocketFactoryASE，见代码（code\java\src\ssl\ **MySSLSocketFactoryASE.java**）：

```
#001 package ssl;
#002
#003 import java.io.IOException;
#004 import java.net.Socket;
#005 import java.net.UnknownHostException;
#006 import java.util.Properties;
#007 import java.util.Vector;
#008 import javax.net.ssl.SSLSocket;
#009 import javax.net.ssl.SSLSocketFactory;
#010 import com.sybase.jdbcx.SybSocketFactory;
#011 public class MySSLSocketFactoryASE extends SSLSocketFactory implements SybSocketFactory
#012 {
#013
#014     public Socket createSocket(String host, int port) throws IOException, UnknownHostException
#015     {
```

```
#016
#017            // Before creating the socket we need to set some system props
#018            setSystemProperties();
#019            SSLSocket s = (SSLSocket) SSLSocketFactory.getDefault().createSocket(host, port);
#020            setProtocol(s);
#021
#022            // Everything is set up properly now. Go through the handshake to
#023            // finalize the ssl connection.
#024            s.startHandshake();
#025            return s;
#026        }
#027
#028        public Socket createSocket(String host, int port, Properties props) throws IOException, UnknownHostException
#029        {
#030            Vector cipherSuites = new Vector();
#031            String cipherSuiteVal = null;
#032            int cipherIndex = 1;
#033
#034            // Loop through possible multiple cipher suites
#035            do
#036            {
#037                if ((cipherSuiteVal = props.getProperty("CIPHER_SUITES_" + cipherIndex++)) == null)
#038                {
#039                    if (cipherIndex <= 2)
#040                    {
#041                        return createSocket(host, port);
#042                    }
#043                    else
#044                    {
#045                        break;
#046                    }
#047                }
#048                else
#049                {
#050                    cipherSuites.addElement(cipherSuiteVal);
#051                }
#052            }
#053            while (true);
#054
#055            String enableThese[] = new String[cipherSuites.size()];
#056            cipherSuites.copyInto(enableThese);
#057
#058            setSystemProperties();
#059            SSLSocket s = (SSLSocket) SSLSocketFactory.getDefault().createSocket(host, port);
```

```
#060        s.setEnabledCipherSuites(enableThese);
#061        setProtocol(s);
#062
#063        s.startHandshake();
#064        return s;
#065    }
#066
#067    protected void setSystemProperties()
#068    {
#069        String trustStoreLocation = "./cacerts";
#070        System.setProperty("javax.net.ssl.trustStore", trustStoreLocation);
#071    }
#072
#073    protected void setProtocol(SSLSocket s)
#074    {
#075        String[] enableTheseProtocols = { "TLSv1" };
#076        s.setEnabledProtocols(enableTheseProtocols);
#077    }
#078
#079    public java.lang.String[] getDefaultCipherSuites()
#080    {
#081        return null;
#082    }
#083
#084    public java.lang.String[] getSupportedCipherSuites()
#085    {
#086        return null;
#087    }
#088
#089    public java.net.Socket createSocket(java.lang.String host, int port, java.net.InetAddress clientAddress,
#090            int clientPort)
#091    {
#092        return null;
#093    }
#094
#095    public java.net.Socket createSocket(java.net.InetAddress host, int port)
#096    {
#097        return null;
#098    }
#099
#100    public java.net.Socket createSocket(java.net.InetAddress host, int port, java.net.InetAddress clientAddress,
#101            int clientPort)
#102    {
#103        return null;
```

```
#104        }
#105
#106        public java.net.Socket createSocket(java.net.Socket s, String host, int port, boolean autoClose)
#107        {
#108            return null;
#109        }
#110 }
```

我们需要注意，最关键的就是系统属性 javax.net.ssl.trustStore，指定 trustStore 的存储位置，本程序中，它是 cacerts 文件，位于当前目录。类 MySSLSocketFactoryASE 主要实现 createSocket 这个重要的方法用于客户端连接。除此之外，其他的 JDBC 接口访问与常规的 Java 应用程序完全一样。

10 Sybase ASE 中的事务

ASE 数据库支持哪几种事务隔离级呢？使用命令 set transaction isolation level <0|1|2|3>，分别用于定义四种事务隔离级：读未提交、读已提交、可重复读、可串行化。通过例子可以对这几种隔离级进行区分。假设有表 t123，其定义为 create table t123(id int primary key not null, col2 varchar(32))，下边的范例将围绕示例表 t123 展开。

10.1 设置事务模式和隔离级

ASE 支持两种事务模式，一种是符合 SQL 标准的**链式**模式，它在执行任何一个数据检索或者修改操作之前都会隐式开启一个事务。这些语句包括：select、delete、insert、update、open、fetch，必须使用 rollback tran[saction] 或者 commit tran[saction]来结束事务。

另一种是**非链式**模式或 Transact-SQL 模式，它是 ASE 事务的默认模式，它要求显式使用 begin tran[saction]和 rollback tran[saction]或 commit tran[saction]语句对来完成事务。

我们可以通过查询变量@@tranchained 来获取当前的事务模式的值（0 表示为非链式模式，1 表示链式模式）：

```
D:\>isql -Uadventure -Sspring_ase
Password:
1> select @@tranchained
2> go

 -----------
           0

(1 row affected)
```

```
1>
1> set chained on
2> commit
3> go
1> select @@tranchained
2> go

 -----------
           1

(1 row affected)
1> commit
2> go
1> set chained off
2> commit
3> go
1> select @@tranchained
2> go

 -----------
           0

(1 row affected)
```

使用语句 set chained on 可以将事务模式改变为链式模式,这时查询得到的@@tranchained 值为 1；set chained off 则会将事务模式设置为非链式模式,这时查询得到的@@tranchained 值为 0。

ASE 中有四种标准的事务隔离级,我们可以使用语句 set transaction isolation level [0|1|2|3]来设置事务隔离级。ASE 分别用 level 0、1、2、3 来表示读未提交、读已提交、可重复读、可串行化四种事务隔离级,完全遵循 SQL 92 标准。

ASE 中默认的事务隔离级是读已提交,即 level 1。我们通过查询全局变量@@isolation 可以得到默认值。

```
D:\>isql -Uadventure –Sspring_ase
Password:
1> select @@isolation
2> go

 -----------
           1

(1 row affected)
1>
1> set transaction isolation level 3
2> go
1> select @@isolation
```

```
2> go

 -----------
    3

(1 row affected)
1> commit
2> go
```

上边的例子里，我们得到默认的@@isolation值为1（读已提交），接着设置隔离级为3（可串行化）。

通常情况下，事务的隔离级越高，系统的加锁操作会越多，效率也会相应有所下降。实际上往往要按照实际应用的需要选择合适的事务隔离级。

使用链式模式时，SQL 标准要求每个 SQL 数据检索和数据修改语句出现在事务中。开始会话之后，或者提交或中止先前事务之后，将自动从第一个数据检索或数据修改语句开始一个事务。这是链式事务模式。

在链式事务中，ASE 在以下SQL语句之前隐式提交一个 begin transaction 操作来开启一个事务：delete、select、insert、update、open。

而非链式事务模式下，除非显式调用 begin trans，否则每个 SQL 语句都会自动提交一个事务，通常 ASE 数据库的 autocommit 就是通过 set chained off 来实现的。

我们看看默认的非链式事务模式下的一个例子，往表 t123 中插入一条记录，接着开启一个事务，删除该条记录，然后回滚该事务，发现插入操作仍然成功执行。

```
D:\>isql -Uadventure -Siihero_ase
Password:
1> select * from t123
2> go
 id          col2
 ----------- --------------
 1           a
 2           b
 3           c
 4           d
 5           e

(5 rows affected)
1> select @@tranchained
2> go

 -----------
    0

(1 row affected)
```

```
1> insert into t123 values(6, 'f')
2> begin tran
3> delete from t123 where id=6
4> rollback tran
5> go
(1 row affected)
(1 row affected)
1> select * from t123
2> go
 id         col2
 ---------- ------------
 1          a
 2          b
 3          c
 4          d
 5          e
 6 f

(6 rows affected)
1>
```

其实，插入操作并没有开启任何事务（@@trancount 值为 0）。因此，后边的回滚事务也只是撤消后边的 delete 操作。

再看看链式模式下的上述操作所产生的结果：

```
D:\>isql -Uadventure -Siihero_ase
Password:
1> set chained on
2> commit
3> go
1> insert into t123 values(7, 'g')
2> select @@trancount
3> go
(1 row affected)

 -----------
 1

(1 row affected)
1> begin tran
2> delete from t123 where id=7
3> select @@trancount
4> go
(1 row affected)

 -----------
 2
```

```
(1 row affected)
1> rollback tran
2> select @@trancount
3> go

 -----------
           1

(1 row affected)
1> select * from t123
2> go
 id          col2
 ----------- --------------
           1 a
           2 b
           3 c
           4 d
           5 e
           6 f

(6 rows affected)
1>
```

这时，你会发现 insert 操作会隐式开启一个事务（这时的@@trancount 值为 1），后边的 rollback 操作会连它一起撤消。所以，插入操作并不成功。

在成功登录数据库时，该会话的@@trancount 值初始化为 0，当以非链式模式运行时，@@trancount 的值将随着不同的事务控制语句而改变，如表 10-1 所示。

表 10-1　@@trancount 值

命令	@@trancount 值变化
begin tran	@@trancount = @@trancount + 1
commit tran	@@trancount = @@trancount – 1
rollback tran	@@trancount = 0（与以前的@@trancount 值无关）
rollback tran <savepoint>	@@trancount = <savepoint>处的@@trancount 值 – 1

10.2　读未提交（level 0）

读未提交（Read uncommitted），当事务在这个隔离级以内，A 会话中的事务可以读到 B 里的事务中还未提交的更新数据，因而也叫读脏数据。时间顺序为 T1<T2<T3，我们开启两个命令窗口，使用 isql 来模拟两个事务。

A 会话（时刻 T1，开启事务并读取数据）：

```
1> set transaction isolation level 0
2> go
1> begin tran
2> select * from t123
3> go
 id          col2
 ----------- ----------------

(0 rows affected)
```

B 会话（时刻 T2，开启事务并插入数据，这时事务仍未提交）：

```
1> set transaction isolation level 0
2> go
1> begin tran
2> insert into t123 values(1, 'h')
3> go
(1 row affected)
```

A 会话（时刻 T3，再次读取数据）：

```
1> select * from t123
2> go
 id          col2
 ----------- ----------------
 1           h

(1 row affected)
```

我们可以看到 A 事务在 T3 时刻居然能看到 B 事务中未提交的插入数据,这就是所谓的脏数据。

10.3　读已提交（level 1）

读已提交（read committed），如果事务的隔离级为读已提交，A 在读取数据时，只能读取到其他事务中已提交的更新数据，它避免了读取脏数据。同样使用刚才的示例进行说明，提交后把表 t123 清空。

A 会话（时刻 T1，开启数据并读取数据）：

```
1> set transaction isolation level 1
2> go
1> select * from t123
2> go
 id          col2
 ----------- ----------------

(0 rows affected)
```

B 会话（时刻 T2，开启事务，插入数据）：

```
1> set transaction isolation level 1
2> go
1> begin tran
2> insert into t123 values(1, 'h')
3> go
(1 row affected)
```

A 会话（时刻 T3，再次读取数据）：

```
1> select * from t123
2> go
```

我们会发现该事务会等待，表明 t123 被 B 事务锁住了。

B 会话（时刻 T4，提交事务）：

```
1> commit
2> go
```

在 B 提交事务以后，我们立刻发现 A 事务读取到 B 刚提交的结果。

```
1> select * from t123
2> go
 id          col2
 ----------- ----------------
 1           h
```

不过有一点值得提醒的是，读已提交隔离级并不能避免"不可重复读"，即在 A 里的同一事务当中，两次读取操作可能得到不同的结果。这是因为，该事务并不对表 t123 进行锁定。

B 会话（时刻 T5，插入新数据并提交）：

```
1> begin tran
2> insert into t123 values(2, 'w')
3> commit
4> go
(1 row affected)
```

A 会话（时刻 T6，同一事务里，再次读取数据）：

```
1> select * from t123
2> go
 id          col2
 ----------- ----------------
 1           h
 2           w

(2 rows affected)
```

我们发现 A 会话在同一读取事务里边，两次读取的结果并不一样。

10.4 可重复读（level 2）

可重复读（repeatable read），可重复读隔离级的事务在执行读取操作时，会锁住所有符合条件的行，当 A 在读取某些记录行时，B 中的事务不能更新那些记录行，直到 A 中的事务结束。但是，A 并不能保证锁住那些符合条件的但并不存在的记录行，这样，B 可以插入不存在的记录行，从而出现"幻影读"现象。

ASE 数据库只对类型为 data-only-locked 的表提供可重复读支持，如果是 allpages-locked 表（APL），它会强制执行 level 3（可串行化）的事务隔离级。而 ASE 默认的锁策略是 allpages 方式，要支持 data-only-locked（DOL）的表，可以使用 datapages 或者 datarows 锁策略。使用 sa 用户登录，运行如下命令来改变默认的锁策略，然后重新启动 ASE 数据库。

```
1> sp_configure "lock scheme", 0, datarows
2> go
Parameter Name     Default       Memory Used Config Value    Run Value      Unit            Type
-----------------  ------------  -------------------------   ------------   -------------   ----------
lock scheme        allpages      0                           datarows       datarows name   dynamic

(1 row affected)
Configuration option changed. ASE need not be rebooted since the option is
dynamic.
Resulting configuration value and memory use have not changed from previous
values: new configuration value 0, previous value 0.
(return status = 0)
```

因为 lock scheme 是全局属性，只针对设置该属性以后的表起作用，因此我们要重新创建表 t123，这里我们还往表里插入三条记录，接着我们看看下边的示例。

```
1> drop table t123
2> go
1> create table t123(id int primary key, col2 varchar(32))
2> insert into t123 values(1, 'a')
3> insert into t123 values(2, 'b')
4> insert into t123 values(3, 'c')
5> commit
6> go
(1 row affected)
(1 row affected)
(1 row affected)
```

T1 时刻在 A 会话里，进行数据读取操作。

A 会话（时刻 T1，开启事务，读取数据）：

```
1> set transaction isolation level 2
2> go
1> begin tran
2> select * from t123
3> go
 id         col2
 ---------- -------------
 1          a
 2          b
 3          c
(3 rows affected)
```

B 会话（时刻 T2，开启事务，试图插入数据）：

```
1> set transaction isolation level 2
2> go
1> begin tran
2> insert into t123 values(4, 'd')
3> go
(1 row affected)
```

我们发现 B 里的 insert 成功了，即 B 可以插入一条符合条件的记录，而 A 并不会察觉。

A 会话（时刻 T3，继续读取数据）：

```
1> select * from t123
2> go
 id         col2
 ---------- -------------
 1          a
 2          b
 3          c

(3 rows affected)
```

确实在同一个事务里可以重复读取，但是，当 B 提交事务以后，A 会话重新读取数据，发现读到了"幻影数据"。

B 会话（时刻 T4，提交事务）：

```
1> commit
2> go
```

A 会话（时刻 T5，再次读取数据）：

```
1> select * from t123
2> go
```

```
id          col2
----------- ---------------
1           a
2           b
3           c
4           d

(4 rows affected)
```

10.5 可串行化（level 3）

可串行化（serializable），可串行化隔离级别除了满足可重复读的特征以外，还有效地杜绝了"幻影读"现象。在这种隔离级别下，在 A 的事务里进行读取操作时，会锁住所有符合条件的行，当 A 在读取某些记录行时，B 中的事务不能更新那些记录行，直到 A 中的事务结束。同时，A 还能确保锁住那些符合条件的但并不存在的记录行，这样，B 将无法插入或者更新不存在的记录行，从而彻底避免了"幻影读"现象。

下边的范例里，首先会清除表 t123 中的数据。

```
1> drop table t123
2> go
1> create table t123(id int primary key, col2 varchar(32))
2> go
1> commit
2> go
1> insert into t123 values(1, 'a')
2> insert into t123 values(2, 'b')
3> insert into t123 values(3, 'c')
4> go
(1 row affected)
(1 row affected)
(1 row affected)
1> commit
2> go
```

A 会话（时刻 T1，开启事务，读数据）：

```
1> set transaction isolation level 3
2> go
1> begin tran
2> begin tran
3> select * from t123 where id>2
4> go
 id          col2
 ----------- ---------------
```

```
3      c

(1 row affected)
```

B 会话（时刻 T2，开启事务，试图插入 id 为 4 的记录）：

```
1> set transaction isolation level 3
2> go
1> begin tran
2> insert into t123 values(4, 'd')
3> go
```

我们发现 B 中的事务会在等待，可以推断 A 中的读操作锁定了符合条件的记录。

A 会话（时刻 T3，提交事务）：

```
1> commit
2> go
1>
```

B 会话（时刻 T4，插入操作成功并提交）：

```
(1 row affected)
1> commit
2> go
1>
```

我们接着做如下实验，试着让 A 查询 id<2 的记录，然后让 B 插入一条!(id<2)，如 id=5 的记录，看看出现什么结果。

A 会话（时刻 T5，开启新事务，读取 id<2 的记录）：

```
1> begin tran
2> select * from t123 where id<2
3> go
 id          col2
 ----------- ----------------
  1          a

(1 row affected)
1>
```

B 会话（时刻 T6，开启新事务，插入 id=5 的记录）：

```
1> begin tran
2> insert into t123 values(5, 'e')
3> go
(1 row affected)
1>
```

这时我们发现 B 中的插入操作立即执行，不会等待，原因是 A 中的事务并没有锁定那些不符合条件的记录。

10.6 如何在事务中允许 DDL 操作

ASE 默认情况下并不允许在事务中执行 DDL 操作。

```
d:\>isql -Sspring_ase -Uspring
Password:
1> begin tran
2> create table t(id int primary key, col2 varchar(32))
3> commit tran
4> go
Msg 2762, Level 16, State 3:
Server 'BJEASPE2900', Line 2:
The 'CREATE TABLE' command is not allowed within a multi-statement transaction
in the 'spring' database.
1>
```

若想支持在事务中执行 DDL 操作，需要将当前数据库的 ddl in tran 属性设置为 true，这需要 DBA 用户来设置，接着执行 checkpoint 命令。下面的例子是先设置当前数据库支持 ddl in tran，然后在一个事务里创建表 t，最后回滚事务，发现表 t 的创建操作确实被撤消。

```
d:\>isql -Sspring_ase -Usa
Password:
1> sp_dboption spring, "ddl in tran", true
2> go
Database option 'ddl in tran' turned ON for database 'spring'.
Running CHECKPOINT on database 'spring' for option 'ddl in tran' to take effect.

(return status = 0)
1> use spring
2> checkpoint
3> go
1> begin tran
2> create table t(id int primary key, col2 varchar(32))
3> rollback tran
4> go
1> select * from t
2> go
Msg 208, Level 16, State 1:
Server 'EAS2K470', Line 1:
t not found. Specify owner.objectname or use sp_help to check whether the object

exists (sp_help may produce lots of output).
1>
```

通过 GUI 方式，也可以方便地设置数据库的一些属性，打开 SybCentral 面板，选择要设定的数据库，然后进入它的属性面板，如图 10-1 所示。

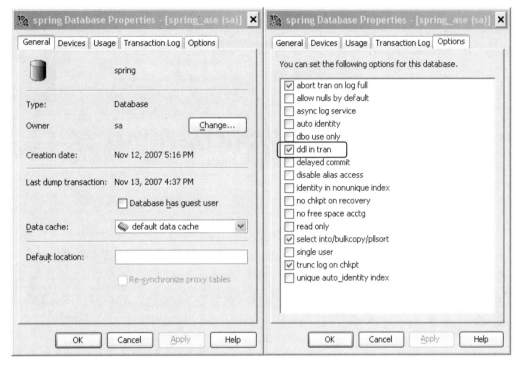

图 10-1　设置数据库的 ddl in tran 选项

11

ASE 数据库的事务日志

恢复（recovery）一词，常被用于描述将数据库从一个已有备份的数据库文件里进行还原的过程。在 ASE 当中，它特指当数据库服务器启动时，将日志同步到物理数据库文件里的过程。那么数据库日志与数据文件是如何进行同步的呢？

为确保数据库始终保持一致，并能重现数据库中数据的所有变化，ASE 数据库内部将这些数据的变化记录下来，形成了事务日志。

11.1 事务

什么是事务？可以用一句话来形容，就是数据库当中最小的工作单元。完整的定义是，事务是一个任务单元，不管它有多大，有多小，也无论它是简单还是非常复杂，它都不能再分割（基于实际业务及应用需求）。在数据库中，事务实际上就是一串原子操作，这些操作作为一个整体，要么全部执行，要么全都不执行，没有任何中间状态。

举例来说，这里有一个简单的银行存取款的例子，用户需要将钱从他的一个账户里取出来，存进他的另一个账户。这两个操作必须定义为一个任务，如果变成两个任务，它们有可能会出错，最终会出现以下几种结果：

1. 钱成功地从一个账户转移到另一个账户中（两个任务都成功）。
2. 钱成功地从一个账户里取出，但是存到另一个账户中时失败了（第二个任务失败）。
3. 钱成功地从一个账户里取出，存到另一个账户时失败了，最终同样数目的钱又返还到前一个账户当中。

显然，第 1 种和第 3 种结果是可以接受的，因为这时没有任何钱的"损失"。而第 2 种结果是不能接受的，因为发生了钱的丢失现象，取出的那部分钱不知去向。

解决这种问题的唯一办法是将这两个操作定义为一个事务，要么这两个操作全部成功，要么其中有操作失败，事务将使两个账户恢复到原有状态，这样，用户没有任何损失。

11.2　事务日志

在 ASE 中，事务日志记录到系统表 syslogs 中。这个表记录了数据库中所有的改变，同时也记录为保持数据库完整性所需要的一些额外信息。

ASE 以预先写入的方式来记录事务日志，数据库的所有改变总是在正式写入数据库之前，先写到事务日志当中。syslogs 表虽然可以使用 Select 语句进行查询，但是它的内容可读性非常差，必须使用专用的工具，如 dbcc log，将日志的内容转换成有用的形式。

11.3　事务提交（commit）

数据库中数据发生改变时，这些改变在存储到具体的数据表之前，都会保存到事务日志里。但是，任意改变在写进磁盘之前，都是在缓冲区里进行处理的，ASE 里使用 cache 来缓存这些改变。

在 ASE 中，每个用户连接都会分配少量内存（默认大小为 2KB，或者是在已经配置了服务器页大小时的服务器页大小）用作用户日志缓存（User Log Cache，ULC）。这些用户日志缓存可以用于收集若干事务及相应的数据改变的语句，这样当需要把事务写进日志里时，不需要多个来回就可以把多个事务及相应的语句写进日志，从而有效地避免访问事务日志时产生竞争现象。

另外，系统有专门的 cache 用于记录数据页和日志页。当读取一个页时，系统首先会将其读到 cache 中，而当一页（数据或日志）发生改变时，这些改变首先会被记录到一个 cache 的副本当中，然后再写回磁盘。而数据和日志两者的 cache 页写回磁盘的具体时机又有所不同。

通常，当一个事务提交的时候，系统会将这个事务的所有日志页写入磁盘；如果系统允许这些日志页保留到缓存而无须写回磁盘时，在系统意外掉电或宕机时，一个已经完成的事务可能会丢失。为了避免这种现象的发生，commit 操作直到系统确认事务已经成功地写入磁盘，才算是成功执行。另外，当 cache 装得太满的时候，系统也会将日志页写入磁盘。这种情况下，系统不能保证整个事务会被完全写入，但是可以确保当事务提交的时候，有关这个事务的所有日志都会记录到磁盘上。

对于数据页而言，它将一直保存到 cache 当中，除非它所在的 cache 空间有其他用途。比如，当这个数据页是"脏"页（即 cache 页中记录了它的改变，但是没有写回磁盘），该页就必须要写回磁盘。

当数据库中的数据发生改变时，系统会执行如下操作：

Step 1　在用户日志 cache 中记录 begin tran（用于开启一个事务）。

Step 2　将数据改变的记录添加到日志里（用户日志 cache）。

Step 3 将数据改变写到数据页（用户日志 cache）。

Step 4 当 commit tran 写入用户日志 cache 时，用户日志 cache 被写到 cache 中的事务日志里。

Step 5 系统将事务的日志页复制，写回磁盘中的事务日志里。

这个过程可以用图 11-1 来描述。

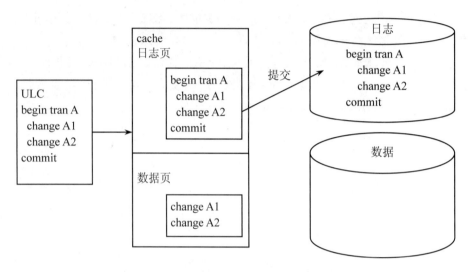

图 11-1 事务日志处理过程

然而上述过程不能避免一种情形：数据改变并不写回磁盘，除非当 cache 满的时候。这样当 cache 未满时，此时如果系统宕机，上述数据改变可能会产生丢失现象。有两个系统进程可以避免这种现象的产生：检查点（checkpoint）进程和恢复（recovery）进程。

11.4 检查点

在 ASE 中，检查点是一个事件，它会通知系统将数据库中所有的脏页（既包括数据页，也包括日志页）从 cache 写回磁盘中。检查点可以由 ASE 服务器触发（时间间隔由系统的行为决定），也可以由数据库的属主用户发出 checkpoint 命令来强制检查点的发生（要注意的是，很多数据库命令只有在检查点发生以后才会生效）。

检查点这个进程的目的，主要是进一步减少服务器在恢复过程中所做的工作，即加速服务器的恢复操作。

当检查点这一事件发生时，系统会执行如下操作：

Step 1 将所有脏日志页写回磁盘。

Step 2 将所有脏数据页写回磁盘。

Step 3 将一条 checkpoint 记录写到日志中。

其过程可以用图 11-2 来表示。

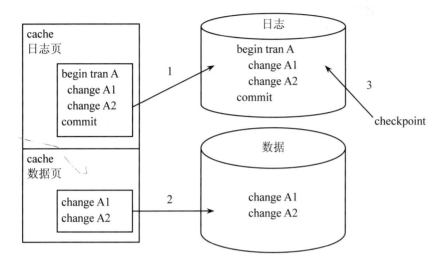

图 11-2　checkpoint

值得注意的是，虽然这时日志和磁盘中的数据已经有了副本，这些数据要到事务提交语句在事务日志里记录下来，即事务提交完成时才算有效。如果还有事务仍然处于打开状态（即在检查点触发时仍未提交），这些事务仍然可以继续执行下去。检查点进程会在事务日志里完整地记录这些未完成的事务信息，这样在恢复进程处理过程当中，可以做相应的处理。

11.5　恢复（recovery）

在 ASE 中，恢复进程用于确保事务日志中存储的数据与数据库中存储的物理数据达成一致，当日志中有数据改变，而在物理数据中没有改变时，最终呈现的数据则要与日志保持一致。如果有的事务已经完成，却没有反映到具体的物理数据当中，系统会执行这些事务以改变数据。如果数据库中有的事务没有完成（由于客户端或者服务器端的失败），恢复进程则会回滚这些事务。服务器每次启动的时候，都会执行这些操作，在服务器装载事务日志的时候，也会执行这类操作。

ASE 中，恢复进程主要执行下述操作：

Step 1 定位事务日志当中最近的一次检查点。

Step 2 构造检查点时刻起，所有活跃的事务列表。

Step 3 执行事务列表中要提交或回滚的事务操作。

Step 4 标识出那些未完成的事务，并将这些事务强制回滚。

Step 5 写一条 checkpoint 记录到日志里，用于记录下次恢复操作时的起始检查点时刻，即一次恢复进程可以从这个新的 checkpoint 记录处开始执行。

这样看来，恢复进程和检查点进程是密切相关的。恢复进程写入一个检查点，表明它不再需要从头开始处理整个事务日志，只需要找到最近的一次检查点即可。同样，当一个检查点将所有的脏数据页和脏日志页写回磁盘时，也等于是从日志中恢复数据的一部分工作。

检查点进程也发生在高强度事务负载时的备份（backup）命令当中，它也有可能发生在系统活动很少的时候。当服务器处于空闲状态时，ASE 会将脏页从内存写回磁盘。

当所有的脏页都写到磁盘上，并且离最近一次检查点至少有 100 条日志记录，系统就会触发一个空闲检查点事件。这个事件是在系统非常空闲时发生，它的触发不会明显地影响系统进程。该空闲检查点进程在 sp_who 命令的输出里标识为 HouseKeeper。通常你会发现该命令的 cmd 列的值为 Checkpoint Sleep，status 列值为 sleeping。注意，到了 ASE 15.0 以后，Sybase 将 HouseKeeper 进程分为三个不同的进程：HK WASH、HK GC、HK CHORES（注：HK 是 HouseKeeper 的缩写）。

11.6 恢复间隔

恢复一个数据库所花时间与事务日志里未完成的事务的数量成正比。如果从最近一个检查点开始，几乎没有什么数据改变，那么数据库瞬间就能完成恢复。如果未恢复日志记录达数千条，那么恢复过程将会耗费很长时间。这种事情如果发生在数据库启动的时刻，确实有些让人感到头疼，因为在数据库完成恢复过程以前，它将会一直处于离线状态。默认情况下，系统会按照数据库 ID 的顺序依次进行恢复，系统管理员 SA 可以进行定制，设定最重要的数据库先行恢复。

ASE 提供了一个可配置的参数，通过设定一个上限值，可以用于限定每个数据库恢复所用的最长时间。到了规定的时间间隔时，服务器会在对应的数据库上自动发出 checkpoint 命令，以保持数据的一致。

检查点任务大约每分钟检查一次服务器上的每个数据库，以便了解自上次进行检查点操作以来，已添加到事务日志中的记录数。如果服务器估算恢复这些事务所需的时间大于数据库的恢复时间间隔，Adaptive Server 会发出一个检查点事件。设置恢复时间间隔的命令，示例如下：

sp_configure "recovery interval in minutes, " 10

它的含义是允许每个数据库在 10 分钟内完成恢复。系统的默认恢复时间间隔是 5 分钟。

通常，间隔值越小，数据库恢复越快。然而，如果间隔值太小了，服务器会频繁执行检查点操作，从而导致过多的 I/O 操作。

> **注意**
>
> 恢复间隔对以下事务没有影响：在 Adaptive Server 发生故障时处于活动状态、长期运行且记入日志内容很少的事务（如 create index）。恢复这些事务所花费的时间可能与运行它们所花费的时间相等。为避免长时间的延迟，需要在数据库的一个表上创建索引后，立即转储每个数据库。

11.7 日志填满

事务日志记录着整个数据库所有的数据改变，非常重要的一点是要确保日志空间不能填满。一旦日志空间用完，日志中再也不能记载数据的变化了。因为要确保成功地记录到事务日志当中，因此，当日志填满时，所有对数据库的改变将会一直挂起，直到有可用的日志空间。当事务日志的空间快用完时，系统的最后机会阈值（lastchange Shreshold）将被激活。该进程会转储整个日志。如果日志空间已经填满，标准的事务日志例程（转储事务、扩展日志）也会失败。系统管理员有必要确保日志有适当的增长空间，并且不会被完全填满。

要避免日志空间被填满，数据库有必要对日志进行合理的截短。通常情况下，这会发生在正常的日志备份的情况下，当日志复制完成以后，完全可恢复的事务会从日志里移除，从而释放了一些可用空间。日志截短的位置主要取决于当前运行的事务。通常情况下，最后机会阈值确保你有足够的日志空间来转储事务日志。

显然，日志文件中记录的最早未完成的事务确定了截短日志的位置，只有在这个最早未完成事务之前的所有事务才可以被截短。

12

ASE 数据库的备份、恢复及数据迁移

对于 DBA 而言，恐怕最重要的工作就是备份与恢复了，对于一个已经运行的数据库系统来说，性能未达标可以进行调优，数据存放位置不对可以进行调整，配置参数不对可以进行修改，可是如果数据没有及时备份而发生数据丢失，那就是失职了。即便是少量数据的丢失，也有可能意味着一场灾难。所以，很难说是承载数据的物理设备更昂贵，还是物理设备里装载的数据更昂贵。

因此，定期对数据库系统的备份非常重要。

12.1 备份权限及周期

12.1.1 备份需要的权限

通常情况下，DBO（数据库属主）有权对自己拥有的数据库进行备份恢复（dump/load）。这些权限并不会直接赋给其他登录用户，除了被添加为 alias 的用户外。

这意味着 sa_role 用户可以对任意一个数据库进行备份与恢复。同时，我们前边也看到，oper_role 也拥有对所有数据库进行备份与恢复的权限，并且专用于此类操作。所以，如果 DBA 想将任务细分，可以为某一个登录用户专门授予此角色。

执行命令：

```
sp_role  "grant", oper_role,  login_name
```

或者

```
grant role oper_role  to login_name [, login_name, …]
```

12.1.2 备份周期（策略）

对数据库进行备份，最终是为了防止表损坏、数据库损坏、存储介质损坏以及运行环境的损坏。虽然备份对数据库即时运行的性能有影响（从统计来看，会使性能下降至少 5%，因此 DBA 不会在系统运行高峰期对数据库进行备份），数据库在运行的时候，依然可以对其进行备份。

在实际的生产环境中，绝大部分都采用对数据库一天一备，对事务日志几小时甚至一小时一备的策略，以尽可能小地减少数据丢失的损失。针对实际情况，有可能有的应用不需要那么频繁地对数据库进行备份，但最保守的做法是，至少每周应该对所有数据库进行一次备份。备份数据库（含日志）是 DBA 最重要的工作之一，也是 DBA 区分于数据库开发人员的一个方面，因为开发人员不会执行或从事任何与备份恢复有关的任务。

事务日志的备份属于增量备份，耗费时间少，同时有另一个好处是，可以让当前的日志被截短，以减少开销。所以，事务日志的备份周期宜短不宜长，是常规备份必需的一个要求。

恢复一个数据库的总的时间是新建一个数据库的时间加上装载数据库的时间，再加上每个事务日志的装载时间。实现这个任务的时候，创建数据库的语句必须加上 load 选项，表示专用于装载，这个选项可以加快装载速度。

设想，如果数据库一年只转储一次，而事务日志每天转储一次，如果数据库在最后一天（假设是第 365 天）损坏了，那么你必须装载转储过的数据库，并且要按顺序装载 364 个事务日志，事务日志的装载量是不是有点多了？如果将完全备份的周期缩短为一个星期，那么就不用装载这么多事务日志了，出错的机率也会小一些。DBA 也不能保证每一次完全备份都没有任何错误，但他有责任对自己所做的备份（完全备份和增量备份）定期进行验证。这里举一个例子，在 2001 年大运会结束的时候，我们的一个 DBA 对大运会官网的数据库进行了一次备份（当时官网硬件设备都是由 Sun 公司捐赠的，他们要立即撤走），比较着急，结果那次备份出了点故障，我们拿回去恢复的时候，很多数据都不能恢复了，非常可惜，可见对备份进行验证是何等的重要。

需要说明的是，对于生产环境，完全备份和增量备份是备份恢复策略中必须要考虑的，比如每天或每两天（至多一星期）一次完全备份，每天一次或若干次增量备份。对于开发环境而言，则没有必要那么频繁地备份和恢复。

12.2 简单备份

通常情况下，备份服务器（Backup Server）和数据库服务器（DB Server）位于同一台机器上。如果只是实现在服务器本机上备份数据，配置起来都比较方便。

ASE 备份的基本体系结构如图 12-1 所示。

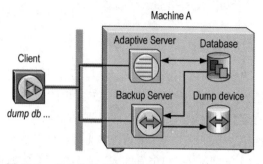

图 12-1　备份的基本体系结构

ASE 备份的基本思想是，用户发送 dump 命令给数据库服务器，数据库服务器解析命令，然后使用 RPC（远程过程调用）把请求转发给备份服务器，备份服务器直接读取库文件，把数据写入最终指定的输出文件里边。

在首次安装完 ASE 后进行备份时，往往出现类似于如下错误：

```
1> dump database pubs2 to 'i:\pubs2.dat'
2> go
Msg 7205, Level 17, State 2:
Server 'BJEASPE1850_', Line 1:
Can't open a connection to site 'SYB_BACKUP'. See the error log file in the ASE
boot directory.
Msg 7205, Level 18, State 3:
Server 'BJEASPE1850_', Line 1:
Can't open a connection to site 'SYB_BACKUP'. See the error log file in the ASE
boot directory.
```

这是因为 SYB_BACKUP 在系统表 master..sysservers 中并没有注册或者注册有误，通过查询 master..sysservers，可以看到备份服务器的服务名 SYB_BACKUP，被注册为默认的网络服务名 SYB_BACKUP，我们应该将其更新为系统安装时的备份服务器名。

```
1> select * from master..sysservers
2> go
 srvid    srvstatus srvname        srvnetname         srvclass  srvsecmech  srvcost
 -------  --------- -------------- ------------------ --------  ----------  -------
 4        8 SYB_BACKUP             SYB_BACKUP         7         NULL        1000
 2        1024 SYB_EJB             EJBServer          10        NULL        NULL
 0        8 BJEASPE1850_           BJEASPE1850_       0         NULL        0
 3        8 BJEASPE1850__XP        BJEASPE1850__XP    11        NULL        1000

(4 rows affected)
1> sp_helpserver
```

```
2> go
 name                       network_name                  class
 status                                                                              id        cost
 --------------------------- -------------------------------- -------------------
 ------------------------------------------------------------------------------ ---------- ------------
  BJEASPE1850_              BJEASPE1850_                 local
                                                                                   0         0
  BJEASPE1850__XP           BJEASPE1850__XP              RPCServer
 timeouts, no net password encryption, writable , rpc security model A           3         1000
  SYB_BACKUP                SYB_BACKUP                   ASEnterprise
 timeouts, no net password encryption, writable , rpc security model A           4         1000
  SYB_EJB                   EJBServer                    ASEJB
 external engine auto start                                2                    NULL
(return status = 0)
```

通过查看%SYBASE%\ini\sql.ini 或者$SYBASE/interfaces 文件,可以找到备份服务器的注册项:

```
[BJEASPE1850_BS]
master=NLWNSCK,bjeaspe1850,5001
query=NLWNSCK,bjeaspe1850,5001
```

于是,我们可以使用存储过程 sp_addserver 覆盖旧的 SYB_BACKUP 注册项:

```
1> sp_addserver SYB_BACKUP, null, BJEASPE1850_BS
2> go
Changing physical name of server 'SYB_BACKUP' from 'SYB_BACKUP' to
'BJEASPE1850_BS'
(return status = 0)
1>
```

现在可以成功地进行备份:

```
1> dump database pubs2 to 'i:\pubs2.dat'
2> go
Backup Server session id is:   38.   Use this value when executing the
'sp_volchanged' system stored procedure after fulfilling any volume change
request from the Backup Server.
Backup Server: 4.41.1.1: Creating new disk file i:\pubs2.dat.
Backup Server: 6.28.1.1: Dumpfile name 'pubs2073200F628   ' section number 1
mounted on disk file 'i:\pubs2.dat'
Backup Server: 4.188.1.1: Database pubs2: 784 kilobytes (59%) DUMPed.
Backup Server: 4.188.1.1: Database pubs2: 1314 kilobytes (100%) DUMPed.
Backup Server: 3.43.1.1: Dump phase number 1 completed.
Backup Server: 3.43.1.1: Dump phase number 2 completed.
Backup Server: 3.43.1.1: Dump phase number 3 completed.
Backup Server: 4.188.1.1: Database pubs2: 1324 kilobytes (100%) DUMPed.
Backup Server: 3.42.1.1: DUMP is complete (database pubs2).
1>
```

12.3 远程备份

有时候我们也需要进行数据库的远程备份,即把远程数据库服务器的数据备份到本地来。通常情况下,这要借助于本地的备份服务器和远程的备份服务器进行合作,才能完成整个备份任务。这里假设要把 Server2 数据库上的数据备份到机器 Server1 上,其基本思想是(如图 12-2 所示):备份请求先发至 Server2 的数据库服务器里,数据库服务器接着将备份请求转发至 Server1 的备份服务器里,然后它再与 Server2 上的备份服务器交互,由 Server2 上的备份服务器处理请求,把备份结果发送给 Server1,Server1 再经由数据库服务器,把结果返回给最终用户。

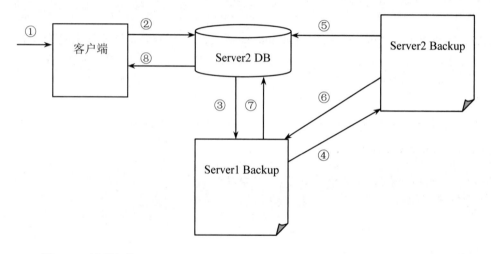

图 12-2 远程备份

前提条件是:本地有一个备份服务器 Server1 Backup,远程数据库服务器 Server2 上同时有备份服务器在运行。

备份方法如下:

Step 1 在 Server2 上的 interface 文件里添加 Server1 备份服务器的注册项(XIONGHEXP_BS):

```
bjeaslinux1_bs
    master tcp ether bjeaslinux1 5001
query tcp ether bjeaslinux1 5001
XIONGHEXP_BS
    master tcp ether xionghe-xp 5001
query tcp ether xionghe-xp 5001
```

Step 2 在本地备份服务器的 interface 文件里添加 Server2 上远程备份服务器的注册项,这个注册项必须与远程服务器安装时的注册项一致。

```
[XIONGHEXP_BS]      #本地注册项
master=NLWNSCK,xionghe-xp,5001
query=NLWNSCK,xionghe-xp,5001
    [BJEASLINUX1_BS]    #远程备份服务器的注册项
    master=tcp,bjeaslinux1,5001
    query=tcp,bjeaslinux1,5001
```

Step 3 在 Server1 里往 sysservers 表中添加远程备份服务器名。

```
1> sp_addserver SYB_BACKUP, null, BJEASLINUX1_BS
2> go
Adding server 'SYB_BACKUP', physical name 'BJEASLINUX1_BS'
Server added.
(return status = 0)
1>
1> sp_helpserver
2> go
 name                network_name              class
status
id          cost
 ------------------------- --------------------------------------- --------------
---------------------------------------------------------------------------------
------------ ------------
  SYB_BACKUP      BJEASLINUX1_BS          ASEnterprise
no timeouts, no net password encryption, writable , rpc security model
3       1000
  SYB_RTMS        SYB_RTMS                TIBCO_JMS
timeouts, no net password encryption, writable , rpc security model A
2       NULL
  bjeaslinux1         bjeaslinux1                local

0        0
(return status = 0)
1>
```

Step 4 使用 sa 用户连接远程数据库 Server2 进行备份。

备份的命令格式如下：

dump database <db> to <本地存储位置> at <本地备份服务器注册项>
```
1> dump database pubs2 to 'e:\t\pubs2.dat' at XIONGHEXP_BS
2> go
Backup Server session id is:    7.   Use this value when executing the
```

'sp_volchanged' system stored procedure after fulfilling any volume change request from the Backup Server.
Backup Server: 4.41.1.1: Creating new disk file e:\t\pubs2.dat.
Backup Server: 6.28.1.1: Dumpfile name 'pubs2073200F125   ' section number 1 mounted on disk file 'e:\t\pubs2.dat'
Backup Server: 4.188.1.1: Database pubs2: 876 kilobytes (1%) DUMPED.
Backup Server: 4.188.1.1: Database pubs2: 1720 kilobytes (4%) DUMPED.
Backup Server: 4.188.1.1: Database pubs2: 2398 kilobytes (11%) DUMPED.
Backup Server: 4.188.1.1: Database pubs2: 2570 kilobytes (18%) DUMPED.
Backup Server: 4.188.1.1: Database pubs2: 2740 kilobytes (25%) DUMPED.
Backup Server: 4.188.1.1: Database pubs2: 2912 kilobytes (32%) DUMPED.
Backup Server: 4.188.1.1: Database pubs2: 3082 kilobytes (40%) DUMPED.
Backup Server: 4.188.1.1: Database pubs2: 3254 kilobytes (47%) DUMPED.
Backup Server: 4.188.1.1: Database pubs2: 3424 kilobytes (54%) DUMPED.
Backup Server: 4.188.1.1: Database pubs2: 3596 kilobytes (61%) DUMPED.
Backup Server: 4.188.1.1: Database pubs2: 3766 kilobytes (68%) DUMPED.
Backup Server: 4.188.1.1: Database pubs2: 3938 kilobytes (75%) DUMPED.
Backup Server: 4.188.1.1: Database pubs2: 4108 kilobytes (82%) DUMPED.
Backup Server: 4.188.1.1: Database pubs2: 4280 kilobytes (89%) DUMPED.
Backup Server: 4.188.1.1: Database pubs2: 4450 kilobytes (96%) DUMPED.
Backup Server: 4.188.1.1: Database pubs2: 4564 kilobytes (100%) DUMPED.
Backup Server: 3.43.1.1: Dump phase number 1 completed.
Backup Server: 3.43.1.1: Dump phase number 2 completed.
Backup Server: 3.43.1.1: Dump phase number 3 completed.
Backup Server: 4.188.1.1: Database pubs2: 4572 kilobytes (100%) DUMPED.
Backup Server: 3.42.1.1: DUMP is complete (database pubs2).
1>

12.4　dump/load 命令的使用

前两节只是分别简单介绍了本地简单备份和远程备份的使用场景。完整的备份不仅包括数据库的转储（dump），还包括对事务日志的转储。在恢复时，需要使用到这两者的装载。

组合起来，共有四个命令供 DBA 使用：

- dump database，转储数据库。
- dump transaction，转储事务日志。
- load database，装载数据库。
- load transaction，装载事务日志。

这几个命令一般包括如下选项：

- 进行备份或装载的数据库名。
- 用于备份或装载的设备名。

- 要使用的备份服务器名。
- 备份或装载活动中，要使用的备份文件名。
- 备份服务器要将信息发送到的位置。

备份一个数据库的过程，就是先进行事务日志的转储，再进行数据库本身的转储。

备份数据库前两节已有介绍，我们来看看事务日志的备份：

dump transaction 命令主要用于转储 ASE 数据库的事务日志，可以删除日志中不活跃的部分。其语法格式为：

dump tran[saction] database_name to <location> [with { … }]

对于日志备份而言，它与数据库备份略有不同。随着数据库事务的提交完成，有很多日志数据逐渐变成不活跃日志，其对应的事务操作已经结束。针对这部分日志，日志备份有几个可用选项：

1. 截断不活跃日志，不生成备份。

    dump tran database_name with trundcate_only

2. 截断不活跃日志，不生成备份，这个命令主要用于日志空间满的情形。

    dump tran database_name with no_log

3. 仅备份日志，不截断日志。

    dump tran database_name with no_truncate

4. ASE 默认方法，生成日志的备份，然后截断不活跃日志。

    dump tran database_name

通常，用得最多的是第 4 种情形。同时，我们需要谨记的是，进行日志备份的数据库不能将 trunc log on chkpt 选项设置为 true。该选项为 true 时，意味着每到检查点，会自动截断日志。

在 ASE 中，恢复一个数据库的基本过程大致如下：

Step 1 创建一个用于装载转储的数据库（create database for load）。

Step 2 从最近的数据库备份中恢复数据库（load database）。

Step 3 从最近的事务日志备份中恢复日志（load transaction）。

Step 4 对数据库执行联机操作（online database），以允许用户访问该数据库。

下面请看几个简单的示例：

1. 对 iihero 数据库日志及数据库进行备份。

```
1> dump database iihero to "e:/temp/iihero.dmp"
2> go
```

备份服务器会话标识是 7。当执行 sp_volchanged 系统存储过程的时候，在备份服务器完成请求改变任一卷之后使用这个值。

```
Backup Server: 6.28.1.1: 转储文件名'iihero12147121C9', 区域号 1 安装在磁盘文件
'e:/temp/iihero.dmp'上。
Backup Server: 4.188.1.1: 数据库 iihero:1672 KB (17%) DUMPED
Backup Server: 4.188.1.1: 数据库 iihero:2294 KB (100%) DUMPED
Backup Server: 3.43.1.1: 转储段号 1 完成
Backup Server: 3.43.1.1: 转储段号 2 完成
Backup Server: 3.43.1.1: 转储段号 3 完成
Backup Server: 4.188.1.1: 数据库 iihero:2304 KB (100%) DUMPED
Backup Server: 3.42.1.1: DUMP 完成(数据库 iihero)
1> dump transaction iihero to "e:/temp/iihero.log.dmp"
2> go
```

备份服务器会话标识是 10。当执行 sp_volchanged 系统存储过程的时候,在备份服务器完成请求改变任一卷之后使用这个值。

```
Backup Server: 6.28.1.1: 转储文件名'iihero12147121D7', 区域号 1 安装在磁盘文件
'E:/temp/iihero.log.dmp'上
Backup Server: 4.58.1.1: 数据库 iihero: 8 千字节 DUMPED
Backup Server: 3.43.1.1: 转储段号 3 完成
Backup Server: 4.58.1.1: 数据库 iihero: 12 千字节 DUMPED
Backup Server: 3.42.1.1: DUMP 完成(数据库 iihero)
```

2. 对 iihero 数据库进行还原。

```
1> drop database iihero
2> go
1> create database iihero on iihero="20M" log on iihero_log="20M" for load
2> go
CREATE DATABASE:分配磁盘 'iihero' 上的 5120 逻辑页 (20.0 MB)
CREATE DATABASE:分配磁盘 'iihero_log' 上的 5120 逻辑页 (20.0 MB)
1> load database iihero from "E:/temp/iihero.dmp"
2> go
```

备份服务器会话标识是 13。当执行 sp_volchanged 系统存储过程的时候,在备份服务器完成请求改变任一卷之后使用这个值。

```
Backup Server: 6.28.1.1: 转储文件名'iihero12147121C9', 区域号 1 安装在磁盘文件
'E:/temp/iihero.dmp'上
Backup Server: 4.188.1.1: 数据库 iihero:6632 KB (16%) LOADED
Backup Server: 4.188.1.1: 数据库 iihero:40966 KB (100%) LOADED
Backup Server: 4.188.1.1: 数据库 iihero:40976 KB (100%) LOADED
Backup Server: 3.42.1.1: LOAD 完成(数据库 iihero)
Started estimating recovery log boundaries for database 'iihero'.
Database 'iihero', checkpoint=(5446, 22), first=(5446, 22), last=(5446, 22).
Completed estimating recovery log boundaries for database 'iihero'.
Started ANALYSIS pass for database 'iihero'.
Completed ANALYSIS pass for database 'iihero'.
```

```
Started REDO pass for database 'iihero'. The total number of log records to
process is 1.
Completed REDO pass for database 'iihero'.
请使用 ONLINE DATABASE 命令使此数据库联机；ASE 不会自动使其联机
1> load tran iihero from "e:/temp/iihero.log.dmp"
2> go
```

备份服务器会话标识是 15。当执行 sp_volchanged 系统存储过程的时候，在备份服务器完成请求改变任一卷之后使用这个值。

```
Backup Server: 6.28.1.1: 转储文件名'iihero12147121D7 ', 区域号 1 安装在磁盘文件
'E:/temp/iihero.log.dmp'上
Backup Server: 4.58.1.1: 数据库 iihero: 12 千字节 LOADED
Backup Server: 3.42.1.1: LOAD 完成(数据库 iihero)
Started estimating recovery log boundaries for database 'iihero'.
Database 'iihero', checkpoint=(5446, 22), first=(5446, 22), last=(5446, 22).
Completed estimating recovery log boundaries for database 'iihero'.
Started ANALYSIS pass for database 'iihero'.
Completed ANALYSIS pass for database 'iihero'.
Started REDO pass for database 'iihero'. The total number of log records to
process is 1.
Completed REDO pass for database 'iihero'.
请使用 ONLINE DATABASE 命令使此数据库联机;ASE 不会自动使其联机
1> online database iihero
2> go
Started estimating recovery log boundaries for database 'iihero'.
Database 'iihero', checkpoint=(5446, 22), first=(5446, 22), last=(5446, 22).
Completed estimating recovery log boundaries for database 'iihero'.
Started ANALYSIS pass for database 'iihero'.
Completed ANALYSIS pass for database 'iihero'.
Recovery of database 'iihero' will undo incomplete nested top actions.
数据库'iihero' 联机.
1>
```

在这里，我们要注意，备份同样存在字符集的问题，我们要配置备份服务器的时候，它所使用的字符集必须与数据库服务器所用的字符集保持一致。

以 Windows 为例，它的字符集参数配置位于%SYBASE%\%SYBASE_ASE%\install\RUN_<SERVER>_BS.bat 中，由-J 选项指定。

备份时指定的数据库名和恢复时指定的数据库名不一定要完全一样。可以使用不同的数据库进行装载，前提是这个数据库不要有同其他用户数据库的关联关系。

考虑到备份文件可能很大，上百 G 的大小也很普遍，这时可以通过指定参数 stripe on 子句来指定多个设备文件进行备份，每个文件就可以平均分配大小。

12.5 用户数据库的备份与恢复

正常的用户数据库备份需要遵循一定的原则，主要有下述原则：

1. 任何数据库结构发生改变，立即备份数据库。这些改变包括表结构、索引的创建与改变，创建与删除用户、授权的改变等。
2. 某些无日志的操作执行后，立即备份数据库。主要是 truncate table、bcp 数据导入操作等。
3. 人为截断日志后，应该立即备份数据库。人为截断日志时，并不会自动备份数据库。这时要想数据库得到后续的完整恢复，必须进行数据库的转储操作才能得到保障。
4. 备份的周期性策略，比如一天一次数据库转储，数据库日志的两小时一次转储等。

对于 DBA 而言，用户数据库备份的工作不只是 dump 和 load 两个操作，应该有一套完整的工作流程，对一个用户数据库进行备份，需要遵循如下步骤：

Step 1 执行一致性检查。

```
1> dbcc checkdb(iihero)
2> go
正在检查 iihero:逻辑页大小为 4096 字节
正在检查表 'sysobjects' (对象 ID 1):逻辑页大小为 4096 字节

正在检查表 'sysobjects' 的分区 'sysobjects_1' (分区 ID 1)。此表的逻辑页大小为 4096 字节
分区 'sysobjects_1' (分区 ID 1)中的数据页总数是 3
分区 'sysobjects_1' (分区 ID 1)有 79 个数据行
… …
正在检查表 't1222' (对象 ID 428525529):逻辑页大小为 4096 字节

正在检查表 't1222' 的分区 't1222_428525529' (分区 ID 428525529)。此表的逻辑页大小为 4096 字节
分区 't1222_428525529' (分区 ID 428525529)中的数据页总数是 1
分区 't1222_428525529' (分区 ID 428525529)有 0 个数据行

该表的数据页总数为 1
表包含 0 数据行
DBCC 运行结束。如果 DBCC 打印出错误消息，请与有系统管理员角色的用户联系
```

Step 2 设置单用户模式。

在备份过程当中，最好设置数据库为单用户模式，以防止在备份期间，其他用户对数据库执行更新操作。

```
1> sp_dboption 'iihero', 'single user', true
2> go
数据库'iihero'的数据库选项'single user'被打开
```

对数据库 'iihero' 运行 checkpoint，以便使选项 'single user' 生效
(return status = 0)

Step 3 执行检查点操作，以同步内存和磁盘中的内容。

```
1> use iihero
2> go
1> checkpoint
2> go
```

Step 4 执行数据库及日志转储。

```
1> dump database iihero to 'e:/temp/iihero.dmp'
2> go
```
备份服务器会话标识是 18。当执行 sp_volchanged 系统存储过程的时候，在备份服务器完成请求改变任一卷之后使用这个值

Backup Server: 6.28.1.1: 转储文件名'iihero12147139C3 '，区域号 1 安装在磁盘文件 'E:/temp/iihero.dmp' 上

Backup Server: 4.188.1.1: 数据库 iihero:1704 KB (17%) DUMPED

Backup Server: 4.188.1.1: 数据库 iihero:2326 KB (100%) DUMPED

Backup Server: 3.43.1.1: 转储段号 1 完成

Backup Server: 3.43.1.1: 转储段号 2 完成

Backup Server: 3.43.1.1: 转储段号 3 完成

Backup Server: 4.188.1.1: 数据库 iihero:2336 KB (100%) DUMPED

Backup Server: 3.42.1.1: DUMP 完成(数据库 iihero)

```
1> dump transaction iihero to 'e:/temp/iihero.log.dmp'
2> go
```
备份服务器会话标识是 21。当执行 sp_volchanged 系统存储过程的时候，在备份服务器完成请求改变任一卷之后使用这个值

Backup Server: 6.28.1.1: 转储文件名'iihero12147139D3 '，区域号 1 安装在磁盘文件 'E:/temp/iihero.log.dmp' 上

Backup Server: 4.58.1.1: 数据库 iihero: 12 千字节 DUMPED

Backup Server: 4.58.1.1: 数据库 iihero: 18 千字节 DUMPED

Backup Server: 3.43.1.1: 转储段号 3 完成

Backup Server: 4.58.1.1: 数据库 iihero: 22 千字节 DUMPED

Backup Server: 3.42.1.1: DUMP 完成(数据库 iihero)

在了解了上述流程之后，DBA 可以为其创建一个按时间点自动执行的任务计划，即在指定的时间点周期性地执行备份脚本。

以 Windows 为例，设这个脚本文件名为 dumpdb.bat，该示例脚本为：

```
@echo off

setlocal
```

```
echo %date%
set DES=e:\temp
set db_file=%DES%\iihero_%date:~0,10%.dmp
set log_file=%DES%\iihero_%date:~0,10%.log.dmp
set err_file=%DES%\error_%date:~0,10%.log

echo dbcc checkdb(iihero)>dump.sql
echo go>>dump.sql
echo sp_dboption 'iihero', 'single user', true>>dump.sql
echo go>>dump.sql
echo use iihero>>dump.sql
echo go>>dump.sql
echo dump database iihero to '%db_file%'>>dump.sql
echo go>>dump.sql
echo dump transaction iihero to '%log_file%'>>dump.sql
echo go>>dump.sql
echo use master>>dump.sql
echo go>>dump.sql

isql -SXIONGHE -Usa -P -idump.sql -o%err_file%
endlocal
```

整个备份将在 e:\temp 目录下生成三个文件：

- iihero_<日期>.dmp：为数据库转储文件。
- iihero_<日期>.log.dmp：为日志转储。
- error_<日期>.log：为 bat 文件执行的输出结果。

将 dumpdb.bat 添加到任务里，即可定制为周期性执行的任务计划，非常方便。或者直接使用命令行为 Windows 平台创建一个任务：

```
D:\MyDocument\MyBooks\ASE\code>at 00:00 /every:M,T,W,Th,F,S,Su D:\MyDocument\MyBooks\ASE\code\dumpdb.bat
Added a new job with job ID = 1
```

即每周的每晚 12 点定时执行这个备份脚本。

UNIX 或者 Linux 平台的备份脚本与其类似，所不同的是，可以使用 crontab 来定制任务周期。与此类似的 dumpdb.sh 脚本内容大致如下：

```
DES=/home/sean/backup
db_file=$DES/iihero_`date +%Y-%m-%d`.dmp
log_file=$DES/iihero_`date +%Y-%m-%d`.log.dmp
err_file=$DES/error_`date +%Y-%m-%d`.log

echo dbcc checkdb(iihero)>dump.sql
echo go>>dump.sql
```

```
echo sp_dboption 'iihero', 'single user', true>>dump.sql
echo go>>dump.sql
echo use iihero>>dump.sql
echo go>>dump.sql
echo dump database iihero to '$db_file'>>dump.sql
echo go>>dump.sql
echo dump transaction iihero to '$log_file'>>dump.sql
echo go>>dump.sql
echo use master>>dump.sql
echo go>>dump.sql

isql -SXIONGHE -Usa -P -idump.sql -o$err_file
```

用 crontab 定制此任务,每晚 12 点执行:

```
crontab -e
0    0   *   *    *    /home/sean/dumpdb.sh
```

13

应用 Open Client 库编程

为方便用户开发客户端和服务器应用，Sybase 为用户提供了一套客户/服务器框架，分别是：Open Client 和 Open Server。Open Client 为客户端应用提供 API，从而与 Sybase 数据库 ASE 和 Open Server 进行交互，实际上 ASE 服务器就是一个 Open Server 的应用。

Open Client 提供了两套核心编程接口用于开发客户端应用，一个是 ctlib，另一个是 dblib。前者提供了底层 API 的完备支持。dblib 则主要是为了支持旧版本的 Open Client，现已不多用。ctlib 则是使用最广泛、最稳定的 Open Client 接口库。如无特殊说明，后边的 Open Client 接口指的就是 ctlib 库接口。

13.1 环境搭建

在 ASE 数据库服务器安装完以后，Open Client 库的根目录就在%SYBASE%\%SYBASE_OCS%（或者 UNIX/Linux 下的$SYBASE/$SYBASE_OCS）。作为独立的客户端，也可以安装 Sybase Open Server/OpenClientSDK 15.0。要将它用作独立的客户端，做法是直接将整个目录%SYBASE%\%SYBASE_OCS%复制到客户端的机器上。在这一点上，Open Client 几乎可以算作是"绿色"客户端，因为它只需要一些环境变量即可工作，不需要对 Windows 上的注册表进行修改。相比之下，Oracle 的 OCI 环境就要复杂得多，需要安装一套 Oracle 客户端，它在后期推出的 Instant Client 并未得到大规模的应用。

13.1.1 Windows 下的环境

在 Windows 平台下，假设 Server 安装目录是 D:\sybase，对应于环境变量%SYBASE%，下边的子目录 OCS-15_0 对应于环境变量%SYBASE_OCS%。测试的时候，我们可以在另一台机器上创建一个完全相同的目录结构 D:\sybase。

我们要依次复制以下目录和文件到目标机器的 D:\sybase：

- charsets（字符集信息）
- collate（排序信息）
- config（如果有，则复制）
- ini（interfaces 及其他配置文件）
- locale（语言配置信息）
- OCS-15_0（OpenClient 库主体）
- SYBASE.bat（环境变量）

原始的 sybase.bat 内容如下：

```
set SYBASE_ASE=ASE-15_0
set PATH=D:\sybase\ASE-15_0\bin;D:\sybase\ASE-15_0\dll;%PATH%
set SYBASE=D:\sybase
set SYBASE_OCS=OCS-15_0
set PATH=D:\sybase\OCS-15_0\bin;D:\sybase\OCS-15_0\dll;D:\sybase\OCS-15_0\lib3p;%PATH%
set INCLUDE=D:\sybase\OCS-15_0\include;%INCLUDE%
set LIB=D:\sybase\OCS-15_0\lib;%LIB%
set SYBASE_JRE=D:\sybase\Shared\Sun\jre142
set SYBASE_SYSAM=SYSAM-2_0
set SYBASE_UA=D:\sybase\ua
set SCROOT=D:\sybase\Shared\Sybase Central 4.3
set PATH=D:\sybase\RPL-15_0\bin;%PATH%
set SYBROOT=D:\sybase
set SYBASE_JRE5_32=D:\sybase\Shared\Sun\JRE-5_0_12_32BIT
set SYBASE_JRE5=D:\sybase\Shared\Sun\JRE-5_0_12_32BIT
set PATH=D:\sybase\UAF-2_6\bin;%PATH%
set SYBASE_UA=D:\sybase\UAF-2_6
set SYBASE_PLATFORM=windows
```

我们只需要在目标机器的用户环境变量（我的电脑→属性→高级→环境变量）里设定上边加粗部分的几个值。

13.1.2　UNIX/Linux 下的环境

对于 UNIX/Linux 平台，我们要在目标机器上建立目录 $SYBASE，并复制如下子目录和文件：

- charsets
- collate
- config
- DataAccess
- locales

- OCS-15_0
- interfaces
- SYBASE.sh 或 SYBASE.csh（取决于你使用的 shell）

然后裁减环境变量，保留如下环境变量值：

```
SYBASE="/opt/sybase/ase1502"
export SYBASE
SYBASE_ASE="ASE-15_0"
export SYBASE_ASE
SYBASE_OCS="OCS-15_0"
export SYBASE_OCS
PATH="/opt/sybase/ase1502/OCS-15_0/bin":$PATH
export PATH
LD_LIBRARY_PATH="/opt/sybase/ase1502/OCS-15_0/lib:/opt/sybase/ase1502/OCS-15_0/lib3p":$LD_LIBRARY_PATH
export LD_LIBRARY_PATH
INCLUDE="/opt/sybase/ase1502/OCS-15_0/include":$INCLUDE
export INCLUDE
LIB="/opt/sybase/ase1502/OCS-15_0/lib":$LIB
export LIB
SYBROOT="/opt/sybase/ase1502"
export SYBROOT
```

当然，如果想图方便，你还可以添加两个有用的环境变量 DSLISTEN 和 DSQUERY。

13.1.3 验证连接

interfaces 文件的配置在第 4 章网络连接里有详细描述，配置好 interfaces 文件，就可以使用 isql 命令行来验证对 ASE 服务器的远程连接是否有效了，它也是最简便有效的验证 Open Client 库是否配置完整的方法，下边是一段在 Windows 下的验证过程。

首先，编辑 ini\sql.ini 文件，内容如下：

```
[BJEASLINUX1]
master=tcp,bjeaslinux1,5000
query=tcp,bjeaslinux1,5000
```

然后，导入环境变量（因为要本机测试），并使用 isql 连接该 Server。

```
D:\shared\openclient-15.0>SYBASE.bat

D:\shared\openclient-15.0>set SYBASE=D:\shared\openclient-15.0
D:\shared\openclient-15.0>set SYBASE_OCS=OCS-15_0
D:\shared\openclient-15.0>… …

D:\shared\openclient-15.0>isql -Uspring -Pspring1 -SBJEASLINUX1
??! ??????????????????????????????????('?')?
1> quit
```

D:\shared\openclient-15.0>set LANG=chs

D:\shared\openclient-15.0>isql -Uspring -Pspring1 -SBJEASLINUX1
1> quit

不过，在上边的测试过程中出现了字符集问题。因为我们在 Windows 平台下，并且没有设置环境变量 LANG 的值，它会默认取值 default，我们再看看 locales\locales.dat 文件中的[NT]这一项下的语言及字符集列表：

[NT]
 locale = enu, us_english, iso_1
 locale = fra, french, iso_1
 locale = deu, german, iso_1
 locale = rus, russian, cp1251
 locale = hun, us_english, cp1250
 locale = ell, us_english, cp1253
 locale = heb, us_english, cp1255
 locale = ara, us_english, cp1256
 locale = trk, us_english, cp1254
 locale = esp, spanish, iso_1
 locale = jpn, japanese, sjis
 locale = japanese, japanese, sjis
 locale = chs, chinese, eucgb
 locale = cht, tchinese, big5
 locale = kor, korean, cp949
 locale = us_english.utf8, us_english, utf8
 locale = default, us_english, iso_1

default 对应的字符集是 ISO_1，与 GB2312 不兼容。与 GB2312（EUCGB）对应的 LANG 值为 chs，所以，当我们设定 LANG 环境变量为 chs 时，一切显示正常。因此，locales.dat 中的字符集问题仍然比较重要。另一个常用的做法是修改 locale=default 项的值，即将最后一行改为 locale = default, chinese, eucgb。

需要牢记的就是，客户端需要根据自己的操作系统类型（这里是 NT）找到 locales.dat 中的对应项去初始化 locales 相关结构，与数据库服务器进行交互。

而上述环境变量对于使用 Open Client ctlib 接口进行编程来说也是必需的。

13.1.4 开发环境

在 Windows 下，ASE 15.0 仍然支持 VC 6.0 进行开发，因为它里边的 Open Client 库全是使用 VC 6.0 的编译器编译链接的。基于兼容的考虑，我们也完全可以基于 VC 7.x（对应于 Visual Studio 2003）、VC 8.0（对应于 Visual Studio 2005）、VC 9.0（对应于 Visual Studio 2008）来进行开发。

在 UNIX/Linux 下，IBM RS6000 使用 AIX 下的编译器 xlC_r，Solaris 使用 Sun Studio 的 cc，HP UNIX 会使用它的编译器 acc。这里要提示的是，这些 UNIX 平台使用的都不是我们通常所用的

gcc 编译器，使用 gcc 基于 Open Client 来开发应用肯定会出错。在 Linux 下，则是统一使用 gcc 编译器。

13.2 编程模型

ASE Adaptive Server 实际上就是 Open Server 的一个实例，它完全基于 Open Server 的框架开发而成，从而 Open Client 与 Adaptive Server 组成经典的 Open Client/Open Server 这种 C/S 访问结构。

Open Client 从 12.5.x 到 15.0.x 有一个重要的变化是，它的动态库的名字发生了改变。其对照表如表 13-1 所示。

表 13-1 动态库名对照表

Open Client12.5.x	Open Client 15.0.x	作用
libct.dll (so)	libsybct.dll (so)	Client 库
libcs.dll (so)	libsybcs.dll (so)	CS 库
libsrv.dll(so)	libsybsrv.dll(so)	Server 库
libblk.dll(so)	libsybblk.dll(so)	Bulk 库

还有一些库，这里不一一列举。其他所有 OCS-12_5/dll 或 lib 下边的库名到了 15.0 版本以后，都在中间加上了 syb 标志串。

基于 Open Client/Open Server 开发，需要一些公共的库：

1. libsybcs.dll (so)：主要用于 Client 和 Server 的公共框架。
2. libsybblk.dll (so)：主要用于 Bulk 操作。

除此以外，Open Client 开发还需要自己的客户端库 libsybct.dll (so)。也就是说，使用 Open Client 开发应用，至少应该包括三个库：libsybcs、libsybblk、libsybct（针对 12.5.x 版本，使用的三个库是 libcs、libblk、libct）。

Open Client 和 Open Server 库之间的关系是：Client-Library(ctlib)库的 API 都是以 ct_打头；CS-Library(cslib)库的 API 都是以 cs_打头；Bulk-Library(blklib)库的 API 都是以 blk_打头，如图 13-1 所示。

一个比较典型的 Open Client 访问 ASE 的流程是：

1. 创建到 Server 的连接。
2. 发送命令到 Server 端。
3. 使用 while 循环处理命令的结果，直到没有结果或者出错。
4. 释放结果。
5. 关闭连接。

下面将结合具体的 SQL 操作，逐步介绍如何使用 Open Client 中的 API 实现上述流程。

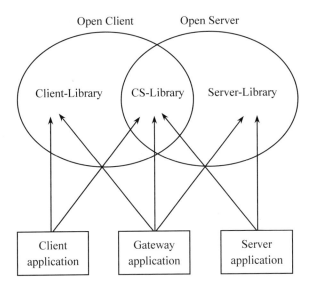

图 13-1 Open Client 和 Open Server 库间的关系

13.3 连接数据库

13.3.1 创建连接

Open Client 创建连接需要做大量的初始化工作。先看一段创建数据库连接的代码清单（完整清单见 openclient\ocs_example1.cpp）：

```
#001 #define CS_NO_LARGE_IDENTIFIERS
#002 #define EX_CTLIB_VERSION CS_VERSION_125
#003
#004 #ifdef WIN32
#005 #pragma comment(lib, "libsybct.lib")
#006 #pragma comment(lib, "libsybcs.lib")
#007 #pragma comment(lib, "libsybdb.lib")
#008 #endif
#009
#010 static CS_CONTEXT*        context = NULL;
#011 static CS_CONNECTION*     connection = NULL;
#012 static CS_COMMAND*        cmd = NULL;
#013
#014 static const char* TEST_INTERFACES = "D:\\SybaseASE\\ini\\sql.ini";
#015
#016 void test_open(const char* server, const char* userName, const char* password,
#017                const char* database, const char* lang = "chinese",
```

```
#018                         const char* charset = "eucgb", const char* ini = TEST_INTERFACES)
#019 {
#020      context = (CS_CONTEXT *)NULL;
#021      CS_RETCODE ret = cs_ctx_alloc(EX_CTLIB_VERSION, &context);
#022      check_error(context, ret, "cs_ctx_alloc failed");
#023
#024      // Initialize Client-Library.
#025      ret = ct_init(context, EX_CTLIB_VERSION);
#026      check_error(context, ret, "ct_init failed");
#027
#028      // Set up error handling
#029      ret = cs_config(context, CS_SET, CS_MESSAGE_CB,
#030           (CS_VOID *)csmsg_callback,
#031           CS_UNUSED, NULL);
#032      check_error(context, ret,
#033           "cs_config(CS_MESSAGE_CB) failed");
#034
#035      ret = ct_callback(context, NULL, CS_SET, CS_CLIENTMSG_CB,
#036           (CS_VOID *)clientmsg_callback);
#037      check_error(context, ret,
#038           "ct_callback for client messages failed");
#039
#040      ret = ct_callback(context, NULL, CS_SET, CS_SERVERMSG_CB,
#041           (CS_VOID *)servermsg_callback);
#042      check_error(context, ret,
#043           "ct_callback for server messages failed");
#044
#045      // set the charset and language
#046      CS_LOCALE* locale;
#047      check_error(context, (ret = cs_loc_alloc(context, &locale)), "Alloc locale handler failed!");
#048      check_error(context, (ret = cs_config(context, CS_SET, CS_LOC_PROP, locale, CS_UNUSED,
#049           NULL)), "CS config locale failed!");
#050
#051      if (lang)
#052      {
#053           check_error(context, (ret = cs_locale(context, CS_SET, locale, CS_SYB_LANG,
#054                (char*)lang, CS_NULLTERM, (CS_INT*)NULL)), "cs_locale language error!");
#055      }
#056      if (charset)
#057      {
#058           check_error(context, (ret = cs_locale(context, CS_SET, locale, CS_SYB_CHARSET,
#059                (char*)charset, CS_NULLTERM, (CS_INT*)NULL)), "cs_locale charset error!");
#060      }
#061
#062      check_error(context, (ret = cs_config(context, CS_SET, CS_LOC_PROP, locale,
#063           CS_UNUSED, NULL)), "CS config locale failed!");
#064      check_error(context, (ret = cs_loc_drop(context, locale)), "CS LOC DROP error!");
#065
```

```
#066        CS_INT logintimeout = 30;
#067        ret = ct_config(context, CS_SET, CS_LOGIN_TIMEOUT, &logintimeout, CS_UNUSED, NULL);
#068        check_error(context, ret, "ct_config login timeout failed");
#069
#070        logintimeout = 60;
#071        ret = ct_config(context, CS_SET, CS_TIMEOUT, &logintimeout, CS_UNUSED, NULL);
#072        check_error(context, ret, "ct_config cs_timeout failed");
#073
#074        // Specify interfaces file
#075        ret = ct_config(context, CS_SET, CS_IFILE, (CS_VOID*)ini, CS_NULLTERM, NULL);
#076        check_error(context, ret, "ct_config interfaces failed");
#077
#078        // Allocate connection handle
#079        ret = ct_con_alloc(context, &connection);
#080        check_error(context, ret, "ct_con_alloc() failed");
#081
#082        ret = ct_con_props(connection, CS_SET, CS_USERNAME,
#083            (char*)userName, CS_NULLTERM, NULL);
#084        check_error(context, ret, "Could not set user name");
#085        ret = ct_con_props(connection, CS_SET, CS_PASSWORD,
#086            (char*)password, CS_NULLTERM, NULL);
#087        check_error(context, ret, "Could not set password");
#088
#089        // Connect and use database
#090        if(server == NULL)
#091            ret = ct_connect(connection, (CS_CHAR *)NULL, 0);
#092        else
#093            ret = ct_connect(connection, (CS_CHAR *)server, strlen(server));
#094        check_error(context, ret, "Could not connect!");
#095
#096        char szUseDB[128]="";
#097        strcat(szUseDB, "use ");
#098        strcat(szUseDB, database);
#099        cmd_exec(connection, szUseDB);
#100
#101        std::cout<<"Connect to : ("<<server<<","<<database<<") successfully!!"<<std::endl;
#102    }
```

首先给定一个宏定义：CS_NO_LARGE_IDENTIFIERS，它是为 Open Client 15.0 兼容 12.5 和更低版本所定义的；如果不考虑兼容性，则不需要这个宏。接着设定初始化的版本号 CS_VERSION_125，即可以使用 15.0 来初始化 12.5 版本的连接上下文。

第 4~8 行定义了 Windows 下链接时需要的三个库文件。UNIX 下则不需要。

函数 test_conn 完整地展示了如何根据服务器名、用户、密码、数据库名以及语言、字符集甚至 interfaces 文件路径来连接后台的 Adaptive Server。

创建连接的流程如下：

Step 1 调用 cs_ctx_alloc 创建 CS_CONTEXT 句柄。以 cs_ 打头的 API 全部来自 CS 库。它是 Client 和 Server 之间的公共库，是一个公共的上下文句柄。

cs_ctx_alloc 调用带两个参数，第一个是 ctlib 的版本，另一个是 CS_CONTEXT**。

接着进行初始化客户端库的操作，调用函数 ct_init(context, EX_CTLIB_VERSION)。

Step 2 错误处理，每一次 API 调用都要严格处理其返回值，这里通过调用 check_error 来统一处理。

```
static void check_error(CS_CONTEXT* context, CS_RETCODE ret, char* str)
{
    if (ret != CS_SUCCEED)
    {
        fprintf(stderr, "Fatal error: %s n", str);
        if (context != (CS_CONTEXT *) NULL)
        {
            (CS_VOID) ct_exit(context, CS_FORCE_EXIT);
            (CS_VOID) cs_ctx_drop(context);
        }
        exit(-1);
    }
}
```

Step 3 为系统设置处理错误消息的回调函数，这个模式比较固定。29～43 行依次设置了 CS 库的回调、Client 端的消息回调、Server 端的消息回调。

CS 库消息回调使用函数 cs_config，设置如下：

```
ret = cs_config(context, CS_SET, CS_MESSAGE_CB,
        (CS_VOID *)csmsg_callback,
        CS_UNUSED, NULL);
CS_RETCODE CS_PUBLIC csmsg_callback PROTOTYPE((
    CS_CONTEXT *context, CS_CLIENTMSG *emsgp ))
{
    fprintf(stderr,
        "CS-Library error:\n");
    fprintf(stderr,
        "\tseverity(%ld) layer(%ld) origin(%ld) number(%ld)",
        (long)CS_SEVERITY(emsgp->msgnumber),
        (long)CS_LAYER(emsgp->msgnumber),
        (long)CS_ORIGIN(emsgp->msgnumber),
        (long)CS_NUMBER(emsgp->msgnumber));

    fprintf(stderr, "\t%s\n", emsgp->msgstring);

    if (emsgp->osstringlen > 0)
    {
        fprintf(stderr, "Operating System Error: %s\n",
```

```
            emsgp->osstring);
    }

    return (CS_SUCCEED);
}
```

函数的原型形式是固定的。在回调函数里，将相应的错误消息打出来。也可以根据需要将错误消息输出到指定文件。

Client 端和 Server 端消息回调设置都调用函数 ct_callback 进行设置，这两个回调函数的实现如下：

```
CS_RETCODE CS_PUBLIC clientmsg_callback PROTOTYPE((
    CS_CONTEXT *context, CS_CONNECTION *conn, CS_CLIENTMSG *emsgp ))
{
    fprintf(stderr,
        "Client Library error:\n\t");
    fprintf(stderr,
        "severity(%ld) number(%ld) origin(%ld) layer(%ld)\n",
        (long)CS_SEVERITY(emsgp->severity),
        (long)CS_NUMBER(emsgp->msgnumber),
        (long)CS_ORIGIN(emsgp->msgnumber),
        (long)CS_LAYER(emsgp->msgnumber));

    fprintf(stderr, "\t%s\n", emsgp->msgstring);

    // OS errors
    if (emsgp->osstringlen > 0)
    {
        fprintf(stderr,
            "Operating system error number(%ld):\n",
            (long)emsgp->osnumber);
        fprintf(stderr, "\t%s\n", emsgp->osstring);
    }
    return (CS_SUCCEED);
}

CS_RETCODE CS_PUBLIC servermsg_callback PROTOTYPE((
    CS_CONTEXT *cp, CS_CONNECTION *chp, CS_SERVERMSG *msgp ))
{
    // ignore 'changed database' or 'changed language to' message
    if ((msgp->msgnumber == 5701) || (msgp->msgnumber == 5703))
    {
        return CS_SUCCEED;
    }

    fprintf(stderr,
        "Server message:\n\t");
    fprintf(stderr,
```

```c
            "number(%ld) severity(%ld) state(%ld) line(%ld)\n",
            (long)msgp->msgnumber, (long)msgp->severity,
            (long)msgp->state, (long)msgp->line);

    if (msgp->svrnlen > 0)
        fprintf(stderr, "\tServer name: %s\n", msgp->svrname);

    if (msgp->proclen > 0)
        fprintf(stderr, "\tProcedure name: %s\n", msgp->proc);

    fprintf(stderr, "\t%s\n", msgp->text);
    return (CS_SUCCEED);
}
```

Step 4 第45～65行设置合适的 locale（即语言和字符集相关选项），这一步是可选操作。在实际开发应用中，出现乱码现象大多与它有关。

首先，调用 cs_loc_alloc 为句柄 CS_LOCALE* locale 分配空间，其参数分别是前边创建的 context 句柄以及一个 CS_LOCALE 的双重指针，最后得到一个默认的 locale，实际上是找到 locales.dat 文件进行初始化的。

然后，调用一次 cs_config 来设置这个默认的 locale。

接着，第 51～61 行，根据参数 lang 的值，设置 locale 中的 CS_SYB_LANG 项；根据参数 charset 的值，设置 locale 中的 CS_SYB_CHARSET 项。

回忆一下 locales.dat 文件的格式，每一个平台下都有如下形式：

    locale = <locale>, <language>, <charset> [, sortorder]

Open Client 加载 locale 的规则如下：

如果是 UNIX/Linux 平台，首先获取环境变量 LC_ALL 的值，如果值不为空，则使用它作为<locale>项，否则试探获取环境变量 LANG 的值，如果值不为空，则直接用它作为<locale>项，根据这一项的值以及当前的平台名，在 locales.dat 文件中加载到对应行，如果 LANG 值没有设定，则设定 locale 值为 default，在对应平台下查找相关行。

CS_SYB_LANG 项对应的就是<language>，用来修改加载 locale 完整项以后的 language 值。CS_SYB_CHARSET 项要覆盖的就是<charset>的值。

如果不设置 language 和 charset 信息，Open Client 就按照默认规则进行 locale 信息的加载。其实，我们还可以在第 51 行之前添加以下代码设置 CS_LC_ALL 项的值：

```c
char lc_msg[80] = "zh_CN";
check_error(context, (ret = cs_locale(context, CS_SET, locale, CS_LC_ALL,
    (char*)lc_msg, CS_NULLTERM, (CS_INT*)NULL)), "cs_locale language error!");
```

可以把 lc_msg 作为函数 test_conn 的一个参数，一旦设定，就不用取环境变量 LC_ALL 或者 LANG 的值了。它的值将直接用作<locale>的值进行检索。

Step 5 设置连接的 timeout（可选）。这里设置最长登录时间为 30 秒。

Step 6 指定 interfaces 文件位置（可选），第 75 行。当你不想使用默认的 sql.ini 或者 interfaces 文件，可以使用该处理。

Step 7 调用 ct_con_alloc，分配连接句柄 CS_CONNECTION*。并为连接句柄指定用户和密码。

Step 8 调用 ct_connect，连接服务器。如果没有指定服务器名，则连接默认的服务器，该服务器由环境变量 DSQUERY 指定，即 ct_connect(connection, (CS_CHAR *)NULL, 0) 会自动连接到 $DSQUERY 指定的服务器。

Step 9 最后，执行命令 "use <database>" 来连接到具体的数据库，第 96~99 行。这里调用了函数 cmd_exec(CS_CONNECTION*, char* sql)，该函数只执行一个 SQL 命令，并忽略返回的任何结果。

Linux 下编译 example1.cpp 的 make 文件如下：

```
DEBUG    = -g
CC       = g++
CFLAGS   = -c $(DEBUG) -fPIC  -Wno-deprecated -D__linux -DLINUX_OS
DLLFLAGS = -c $(DEBUG) -fPIC -Wno-deprecated -D__linux -DLINUX_OS
INCLUDE = $(SYBASE)/$(SYBASE_OCS)/include
LIBPATH = $(SYBASE)/$(SYBASE_OCS)/lib


#example1
OBJS = example1.o

${OBJS}:
   $(CC) $(CFLAGS) -I$(INCLUDE) example1.cpp
exe:   ${OBJS}
   $(CC) ${OBJS} -L$(LIBPATH) -lsybct -lsybcs -lsybblk -o example1
clean:
   rm -f *.o *.exe
```

这里需要提醒的是，需要有三个库的链接：sybct、sybcs、sybblk，同时需要设定好链接的路径 libpath。

13.3.2 处理命令

本节里的处理命令是指，处理一个 DDL 型的 SQL 语句或者完全命令型的 SQL 语句，比如 use database、set tranchained off 等。

它的处理流程如下：

- 调用 ct_cmd_alloc 创建一个 CS_COMMAND* 句柄。
- 调用 ct_command 创建一个 CS_LANG_CMD 型的命令。
- 调用 ct_send 发送命令。

- 在 while 循环里调用 ct_results 处理该命令的结果。
- 根据返回值调用 ct_cmd_drop 来释放 CS_COMMAND*句柄。

我们来看看前边处理命令的函数 cmd_exec(CS_CONNECTION* conn, char* ddl)的实现，如下所示：

```cpp
static CS_RETCODE cmd_exec(CS_CONNECTION* conn, char* ddl)
{
    CS_RETCODE          retcode;
    CS_INT              restype;
    CS_COMMAND          *cmd;
    CS_RETCODE          query_code;

    CS_RETCODE ret = ct_cmd_alloc(conn, &cmd);
    check_error(context, ret, "ct_cmd_alloc() failed");

    if ( (retcode=ct_command(cmd, CS_LANG_CMD, ddl, CS_NULLTERM, CS_UNUSED))
        != CS_SUCCEED )
    {
        std::cerr<<"ex_execute_cmd: ct_command() failed"<<std::endl;
        cmd_close(&cmd);
        return retcode;
    }
    if ((retcode = ct_send(cmd)) != CS_SUCCEED)
    {
        std::cerr<<"ex_execute_cmd: ct_send() failed"<<std::endl;
        cmd_close(&cmd);
        return retcode;
    }

    query_code = CS_SUCCEED;
    while ((retcode = ct_results(cmd, &restype)) == CS_SUCCEED)
    {
        switch((int)restype)
        {
        case CS_CMD_SUCCEED:
        case CS_CMD_DONE:
            break;

        case CS_CMD_FAIL:
            query_code = CS_FAIL;
            break;

        case CS_STATUS_RESULT:
            retcode = ct_cancel(NULL, cmd, CS_CANCEL_CURRENT);
            if (retcode != CS_SUCCEED)
            {
                std::cerr<<("ex_execute_cmd: ct_cancel() failed"
```

```
                    <<std::endl;
                query_code = CS_FAIL;
            }
            break;

        default:

            // Unexpected result type.
            query_code = CS_FAIL;
            break;
    }
    if (query_code == CS_FAIL)
    {
        // Terminate results processing and break out of
        retcode = ct_cancel(NULL, cmd, CS_CANCEL_ALL);
        if (retcode != CS_SUCCEED)
        {
            std::cerr<<("ex_execute_cmd: ct_cancel() failed")
                    <<std::endl;
        }
        break;
    }
}

// Clean up the command handle used
if (retcode == CS_END_RESULTS)
{
    retcode = ct_cmd_drop(cmd);
    if (retcode != CS_SUCCEED)
    {
        query_code = CS_FAIL;
    }
}
else
{
    (void)ct_cmd_drop(cmd);
    query_code = CS_FAIL;
}
return query_code;
}
```

这也是处理一个 CS_LANG_CMD 型命令的常用流程。与 COMMAND 相关的几个函数用法如下：

1. CS_RETCODE ct_cmd_alloc(connection, cmd_pointer)

CS_CONNECTION* connection：指向 CS_CONNECTION 结构句柄。

CS_COMMAND** cmd_pointer：指向 CS_COMMAND 结构句柄的指针，用于分配这个句柄。

返回值是 CS_RETCODE 型，通常它是 long 型。几乎所有的库函数都有返回值。成功，则返回 CS_SUCCEED；否则返回 CS_FAIL 或其他错误码。

2．ct_command 函数，用于初始化一个语句命令、包命令、RPC 命令、消息命令或者发送数据的命令。其原型如下：

```
CS_RETCODE ct_command(cmd, type, buffer, buflen,option)
CS_COMMAND *cmd;
CS_INT type;
CS_VOID *buffer;
CS_INT buflen;
CS_INT option;
```

- cmd：一个 COMMAND 句柄。
- type：命令的类型。
- buffer：缓冲区。
- buflen：缓冲区大小。
- option：该命令相关的选项。

函数 ct_command 中 type 参数值与其他几个参数的有效值的对应关系如表 13-2 所示。

表 13-2　ct_command 中的 type 参数值与其他参数有效值的对应关系

type 值	初始化的 command	buffer 用途	buflen
CS_LANG_CMD	一个语句命令	一个字符串	CS_NULLTERM 或者字符串实际长度
CS_MSG_CMD	一个消息命令	指向 int 变量的指针，该变量保存着消息的 ID	CS_UNUSED
CS_PACKAGE_CMD	包命令	字符串，描述包名	CS_NULLTERM 或者字符串实际长度
CS_RPC_CMD	远程过程调用命令	字符串，描述过程名	CS_NULLTERM 或者字符串实际长度
CS_SEND_DATA_CMD	发送数据的命令	NULL	CS_UNUSED
CS_SEND_BULK_CMD	内部发送批量数据的命令	字符串，描述表名	CS_NULLTERM 或者字符串实际长度

函数 ct_command 中的 type 参数值与最后一个参数 option 有效值的对应关系如表 13-3 所示。

表 13-3　type 与 option 值间的关系

type	option 值	作用
CS_LANG_CMD	CS_MORE	buffer 中的串只是命令的一部分
	CS_END	buffer 中的串是命令的最后一部分
	CS_UNUSED	等价于 CS_END

续表

type	option 值	作用
CS_RPC_END	CS_RECOMPILE	在执行存储过程之前重编译
	CS_NO_RECOMPILE	在执行存储过程之前不进行重编译
	CS_UNUSED	等同于 CS_NO_RECOMPILE
CS_SEND_DATA_CMD	CS_COLUMN_DATA	数据发送给 text 或 image 列的更新
	CS_BULK_DATA	内部使用，用于 bulk 批量复制操作
CS_SEND_BULK_CMD	CS_BULK_INIT	初始化一个批量复制操作
	CS_BULK_CONT	继续执行一个批量复制

3. ct_send 函数，发送一个命令给 Server 端。

4. ct_results 函数，用于处理一个命令的执行结果，我们注意到这里使用了一个 while 循环，直到 ct_results 函数返回失败结束。这是因为在 ASE 数据库中，发送一个命令可以包含多个 SQL 语句，因而可以返回多个结果集。甚至单条 SQL 语句也可以返回多个结果集。

```
CS_RETCODE ct_results(cmd, result_type)
CS_COMMAND *cmd;
CS_INT *result_type;
```

- cmd：CS_COMMAND 句柄。
- result_type：指向一个整数的指针，用于返回结果集类型。

该函数是 Open Client 中非常重要的一个函数，几乎绝大部分处理结果集的实现都需要它。

result_type 的有效值如表 13-4 所示。

表 13-4 result_type 值的含义

范围	result_type	意义	结果集内容
表示命令状态	CS_CMD_DONE	命令已经处理完	不可用
	CS_CMD_FAIL	Server 执行命令时出错	无结果
	CS_CMD_SUCCEED	命令成功执行	无结果
表示结果集	CS_COMPUTE_RESULT	计算行结果集	单行计算结果
	CS_CURSOR_RESULT	由 ct_cursor 命令得到的游标行结果集	0 或多行表数据
	CS_PARAM_RESULT	参数返回值	单行参数返回值
	CS_ROW_RESULT	普通的行结果集	0 或多行表数据
	CS_STATUS_RESULT	存储过程返回值	单行单个状态值
获取元信息	CS_COMPUTEFMT_RESULT	计算值的格式信息	没有结果集返回，但会得到将要返回的计算结果的格式信息，该格式可以通过 ct_res_info、ct_describe、ct_compute_info 得到
	CS_ROWFMT_RESULT	行格式信息	没有结果集，但可以通过 ct_describe 和 ct_res_info 得到格式信息

续表

范围	result_type	意义	结果集内容
获取元信息	CS_MSG_RESULT	表示有消息到达	可以通过 ct_res_info 得到消息的 ID
	CS_DESCRIBE_RESULT	动态 SQL 的描述信息，有 in 和 out 两部分	没有结果集，下边几种操作都可以得到结果集： 调用 ct_res_info 得到各列，ct_describe 得到具体列的元信息 多次调用 ct_dyndesc，得到列数和各列的元信息 调用 ct_res_info 得到列数，ct_dynsqlda 得到各列的元信息

5. ct_cancel 函数，取消一个命令或者放弃一个命令的结果。

```
CS_RETCODE ct_cancel(connection, cmd, type)
CS_CONNECTION *connection;
CS_COMMAND *cmd;
CS_INT type;
```

- connection：一个 CS_CONNECTION 句柄。
- cmd：一个 CS_COMMAND 句柄。
- type：要取消的类型。

这也是很重要的一个处理函数。当 type 值或者说要取消类型为 CS_CANCEL_CURRENT 时，connection 参数值必须为 NULL。当取消类型为 CS_CANCEL_ATTN 或 CS_CANCEL_ALL 时，connection 和 cmd 两者必须有一个为 NULL。如果 cmd 为 NULL，而 connection 非空时，ct_cancel 将把该连接的所有延迟操作都取消。CS_CANCEL_ATTN 类型是指将发送一个通知给 Server，告诉它要取消当前命令。CS_CANCEL_CURRENT 则指放弃当前的结果集。

6. ct_cmd_drop 函数，用于释放一个 CS_COMMAND 句柄。

我们注意到，在 cmd_exec 函数中，直接使用 ct_cancel 调用忽略了除 CS_STATUS_RESULT 类型以外的所有结果集类型。因为它的目的只是执行一个 SQL 命令。

13.3.3 关闭连接

关闭连接的过程实际上是释放已经创建的句柄的过程。其处理过程如下所示：

```
void test_close()
{
    CS_RETCODE ret = ct_close(connection, CS_UNUSED);
    check_error(context, ret, "ct_close failed");
    ret = ct_con_drop(connection);
    check_error(context, ret, "ct_con_drop failed");
```

```
            // ct_exit tells Client-Library that we are done.
            ret = ct_exit(context, CS_UNUSED);
            check_error(context, ret, "ct_exit failed");

            //Drop the context structure.
            ret = cs_ctx_drop(context);
            check_error(context, ret, "cs_ctx_drop failed");

            std::cout<<"Close the connection to "<<TEST_SERVER_NAME<<"!!"<<std::endl;
}
```

1. 调用 ct_close，关闭当前连接。
2. 调用 ct_con_drop，释放 CS_CONNECTION*句柄。
3. 调用 ct_exit，退出客户端环境。
4. 调用 cs_ctx_drop，释放 CS_CONTEXT*公共上下文公柄。

13.4 SQL 中的 DDL 操作

SQL 中的 DDL 操作（Create、Drop、Alter）有一个共同的特点，就是不带返回值。于是可以直接使用上节的 cmd_exec 来处理普通的 DDL 或其他纯命令操作。

最典型的是事务的开始、提交、回滚，甚至设置事务的隔离级，都可以使用它来实现。

有一个比较典型的实现，是关于事务是否自动提交：在 ASE 和 SQL Server 中，系统默认情况下都是事务自动提交，我们可以通过 set chained on 的命令来将事务设置成非自动提交。实际上，在它们的内部实现（jdbc、odbc）中也是这么实现的。

我们可以参照一下 example2.cpp 中使用 cmd_exec 的代码片段，都是通过 cmd_exec 的调用来实现一些常规的功能：

```
bool begin()
{
    return cmd_exec(connection, "begin tran") == CS_SUCCEED;
}

bool rollback()
{
    return cmd_exec(connection, "rollback tran") == CS_SUCCEED;
}

bool commit()
{
    return cmd_exec(connection, "commit") == CS_SUCCEED;
}

bool setAutoCommit(bool autoCommit)
```

```cpp
{
    CS_RETCODE ret = CS_FAIL;
    if (autoCommit)
    {
        ret = cmd_exec(connection, "set chained off");
    }
    else
    {
        ret = cmd_exec(connection, "set chained on");
    }
    return ret == CS_SUCCEED;
}

int main()
{
    test_open(TEST_SERVER_NAME, TEST_USER_NAME, TEST_PASSWORD, TEST_DB_NAME);
    setAutoCommit(false);

    if (!cmd_exec("drop table t12345"))
    {
        std::cout<<"t12345 not exist!"<<std::endl;
    }
    cmd_exec("create table t12345(id int primary key, col2 varchar(32))");

    begin();
    cmd_exec("insert into t12345 values(1, '测试')");
    rollback();

    begin();
    cmd_exec("insert into t12345 values(1, '测试')");
    commit();
    setAutoCommit(true);
    test_close();
    return 0;
}
```

总结一下，可以通过类似于 cmd_exec 来处理的命令包括：

- DDL 操作（Create、Drop、Alter）。
- 单条 Insert/Update/Delete 语句的操作。
- 事务处理命令（begin tran、rollback tran、commit）。
- 权限控制命令（grant、revoke）。
- 数据库选项命令（set ...）。

Linux 下编译链接 example2.cpp 的 make 文件，如下所示：

```
DEBUG     = -g
CC        = g++
CFLAGS    = -c $(DEBUG) -fPIC   -Wno-deprecated -D_linux -DLINUX_OS
```

```
DLLFLAGS = -c $(DEBUG) -fPIC -Wno-deprecated -D_linux -DLINUX_OS
INCLUDE = $(SYBASE)/$(SYBASE_OCS)/include
LIBPATH = $(SYBASE)/$(SYBASE_OCS)/lib

#example2
OBJS = example2.o

${OBJS}:
    $(CC) $(CFLAGS) -I$(INCLUDE) example2.cpp
exe:    ${OBJS}
    $(CC) ${OBJS} -L$(LIBPATH) -lsybct -lsybcs -lsybblk -o example2.exe
clean:
    rm -f *.o *.exe
```

13.5 获取 SQL 查询结果集

13.5.1 简单结果集获取

我们就以样例数据库 iihero 中的表 student 为例，使用 Open Client 获取 student 表中的数据。student 表的表结构如下：

```
create table student
(
    sno int not null primary key,
    sname varchar(32) not null,
    sgender char(1) not null,
    sbirth datetime not null,
    sage numeric(2) null,
    sdept varchar(128) null
)
```

由于表结构预先知道，我们可以通过 ct_results 得到结果之后，由上两节中的内容知道它返回的应该是 CS_ROW_RESULT 型结果，然后通过对这些列进行相关值的绑定，最终可以得到相应结果，详见示例代码中 Open Client 下的 example3.cpp，这里列出函数 test_select_simple 中的实现：

```
#001 void test_select_simple()
#002 {
#003     static const int MAXCOLUMNS = 15;
#004     static const int MAXSTRING = 128;
#005     CS_DATAFMT         columns[MAXCOLUMNS];
#006     CS_INT             datalength[MAXCOLUMNS];
#007     CS_SMALLINT        indicator[MAXCOLUMNS];
#008     int sno;
#009     char   sname[MAXSTRING];
```

```
#010    int sage;
#011    CS_DATETIME sbirth;
#012    CS_DATEREC sb;
#013
#014    cmd_open(&cmd);
#015    CS_RETCODE ret = ct_command(cmd, CS_LANG_CMD,
#016         "select sno, sname, sbirth, sage from student",
#017         CS_NULLTERM, CS_UNUSED);
#018    check_error(context, ret, "ct_command() failed");
#019    check_error(context, ct_send(cmd), "ct_send() failed");
#020
#021    CS_RETCODE results_ret;
#022    CS_INT result_type;
#023    int i=0;
#024    while ( (results_ret = ct_results(cmd, &result_type)) == CS_SUCCEED )
#025    {
#026         switch ((int)result_type)
#027         {
#028         case CS_ROW_RESULT:
#029             {
#030                 columns[0].datatype = CS_INT_TYPE;
#031                 columns[0].format = CS_FMT_UNUSED;
#032                 //columns[0].maxlength = MAXSTRING;
#033                 columns[0].count = 1;
#034                 columns[0].locale = NULL;
#035                 check_error(context,   ct_bind(cmd, 1, &columns[0],
#036                      &sno, &datalength[0],
#037                      &indicator[0]), "ct_bind() for sno failed");
#038
#039                 columns[1].datatype = CS_CHAR_TYPE;
#040                 columns[1].format = CS_FMT_NULLTERM;
#041                 columns[1].maxlength = MAXSTRING;
#042                 columns[1].count = 1;
#043                 columns[1].locale = NULL;
#044                 check_error(context,   ct_bind(cmd, 2, &columns[1],
#045                      sname, &datalength[1],
#046                      &indicator[1]), "ct_bind() for sname failed");
#047
#048                 columns[2].datatype = CS_DATETIME_TYPE;
#049                 columns[2].format = CS_FMT_UNUSED;
#050                 columns[2].count = 1;
#051                 columns[2].locale = NULL;
#052                 check_error(context,   ct_bind(cmd, 3, &columns[2],
#053                      &sbirth, &datalength[2],
#054                      &indicator[2]), "ct_bind() for sbirth failed");
#055
#056                 columns[3].datatype = CS_INT_TYPE;
#057                 columns[3].format = CS_FMT_UNUSED;
```

```
#058                    columns[3].count = 1;
#059                    columns[3].locale = NULL;
#060                    check_error(context,  ct_bind(cmd, 4, &columns[3],
#061                        &sage, &datalength[3],
#062                        &indicator[3]), "ct_bind() for sage failed");
#063
#064                    CS_INT count = 0;
#065                    while (((ret = ct_fetch(cmd, CS_UNUSED, CS_UNUSED,
#066                        CS_UNUSED, &count))== CS_SUCCEED)|| (ret == CS_ROW_FAIL))
#067                    {
#068                        if (ret == CS_ROW_FAIL)
#069                        {
#070                            fprintf(stderr, "Error on row %ld.\n", (long)(count + 1));
#071                        }
#072                            check_error(context, cs_dt_crack(context, CS_DATETIME_TYPE, &sbirth, &sb),
"cs_dt_crack() error!");
#073                        fprintf(stdout, "%d row==> (%d, %s, %d-%d-%d %d:%d:%d", ++i, sno, sname,
#074                            sb.dateyear, sb.datemonth+1, sb.datedmonth, sb.datehour, sb.dateminute, sb.datesecond);
#075                        if (indicator[3] == -1)
#076                        {
#077                            fprintf(stdout, ", NULL)\n");
#078                        }
#079                        else
#080                        {
#081                            fprintf(stdout, ", %d)\n", sage);
#082                        }
#083                    }
#084                    if (ret == CS_END_DATA)
#085                    {
#086                        fprintf(stdout, "\nAll done processing rows.\n");
#087                    }
#088                    else
#089                    {
#090                        check_error(context, CS_FAIL, "ct_fetch failed");
#091                    }
#092
#093                }
#094                break;
#095            case CS_CMD_SUCCEED:
#096                std::cerr<<"No rows returned."<<std::endl;
#097                break;
#098            case CS_CMD_FAIL:
#099                break;
#100            case CS_CMD_DONE:
#101                break;
#102            default:
```

```
#103              check_error(context, CS_FAIL, "ct_results returned unexpected result type");
#104              break;
#105          }
#106      }
#107
#108      switch ((int)results_ret)
#109      {
#110      case CS_END_RESULTS:
#111          break;
#112
#113      case CS_FAIL:
#114          check_error(context, CS_FAIL,
#115              "ct_results() returned CS_FAIL.");
#116          break;
#117      default:
#118          check_error(context, CS_FAIL,
#119              "ct_results returned unexpected return code");
#120          break;
#121      }
#122
#123      cmd_close(&cmd);
#124 }
```

该程序的最终返回值如图 13-2 所示。

```
F:\sharedroot\openclient\example3\Debug\example
Connect to : (DemoServer,iihero) successfully!!
1 row==> (2007001, 李勇, 1987-9-1 0:0:0, NULL)
2 row==> (2007002, 刘晨, 1988-10-22 0:0:0, NULL)
3 row==> (2007003, 王敏, 1990-2-3 0:0:0, NULL)
4 row==> (2007004, 张铁林, 1989-4-1 0:0:0, NULL)

All done processing rows.
Close the connection!!
Press any key to continue
```

图 13-2 example3 运行结果

该函数的处理流程如下：

首先发送 SQL 命令，类型为 CS_LANG_CMD。注意到所用的 SQL 语句为 select sno、sname、sbirth、sage from student，分别为 int、varchar、datetime、numeric 类型。因此会采用不同的绑定方式。

在 while 循环调用 ct_results 函数的处理中，如果有结果，首先返回的就是 CS_ROW_RESULT 类型；如果无结果，results_ret 值将为 CS_END_RESULTS。

这里使用一个 CS_DATAFMT 结构体数组，用其中的四个元素分别与 student 表中的四列调用 ct_bind 函数进行绑定。

ct_bind：将 Server 端结果与程序变量相绑定。

CS_RETCODE ct_bind(cmd, item, datafmt, buffer, copied, indicator)
CS_COMMAND *cmd;
CS_INT item;
CS_DATAFMT *datafmt;
CS_VOID *buffer;
CS_INT *copied;
CS_SMALLINT *indicator;

- cmd：CS_COMMAND 句柄。
- item：表示列、返回参数或状态码的序号，即表示第几列、第几个参数或第几个状态码。第一列的序号是 1，依次递增。对存储过程的参数进行绑定时，只能绑定那些有返回值的参数。而且序号也是只从那些有返回值的参数计起。比如，某存储过程带有两个参数，如果只有第 2 个存储参数带有返回值，那么 item 值只能传 1，而不是 2。

当与存储过程的状态码返回值进行绑定时，item 值只能是 1，因为只可能有一个状态码。要清除所有绑定，只需要向 item 传入值 CS_UNUSED，并将 datafmt、buffer、copied、indicator 都设为 NULL 即可。

- datafmt：CS_DATAFMT 结构体指针，用于描述最终变量或变量数组的格式信息。ct_bind 在返回之前，会复制该结构体中的内容，也就是说，它在返回之后不再引用该地址里的内容。CS_DATAFMT 结构体中各成员的用法如表 13-5 所示。

表 13-5　CS_DATAFMT 结构体的成员及用法

成员名	何时使用	如何赋值
name	保留	无须赋值
namelen	保留	无须赋值
datatype	绑定结果集时	使用常量 CS_xxx_TYPE 来表示最终变量的类型，具体的变量类型在后边将有介绍。并且，Client 端的类型可以与 Server 端的某些类型进行自动转换
format	如果是字符串、binary、text、image 类型，则需要使用该值，其他类型一律为 CS_FMT_UNUSED	对于 text 或者 varchar 等文本串类型：CS_FMT_NULLTERM 用于以 NULL 结束的串，或者使用 CS_FMT_PADBLANK 表示以空格填充定长串中余下的空间
maxlength	用于非定长类型，对于定长类型，此成员值会被忽略	*buffer 的长度。当 buffer 指向一个数组时，maxlength 指的是数组中一个元素的长度。当绑定字符串时，该长度应将结束符计算在内。当 maxlength 值设得偏小，导致*buffer 不足以容纳结果时，ct_fetch 调用会放弃那些超长的结果，并且会返回 CS_ROW_FAIL。这时，*buffer 中的值是不确定的
scale	当绑定 numeric 或 decimal 类型时使用	小数点右边最大位数
precision	同上	最大数字长度，即精度
status	保留	无须赋值

续表

成员名	何时使用	如何赋值
Count	绑定所有类型的结果时	要复制到 ct_fetch 或 ct_scroll_fetch 的变量里的结果的行数,如果它的值比实际的行数要大,那只会复制实际的行数(只有普通查询行或者 cursor 行才允许有多行结果复制)。如果 count 值为 0,那么会取 1 行记录。对于 ct_scroll_fetch,count 值不能小于 CS_CURSOR_ROWS,否则结果不能确定。我们可以利用 count 值,一次取多行结果
usertype	保留	无须赋值
locale	绑定所有类型时	赋给一个具体的 CS_LOCALE* 名柄。默认情况下可以直接赋 NULL

- buffer: 是 datafmt→count 个连续元素的缓冲区地址,每个元素的长度为 datafmt→maxlength。每次当 ct_fetch 被调用时,实际结果就会复制到该缓冲区当中。buffer 必须一直保持有效,直到 CS_COMMAND 句柄被释放时为止。
- copied: datafmt→count 个连续整数数组的地址,在 ct_fetch 被调用时,会将实际复制的字节数放到这个地址里头,它是可选参数,可以传入 NULL。
- indicator: datafmt→count 个 CS_SMALLINT(即 short)数组的地址。主要用于表示该变量值是否为空,它的取值函数如下:
- -1: 获取的结果为 NULL,这时没有值复制到*buffer。
- 0: 取值正常。
- 大于 0 的任意值: 实际的数据值的长度,发生于数据被截短时。

ct_bind 返回如下可能值:

- CS_SUCCEED: 成功返回。
- CS_FAIL: 出错。
- CS_BUSY: 绑定延迟。

通常,绑定失败的原因有可能是:

- 使用了无效的 datatype 值。
- 在不使用的时候,可以将无效的 locale 指针初始化为 NULL。
- 要请求的类型转换不存在。

我们再看看 ct_fetch 函数的语法:

```
CS_RETCODE ct_fetch(cmd, type, offset, option, rows_read)
CS_COMMAND *cmd;
CS_INT type;
CS_INT offset;
CS_INT option;
CS_INT *rows_read;
```

- cmd：CS_COMMAND 句柄。
- type：保留给将来的 ctlib 使用，直接赋值为 CS_UNUSED。
- offset：保留，必须赋值为 CS_UNUSED。
- option：同上。
- rows_read：指向一个整数的指针，ct_fetch 将实际获取的行数赋给该值。

ct_fetch 的返回值如表 13-6 所示。

表 13-6 ct_fetch 的返回值

返回值	意义
CS_SUCCEED	成功，但是应用必须继续调用 ct_fetch，因为整个流程未完
CS_END_DATA	结果集所有的结果都获取完毕，这时应用需要调用 ct_results 得到另一个结果集，直到没有新的结果集
CS_ROW_FAIL	获取记录行时出错，该错可恢复。程序仍要调用 ct_fetch 获取余下的结果，或者调用 ct_cancel 取消余下的结果获取。注意，使用数组绑定时，该返回值意味着只得到部分结果，这时*row_count 告诉你实际获取记录的行数
CS_FAIL	调用失败，*row_count 会告诉你失败的行数。出现这种情况时，应用必须调用 ct_cancel，cancel 类型为 CS_CANCEL_ALL，来取消结果获取
CS_CANCELLED	取消结果获取
CS_PENDING	出现于异步调用
CS_BUSY	出现于异步调用

13.5.2 类型绑定

Open Client 支持多种 Client 端类型到 Server 端类型的自动转换，具体类型如表 13-7 所示。

表 13-7 Open Client 中 Client 端类型到 Server 端类型的转换关系

字段类型	Open Client 类型常量	说明	对应的 C 类型绑定
binary, varbinary	CS_BINARY_TYPE	二进制类型	CS_BINARY
无	CS_LONGBINARY_TYPE	长二进制类型	CS_LONGBINARY
无	CS_VARBINARY_TYPE	变长二进制	CS_VARBINARY
bit	CS_BIT_TYPE	二进制位类型	CS_BIT
char, varchar	CS_CHAR_TYPE	字符串	CS_CHAR
无	CS_LONGCHAR_TYPE	长字符串	CS_LONGCHAR
无	CS_VARCHAR_TYPE	变长字符串	CS_VARCHAR
unichar, univarchar	CS_UNICHAR_TYPE	定长或变长串	CS_UNICHAR
xml	CS_XML_TYPE	变长字符串	CS_XML
date	CS_DATE_TYPE	4 字节日期类型	CS_DATE
time	CS_TIME_TYPE	4 字节时间类型	CS_TIME
datetime	CS_DATETIME_TYPE	8 字节日期时间	CS_DATETIME

续表

字段类型	Open Client 类型常量	说明	对应的 C 类型绑定
smalldatetime	CS_DATETIME4_TYPE	4 字节日期时间	CS_DATETIME4
tinyint	CS_TINYINT_TYPE	unsigned byte	CS_TINYINT
smallint	CS_SMALLINT_TYPE	2 字节整数	CS_SMALLINT
int	CS_INT_TYPE	4 字节整数	CS_INT
bigint	CS_BIGINT_TYPE	8 字节整数	CS_BIGINT
usmallint	CS_USMALLINT_TYPE	2 字节无符短整	CS_USMALLINT
uint	CS_UINT_TYPE	4 字节无符整数	CS_UINT
ubigint	CS_UBIGINT_TYPE	8 字节无符整数	CS_UBIGINT
decimal	CS_DECIMAL_TYPE	decimal 类型	CS_DECIMAL
numeric	CS_NUMERIC_TYPE	numeric 类型	CS_NUMERIC
float	CS_FLOAT_TYPE	8 字节 float 类型，相当于 double	CS_FLOAT
real	CS_REAL_TYPE	4 字节 float 类型	CS_REAL
money	CS_MONEY_TYPE	8 字节 money 类型	CS_MONEY
smallmoney	CS_MONEY4_TYPE	4 字节 money 类型	CS_MONEY4
text	CS_TEXT_TYPE	文本类型	CS_TEXT
unitext	CS_UNITEXT_TYPE	变长文本类型	CS_UNITEXT
image	CS_IMAGE_TYPE	image 类型	CS_IMAGE

我们看到 datetime 类型对应的 C 类型绑定为 CS_DATETIME，这在上一节的例子中得到了应用，即 student 表的 sbirth 列。

CS_DATETIME 的定义如下，

```
typedef struct _cs_datetime
{
CS_INT dtdays;
CS_INT dttime;
} CS_DATETIME;
```

这里，dtdays 是从 1900 年 1 月 1 日算起的天数，dttime 是从半夜 0 点算起 1/300 秒的倍数，所以，即使得到了这个结果，也还需要进一步处理，才能得到用户可以直接认可的日期。

于是在第 72 行用到了相应的类型转换：

check_error(context, cs_dt_crack(context, CS_DATETIME_TYPE, &sbirth, &sb), "cs_dt_crack() error!");

这里，我们将 CS_DATETIME 类型转换成 sb 对应的类型：CS_DATEREC。它的定义就好理解得多：

```
typedef struct _cs_daterec
{
   CS_INT    dateyear;          /* 1900 to the future */
```

```
    CS_INT      datemonth;          /* 0 - 11 */
    CS_INT      datedmonth;         /* 1 - 31 */
    CS_INT      datedyear;          /* 1 - 366 */
    CS_INT      datedweek;          /* 0 - 6 (Mon. - Sun.) */
    CS_INT      datehour;           /* 0 - 23 */
    CS_INT      dateminute;         /* 0 - 59 */
    CS_INT      datesecond;         /* 0 - 59 */
    CS_INT      datemsecond;        /* 0 - 997 */
    CS_INT      datetzone;          /* 0 - 127 */
} CS_DATEREC;
```

这里要注意的是，1月到12月分别对应0到11，星期一到星期天分别对应0到6。

cs_dt_crack专门用于这两种日期类型的转换。

另几个重要的类型辅助函数有：

- cs_calc，用于对decimal、numeric、money等类型进行数值计算。
- cs_cmp，用于datetime、decimal、numeric、money类型的大小比较。
- cs_convert，用于将一种数据类型转换成另一种数据类型。
- cs_dt_crack，将对机器可读的日期类型转换成对人可读的日期类型。
- cs_dt_info，获取与语言有关的日期信息。
- cs_strcmp，比较两个字符串。

表13-7中有些绑定无法进行，因为ASE中没有对应的类型（在上表中标为"无"），千万不要混淆。

13.5.3 获取表的元信息

在实际应用中，经常需要获取表中各个字段的详细定义信息，即表的元信息，甚至是一个查询的各个字段的元信息。

以iihero库中的student表为例：

```
create table student
(
    sno int not null primary key,
    sname varchar(32) not null,
    sgender char(1) not null,
    sbirth datetime not null,
    sage numeric(2) null,
    sdept varchar(128) null
)
```

这里各个字段的定义就是student表的元信息，那么如何使用Open Client调用来实现这一功能呢？其基本流程如图13-3所示。

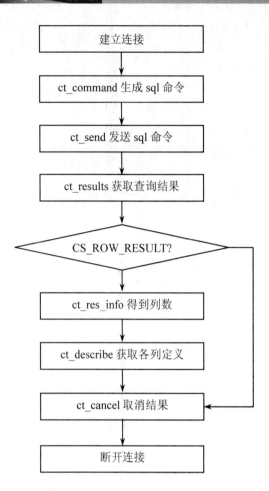

图 13-3 获取表元信息的基本流程

函数 ct_describe 的定义如下：

```
CS_RETCODE ct_describe(cmd, item, datafmt)
CS_COMMAND *cmd;
CS_INT item;
CS_DATAFMT *datafmt;
```

- cmd：CS_COMMAND 指针，表示一个命令。
- item：表示取最终结果（结果集，行集）的第几列；当获取普通列定义时，item 就表示列号，一个 Select 语句中的第 1 列的列号为 1，依此类推。

当获取一个计算列列定义时，item 指的是计算列的列序号。所谓计算列，是指计算子句中的计算列，也是从 1 计起。

当获取存储过程中的返回值定义时，item 值的规则须谨慎。只是对存储过程的返回值参数计数，并且严格按照存储过程的原始定义中的参数顺序进行。第一个输出参数为1，第二个输出参数为2。而输入参数并不计算在内。

当获取存储过程返回的状态码描述信息时，item 始终是 1，因为状态码始终只有一列。

有关获取 student 表的元信息的处理函数，在 openclient\example4.cpp 中的 getMetaData 里，其内容如下：

```
#001  static void outputFieldMetaData(CS_DATAFMT* column)
#002  {
#003      std::cout<<column->name<<",\t"<<column->datatype<<",\t"
#004          <<column->maxlength<<",\t"<<column->status<<std::endl;
#005  }
#006
#007  static void getMetaData()
#008  {
#009      static const int MAXCOLUMNS = 33;
#010      static const int MAXSTRING = 128;
#011      CS_DATAFMT          column;
#012
#013      cmd_open(&cmd);
#014      CS_RETCODE ret = ct_command(cmd, CS_LANG_CMD,
#015          "select * from student",
#016          CS_NULLTERM, CS_UNUSED);
#017      check_error(context, ret, "ct_command() failed");
#018      check_error(context, ct_send(cmd), "ct_send() failed");
#019
#020      CS_RETCODE results_ret;
#021      CS_INT result_type;
#022      int i=0;
#023      int nCols = 0;
#024      while ( (results_ret = ct_results(cmd, &result_type)) == CS_SUCCEED )
#025      {
#026          switch ((int)result_type)
#027          {
#028          case CS_ROW_RESULT:
#029              {
#030                  // 得到各列的元信息，并输出
#031                  check_error(context, ct_res_info(cmd, CS_NUMDATA, &nCols, CS_UNUSED, NULL), "ct_res_info() call failed");
#032                  for (i=0; i<nCols; i++)
#033                  {
#034                      check_error(context, ct_describe(cmd, i+1, &column), "ct_describe failed");
#035                      // 输出第 i+1 列的元信息
#036                      outputFieldMetaData(&column);
#037                  }
#038                  check_error(context, ct_cancel(NULL, cmd, CS_CANCEL_CURRENT), "ct_cancel failed");
#039              }
```

```
#040                break;
#041            case CS_CMD_SUCCEED:
#042                std::cerr<<"No rows returned."<<std::endl;
#043                break;
#044            case CS_CMD_FAIL:
#045                break;
#046            case CS_CMD_DONE:
#047                break;
#048            default:
#049                check_error(context, CS_FAIL, "ct_results returned unexpected result type");
#050                break;
#051            }
#052        }
#053
#054        switch ((int)results_ret)
#055        {
#056        case CS_END_RESULTS:
#057            break;
#058
#059        case CS_FAIL:
#060            check_error(context, CS_FAIL,
#061                "ct_results() returned CS_FAIL.");
#062            break;
#063        default:
#064            check_error(context, CS_FAIL,
#065                "ct_results returned unexpected return code");
#066            break;
#067        }
#068        cmd_close(&cmd);
#069 }
```

在发送 sql 命令以后，得到 CS_ROW_RESULT 类型的结果，直接调用 ct_describe 函数获取各列信息，并调用 outputFieldMetaData 作为一个简单的输出。

按照正常逻辑，必须调用 ct_fetch 取出所有查询结果才算完毕，但这里我们并不需要查询的结果，所以调用 ct_cancel(NULL, cmd, CS_CANCEL_CURRENT)以退出当前的查询结果处理，以免出错。

由于函数 outputFieldMetaData 的输出比较简单，程序的输出结果为：

```
Connect to : (XIONGHE,iihero) successfully!!
sno,        8,    4,    16
sname,      0,    32,   16
sgender,    0,    1,    16
sbirth,     12,   8,    16
sage,       16,   35,   48
sdept,      0,    128,  48
Close the connection!!
```

该函数略去了对 datatype（整型）到具体的数据库列类型名的转换，详细的列类型宏定义如下：

```
#define CS_CHAR_TYPE           (CS_INT)0
#define CS_BINARY_TYPE         (CS_INT)1
#define CS_LONGCHAR_TYPE       (CS_INT)2
#define CS_LONGBINARY_TYPE     (CS_INT)3
#define CS_TEXT_TYPE           (CS_INT)4
#define CS_IMAGE_TYPE          (CS_INT)5
#define CS_TINYINT_TYPE        (CS_INT)6
#define CS_SMALLINT_TYPE       (CS_INT)7
#define CS_INT_TYPE            (CS_INT)8
#define CS_REAL_TYPE           (CS_INT)9
#define CS_FLOAT_TYPE          (CS_INT)10
#define CS_BIT_TYPE            (CS_INT)11
#define CS_DATETIME_TYPE       (CS_INT)12
#define CS_DATETIME4_TYPE      (CS_INT)13
#define CS_MONEY_TYPE          (CS_INT)14
#define CS_MONEY4_TYPE         (CS_INT)15
#define CS_NUMERIC_TYPE        (CS_INT)16
#define CS_DECIMAL_TYPE        (CS_INT)17
#define CS_VARCHAR_TYPE        (CS_INT)18
#define CS_VARBINARY_TYPE      (CS_INT)19
#define CS_LONG_TYPE           (CS_INT)20
#define CS_SENSITIVITY_TYPE    (CS_INT)21
#define CS_BOUNDARY_TYPE       (CS_INT)22
#define CS_VOID_TYPE           (CS_INT)23
#define CS_USHORT_TYPE         (CS_INT)24
#define CS_UNICHAR_TYPE        (CS_INT)25
#define CS_BLOB_TYPE           (CS_INT)26
#define CS_DATE_TYPE           (CS_INT)27
#define CS_TIME_TYPE           (CS_INT)28
#define CS_UNITEXT_TYPE        (CS_INT)29
#define CS_BIGINT_TYPE         (CS_INT)30
#define CS_USMALLINT_TYPE      (CS_INT)31
#define CS_UINT_TYPE           (CS_INT)32
#define CS_UBIGINT_TYPE        (CS_INT)33
#define CS_XML_TYPE            (CS_INT)34
```

可以另外定义一个函数自行转换，另外 CS_DATAFMT 结构体有一个成员 status，它的值在这里也输出了，实际上它是一个掩码，表明该列是什么样的列。只要将该值与下述宏进行"与"（&）运算，如果与某一宏相等，表明它就是该宏对应类型的列。

Open Client 定义了如下类型的宏：

```
#define CS_HIDDEN              (CS_INT)0x1
#define CS_KEY                 (CS_INT)0x2
#define CS_VERSION_KEY         (CS_INT)0x4
#define CS_NODATA              (CS_INT)0x8
```

```
#define CS_UPDATABLE           (CS_INT)0x10
#define CS_CANBENULL           (CS_INT)0x20
#define CS_DESCIN              (CS_INT)0x40
#define CS_DESCOUT             (CS_INT)0x80
#define CS_INPUTVALUE          (CS_INT)0x100
#define CS_UPDATECOL           (CS_INT)0x200
#define CS_RETURN              (CS_INT)0x400
#define CS_RETURN_CANBENULL    (CS_INT) 0x420
#define CS_TIMESTAMP           (CS_INT)0x2000
#define CS_NODEFAULT           (CS_INT)0x4000
#define CS_IDENTITY            (CS_INT)0x8000
```

以列 sdept 为例，它的 status 值是 48，对应于 0x30，与上述宏作&运算：

0x30 & 0x20 = 0x20，所以它符合 CS_CANBENULL，该列允许有空值。

0x30 & 0x10 = 0x10 = CS_UPDATABLE，表明该列是可更新的列。

13.6 数据的插入、更新与删除操作

13.6.1 不带任何参数的 CUD 操作

普通的 CUD 操作（Insert、Update、Delete）如果不包含任何动态参数，可以通过调用 13.4 节中的 DDL 操作函数 cmd_exec(char* sql)直接实现，只需要传入合适的 SQL 语句就可以。

不过，CUD 操作有一个缺点，就是它并没有返回受影响的行数，我们可以对函数 cmd_exec(char* sql)稍作修改，让其返回受影响的行数。这里给出新的函数实现：

```
#001 // 执行一个 sql, 返回受影响的行数
#002 static int execute_command(char* sql)
#003 {
#004      CS_RETCODE        retcode;
#005      CS_INT            restype;
#006      CS_COMMAND        *cmd;
#007      CS_RETCODE        query_code;
#008
#009      CS_RETCODE ret = ct_cmd_alloc(connection, &cmd);
#010      check_error(context, ret, "ct_cmd_alloc() failed");
#011
#012      if( (retcode=ct_command(cmd, CS_LANG_CMD, sql, CS_NULLTERM, CS_UNUSED))
#013              != CS_SUCCEED )
#014      {
#015          std::cerr<<"ex_execute_cmd: ct_command() failed"<<std::endl;
#016          cmd_close(&cmd);
#017          return retcode;
#018      }
#019
```

```
#020        if ((retcode = ct_send(cmd)) != CS_SUCCEED)
#021        {
#022            std::cerr<<"ex_execute_cmd: ct_send() failed"<<std::endl;
#023            cmd_close(&cmd);
#024            return retcode;
#025        }
#026
#027        query_code = CS_SUCCEED;
#028        while ((retcode = ct_results(cmd, &restype)) == CS_SUCCEED)
#029        {
#030            switch((int)restype)
#031            {
#032            case CS_CMD_SUCCEED:
#033            case CS_CMD_DONE:{
#034                    CS_INT row_count = 0;
#035                    if (ct_res_info(cmd, CS_ROW_COUNT, &row_count, CS_UNUSED, NULL)==CS_SUCCEED)
#036                    {
                            retcode = row_count;
#037                    }
#038                }
#039                break;
#040
#041            case CS_CMD_FAIL:
#042                query_code = CS_FAIL;
#043                break;
#044
#045            case CS_STATUS_RESULT:
#046                retcode = ct_cancel(NULL, cmd, CS_CANCEL_CURRENT);
#047                if (retcode != CS_SUCCEED)
#048                {
#049                    std::cerr<<("ex_execute_cmd: ct_cancel() failed")
#050                        <<std::endl;
#051                    query_code = CS_FAIL;
#052                }
#053                break;
#054
#055            default:
#056
#057                // Unexpected result type.
#058                query_code = CS_FAIL;
#059                break;
#060            }
#061            if (query_code == CS_FAIL)
#062            {
#063                // Terminate results processing and break out of
#064                retcode = ct_cancel(NULL, cmd, CS_CANCEL_ALL);
#065                if (retcode != CS_SUCCEED)
#066                {
#067                    std::cerr<<("ex_execute_cmd: ct_cancel() failed")
```

```
#068                            <<std::endl;
#069                    }
#070                    break;
#071               }
#072         }
#073
#074         // Clean up the command handle used
#075         if (retcode == CS_END_RESULTS)
#076         {
#077               retcode = ct_cmd_drop(cmd);
#078               if (retcode != CS_SUCCEED)
#079               {
#080                    query_code = CS_FAIL;
#081               }
#082         }
#083         else
#084         {
#085               (void)ct_cmd_drop(cmd);
#086               query_code = CS_FAIL;
#087         }
#088         return query_code;
#089 }
```

主要区别在第33～38行,在遇到命令执行结束时,通过调用ct_res_info,获取CS_ROW_COUNT属性值,从而得到受影响的行数。

有了这个函数,我们就可以执行下边的调用(详见代码:openclient\example4\example4.cpp):

```
test_open(TEST_SERVER_NAME, TEST_USER_NAME, TEST_PASSWORD, TEST_DB_NAME);
setAutoCommit(false);
char sql[256] = "create table t12345(id int primary key, col2 varchar(32))";
std::cout<<"create table: "<<sql<<std::endl;
std::cout<<"affected count: "<<execute_command(sql)<<std::endl;
strcpy(sql, "insert into t12345 values(1, 'wang')");
std::cout<<"insert data: "<<sql<<std::endl;
std::cout<<"affected count: "<<execute_command(sql)<<std::endl;
strcpy(sql, "delete from t12345");
std::cout<<"delete data: "<<sql<<std::endl;
std::cout<<"affected count: "<<execute_command(sql)<<std::endl;
commit();
strcpy(sql, "drop table t12345");
std::cout<<"drop table: "<<sql<<std::endl;
std::cout<<"affected count: "<<execute_command(sql)<<std::endl;
commit();
setAutoCommit(true);
test_close();
return 0;
```

13.6.2 带动态参数的 CUD 操作

在上一节，我们知道，绑定输出结果使用 ct_bind 函数即可。那么，如果一个 SQL 语句需要动态绑定输入参数，使用 Open Client 如何进行绑定呢？

动态绑定的正常处理流程是：

Step 1 ct_dynamic()，准备一个 SQL 语句。

Step 2 ct_send()，发送准备语句的命令。

Step 3 ct_dynamic()，用于执行一个 SQL 语句。

Step 4 ct_setparam()，用于为动态参数设定值，每个动态参数调用一次。

Step 5 ct_send()，发送 SQL 语句给服务器端执行。

Step 6 ct_results()，处理来自服务器端的结果。

这里有一个动态 SQL 语句：update student set sname=(?) where sno=(?)，根据参数 sname 和 sno 的值进行更新，代码见 openclient\example5\example5.cpp 中的函数 int update_student(int sno, char* name)，该函数详细内容如下：

```
// 示例，按照学号更改某学生的姓名
#001  static int update_student(int sno, char* name)
#002  {
#003      char sql[128] = "update student set sname=(?) where sno=(?)";
#004      CS_RETCODE      retcode;
#005      CS_INT          restype;
#006      CS_COMMAND      *cmd;
#007      CS_RETCODE      query_code;
#008
#009      CS_DATAFMT      col_sno;
#010      CS_DATAFMT      col_sname;
#011      long            len_sno = sizeof(sno);
#012      long            len_name = strlen(name);
#013
#014      memset(&col_sname, 0,  sizeof(col_sname));
#015      strcpy(col_sname.name, "@sname");
#016      col_sname.datatype=CS_VARCHAR_TYPE;
#017      col_sname.maxlength = 32;
#018      col_sname.namelen=strlen("@name");
#019      col_sname.status = CS_INPUTVALUE;
#020
#021      memset(&col_sno, 0,  sizeof(col_sno));
#022      strcpy(col_sno.name, "@sno");
#023      col_sno.datatype=CS_INT_TYPE;
#024      col_sno.maxlength=4;
#025      col_sno.namelen=strlen("@sno");
```

```cpp
#026        col_sno.status = CS_INPUTVALUE;
#027
#028        CS_RETCODE ret = ct_cmd_alloc(connection, &cmd);
#029        check_error(context, ret, "ct_cmd_alloc() failed");
#030
#031        retcode = ct_dynamic(cmd, CS_PREPARE, "student_cursor", CS_NULLTERM, sql, CS_NULLTERM);
#032        retcode = ct_send(cmd);
#033        while ((retcode = ct_results(cmd, &restype)) == CS_SUCCEED){}
#034
#035
#036        retcode = ct_dynamic(cmd, CS_DESCRIBE_INPUT, "student_cursor", CS_NULLTERM, NULL, CS_UNUSED);
#037        if ((retcode = ct_send(cmd)) != CS_SUCCEED)
#038        {
#039            std::cerr<<"ex_execute_cmd: ct_send() failed"<<std::endl;
#040            cmd_close(&cmd);
#041            return retcode;
#042        }
#043
#044        while ((retcode = ct_results(cmd, &restype)) == CS_SUCCEED)
#045        {
#046            switch (restype)
#047            {
#048                case CS_DESCRIBE_RESULT:
#049                    {
#050                        ct_describe(cmd, 1, &col_sname);
#051                        ct_describe(cmd, 2, &col_sno);
#052                    }
#053                    break;
#054                case CS_ROW_RESULT:
#055                    std::cout<<"Row result after dynamic CS_PREPARE"<<std::endl;
#056                    break;
#057                case CS_CMD_SUCCEED:
#058                case CS_CMD_DONE:
#059                    break;
#060                case CS_CMD_FAIL:
#061                    // return CS_FAIL;
#062                    std::cout<<"CS_FAIL..."<<std::endl;
#063                case CS_PARAM_RESULT:
#064                    std::cout<<"Param result after dynamic CS_PREPARE"<<std::endl;
#065                    break;
#066                case CS_CURSOR_RESULT:
#067                    std::cout<<"Cursor result after dynamic CS_PREPARE"<<std::endl;
#068                    break;
#069                default:
#070                    std::cout<<"other result after dynamic CS_PREPARE"<<std::endl;
#071                    break;
#072            }
#073        }
```

```
#074
#075        retcode = ct_dynamic(cmd, CS_EXECUTE, "student_cursor", CS_NULLTERM, NULL, CS_UNUSED);
#076
#077        ct_setparam(cmd, &col_sname, name, &len_name, NULL);
#078        ct_setparam(cmd, &col_sno, &sno, &len_sno, NULL);
#079
#080        if ((retcode = ct_send(cmd)) != CS_SUCCEED)
#081        {
#082            std::cerr<<"ex_execute_cmd: ct_send() failed"<<std::endl;
#083            cmd_close(&cmd);
#084            return retcode;
#085        }
#086
#087        query_code = CS_SUCCEED;
#088        while ((retcode = ct_results(cmd, &restype)) == CS_SUCCEED)
#089        {
#090            switch((int)restype)
#091            {
#092            case CS_CMD_SUCCEED:
#093            case CS_CMD_DONE:{
#094                    CS_INT row_count = 0;
#095                    if (ct_res_info(cmd, CS_ROW_COUNT, &row_count, CS_UNUSED, NULL)==CS_SUCCEED)
#096                    {
#097                        retcode = row_count;
#098                    }
#099                }
#099                break;
#100
#101            case CS_CMD_FAIL:
#102                query_code = CS_FAIL;
#103                break;
#104
#105            case CS_STATUS_RESULT:
#106                retcode = ct_cancel(NULL, cmd, CS_CANCEL_CURRENT);
#107                if (retcode != CS_SUCCEED)
#108                {
#109                    std::cerr<<("ex_execute_cmd: ct_cancel() failed")
#110                        <<std::endl;
#111                    query_code = CS_FAIL;
#112                }
#113                break;
#114
#115            default:
#116
#117                // Unexpected result type.
#118                query_code = CS_FAIL;
#119                break;
#120            }
#121            if (query_code == CS_FAIL)
```

```
#122            {
#123                // Terminate results processing and break out of
#124                retcode = ct_cancel(NULL, cmd, CS_CANCEL_ALL);
#125                if (retcode != CS_SUCCEED)
#126                {
#127                    std::cerr<<("ex_execute_cmd: ct_cancel() failed")
#128                        <<std::endl;
#129                }
#130                break;
#131            }
#132        }
#133
#134        // Clean up the command handle used
#135        if (retcode == CS_END_RESULTS)
#136        {
#137            retcode = ct_cmd_drop(cmd);
#138            if (retcode != CS_SUCCEED)
#139            {
#140                query_code = CS_FAIL;
#141            }
#142        }
#143        else
#144        {
#145            (void)ct_cmd_drop(cmd);
#146            query_code = CS_FAIL;
#147        }
#148        return query_code;
#149 }
```

在不确定如何设定ct_setparam()函数中的CS_DATAFMT参数值时，有必要调用一次"retcode = ct_dynamic(cmd, CS_DESCRIBE_INPUT, "student_cursor", CS_NULLTERM, NULL, CS_UNUSED);"，然后获取CS_DESCRIBE_RESULT类型的结果（见第48～52行），对两个参数的CS_DATAFMT进行有效的初始化，使得后边的ct_setparam能够正常工作。

13.6.3　BLOB/CLOB值的读写

虽然前一小节介绍了普通的字段类型的CUD操作，但是那种方法并不适用于BLOB/CLOB字段（在ASE中，即IMAGE/TEXT类型字段，尤其是遇到长度很长的BLOB/CLOB字段。

ASE的Open Client ct-lib库并不明显区分IMAGE/TEXT字段，它操纵这两个字段的方式几乎是一样的。

BLOB/CLOB写操作的基本处理过程是：

Step 1　如果是新数据，需要先插入一条记录。

Step 2　执行select查询，定位到刚插入的那条记录或者是要更新的那条记录。

Step 3 获取完刚 select 出来的结果，对要写入数据的 BLOB/CLOB，通过调用 ct_data_info，更新与其相关联的 CS_IODESC 结构体内容。

Step 4 对要更新的 BLOB/CLOB 字段发送 CS_SEND_DATA_CMD 命令，不断地传送数据，直至预定长度的数据发送完毕。

Step 5 发送数据完毕之后，调用 ct_results 函数处理其结果，并更改 IODESC 结构体里边的 timestamp，完成 BLOB/CLOB 的最终更新。对每个 BLOB/CLOB 的发送数据命令都需要作这样的处理。

详细的写操作处理过程见示例代码 Openclient\example6\example6.cpp 中的 write_lob 函数：write_lob("C:\\jconn3.read.jar", "C:\\tmp.txt", 101);，是把这两个文件分别写入 multitype_t 中 ID 值为 101 的 col11 和 col12 两列当中。

实现 write_lob 函数要调用 fetchResults() 和 update_image_text_data() 两个子函数。相应代码如下：

```
#001 #define EX_MAX_TEXT      8192
#002 #define BLOCK_LEN        1024
#003
#004 typedef struct _text_data
#005 {
#006     CS_IODESC   iodesc;      /* 与 LOB 相关联的 IODESC 结构 */
#007     CS_TEXT     textbuf[EX_MAX_TEXT];    /* holds the value */
#008     CS_INT      textlen;     /* number of bytes in textbuf */
#009 } TEXT_DATA;
#010
#011
#012 static int fetchResults(CS_COMMAND* cmd, TEXT_DATA* textdata11, TEXT_DATA* textdata12)
#013 {
#014
#015     CS_RETCODE  retcode;
#016
#017     CS_TEXT     *txtptr11;
#018     CS_TEXT     *txtptr12;
#019
#020     CS_INT      count;
#021     CS_INT      len;
#022
#023
#024     while(((retcode = ct_fetch(cmd, CS_UNUSED, CS_UNUSED, CS_UNUSED,
#025         &count)) == CS_SUCCEED) || (retcode == CS_ROW_FAIL))
#026     {
#027         if (retcode == CS_ROW_FAIL)
#028         {
#029             ex_error("FetchResults: ct_fetch() returned CS_ROW_FAIL");
#030             continue;
#031         }
```

```
#032
#033            txtptr11 = textdata11->textbuf;
#034            txtptr12 = textdata12->textbuf;
#035            textdata11->textlen = 0;
#036            textdata12->textlen = 0;
#037
#038            do
#039            {
#040                retcode = ct_get_data(cmd, 1, txtptr11, BLOCK_LEN, &len);
#041                textdata11->textlen += len;
#042                /*
#043                ** Protect against overflowing the string buffer.
#044                */
#045                if ((textdata11->textlen + BLOCK_LEN) > (EX_MAX_TEXT - 1))
#046                {
#047                    break;
#048                }
#049                txtptr11 += len;
#050            } while (retcode == CS_SUCCEED);
#051
#052            if (retcode != CS_END_ITEM)
#053            {
#054                ex_error("FetchResults: ct_get_data() failed");
#055                return retcode;
#056            }
#057            std::cout<<"txtptr1: "<<(char*)(textdata11->textbuf)<<std::endl;
#058
#059            retcode = ct_data_info(cmd, CS_GET,   1, &textdata11->iodesc);
#060
#061            do
#062            {
#063                retcode = ct_get_data(cmd, 2, txtptr12, BLOCK_LEN, &len);
#064                textdata12->textlen += len;
#065                /*
#066                ** Protect against overflowing the string buffer.
#067                */
#068                if ((textdata12->textlen + BLOCK_LEN) > (EX_MAX_TEXT - 1))
#069                {
#070                    break;
#071                }
#072                txtptr12 += len;
#073            } while (retcode == CS_SUCCEED);
#074
#075            if (retcode != CS_END_ITEM && retcode != CS_END_DATA)
#076            {
#077                ex_error("FetchResults: ct_get_data() failed");
#078                return retcode;
#079            }
```

```
#080
#081            std::cout<<"txtptr2: "<<(char*)(textdata12->textbuf)<<std::endl;
#082            retcode = ct_data_info(cmd, CS_GET,    2, &textdata12->iodesc);
#083       }
#084
#085       if (retcode == CS_END_DATA)
#086       {
#087            retcode = CS_SUCCEED;
#088       }
#089       else
#090       {
#091            ex_error("FetchResults: ct_fetch() failed");
#092       }
#093
#094       return retcode;
#095 }
#096
#097 int static processTimestamp(CS_COMMAND* cmd, TEXT_DATA* textdata)
#098 {
#099       CS_RETCODE       retcode;
#100       CS_INT           count;
#101       CS_DATAFMT       datafmt;
#102
#103       retcode = ct_describe(cmd, 1, &datafmt);
#104       if (retcode != CS_SUCCEED)
#105       {
#106                ex_error("ProcessTimestamp: ct_describe() failed");
#107                return retcode;
#108       }
#109
#110       // 检查是否是 TIMESTAMP 类型
#111       if (!(datafmt.status & CS_TIMESTAMP))
#112       {
#113           /*
#114           ** Unexpected parameter data was received.
#115           */
#116                ex_error("ProcessTimestamp: unexpected parameter data received");
#117                return CS_FAIL;
#118       }
#119
#120       // 生成新的 timestamp
#121       datafmt.maxlength = sizeof(textdata->iodesc.timestamp);
#122       datafmt.format    = CS_FMT_UNUSED;
#123       if ((retcode = ct_bind(cmd, 1, &datafmt, (CS_VOID *)textdata->iodesc.timestamp,
#124                        &textdata->iodesc.timestamplen,
#125                        NULL)) != CS_SUCCEED)
#126       {
#127                ex_error("ProcessTimestamp: ct_bind() failed");
```

```
#128                return retcode;
#129        }
#130
#131        retcode = ct_fetch(cmd, CS_UNUSED, CS_UNUSED, CS_UNUSED, &count);
#132        if (retcode != CS_SUCCEED)
#133        {
#134            ex_error("ProcessTimestamp: ct_fetch() failed");
#135            return retcode;
#136        }
#137
#138        retcode = ct_cancel(NULL, cmd, CS_CANCEL_CURRENT);
#139        if (retcode != CS_SUCCEED)
#140        {
#141            ex_error("ProcessTimestamp: ct_cancel() failed");
#142        }
#143        return retcode;
#144 }
#145
#146    CS_RETCODE update_image_text_data(TEXT_DATA* textdata11, char* col11, TEXT_DATA* textdata12, char* col12)
#147 {
#148        CS_RETCODE      retcode;
#149        CS_INT          res_type;
#150        CS_COMMAND      *cmd;
#151        CS_TEXT         *txtptr;
#152        CS_INT          txtlen;
#153
#154
#155        if ((retcode = ct_cmd_alloc(connection, &cmd)) != CS_SUCCEED)
#156        {
#157            ex_error("UpdateTextData: ct_cmd_alloc() failed");
#158            return retcode;
#159        }
#160
#161        if ((retcode = ct_command(cmd, CS_SEND_DATA_CMD, NULL, CS_UNUSED,
#162                CS_COLUMN_DATA)) != CS_SUCCEED)
#163        {
#164            ex_error("UpdateTextData: ct_command() failed");
#165            return retcode;
#166        }
#167
#168        txtptr = (CS_TEXT *)col11;
#169        std::ifstream is;
#170        is.open((char*)txtptr, std::ios::binary);
#171        is.seekg(0, std::ios::end);
#172        txtlen = is.tellg();
#173        is.seekg(0, std::ios::beg);
#174
```

```
#175
#176        textdata11->iodesc.total_txtlen = txtlen;
#177        textdata11->iodesc.log_on_update = CS_TRUE;
#178        retcode = ct_data_info(cmd, CS_SET, CS_UNUSED, &textdata11->iodesc);
#179        if (retcode != CS_SUCCEED)
#180        {
#181            ex_error("UpdateTextData: ct_data_info() failed");
#182            return retcode;
#183        }
#184        int offset = 0;
#185        int readsize = 0;
#186        // for (i = 0; i <txtlen; i++)
#187        while (!is.eof())
#188        {
#189            readsize = offset + BLOCK_LEN < txtlen ? BLOCK_LEN : txtlen - offset;
#190            if (readsize <= 0) break;
#191            is.read((char*)textdata11->textbuf, readsize);
#192            offset += readsize;
#193
#194            retcode = ct_send_data(cmd, textdata11->textbuf, readsize);
#195            if (retcode != CS_SUCCEED)
#196            {
#197                ex_error("UpdateTextData: ct_send_data() failed");
#198                return retcode;
#199            }
#200        }
#201        is.close();
#202
#203        if ((retcode = ct_send(cmd)) != CS_SUCCEED)
#204        {
#205            ex_error("UpdateTextData: ct_send() failed");
#206            return retcode;
#207        }
#208
#209        while ((retcode = ct_results(cmd, &res_type)) == CS_SUCCEED)
#210        {
#211            switch ((int)res_type)
#212            {
#213                case CS_PARAM_RESULT:
#214                    retcode = processTimestamp(cmd, textdata11);
#215                    if (retcode != CS_SUCCEED)
#216                    {
#217                        ex_error("UpdateTextData: ProcessTimestamp() failed");
#218                        ct_cancel(NULL, cmd, CS_CANCEL_ALL);
#219                        return retcode;
#220                    }
#221                    break;
#222
```

```
#223            case CS_STATUS_RESULT:
#224                retcode = ct_cancel(NULL, cmd, CS_CANCEL_CURRENT);
#225                if (retcode != CS_SUCCEED)
#226                {
#227                    ex_error("UpdateTextData: ct_cancel() failed");
#228                    return retcode;
#229                }
#230                break;
#231
#232            case CS_CMD_SUCCEED:
#233            case CS_CMD_DONE:
#234                break;
#235
#236            case CS_CMD_FAIL:
#237                ex_error("UpdateTextData: ct_results() returned CS_CMD_FAIL");
#238                break;
#239
#240            default:
#241                ex_error("UpdateTextData: ct_results() returned unexpected result typ");
#242                ct_cancel(NULL, cmd, CS_CANCEL_ALL);
#243                break;
#244        }
#245    }
#246    switch ((int)retcode)
#247    {
#248        case CS_END_RESULTS:
#249            retcode = CS_SUCCEED;
#250            break;
#251        case CS_FAIL:
#252            ex_error("UpdateTextData: ct_results() failed");
#253            break;
#254            ex_error("UpdateTextData: ct_results() returned unexpected result");
#255            break;
#256    }
#257
#258    // 再次发送 send data 命令
#259    if ((retcode = ct_command(cmd, CS_SEND_DATA_CMD, NULL, CS_UNUSED,
#260            CS_COLUMN_DATA)) != CS_SUCCEED)
#261    {
#262        ex_error("UpdateTextData: ct_command() failed");
#263        return retcode;
#264    }
#265
#266    // col12
#267    txtptr = (CS_TEXT *)col12;
#268    std::ifstream is2;
#269    is2.open((char*)txtptr, std::ios::binary);
#270    is2.seekg(0, std::ios::end);
```

```
#271        txtlen = is2.tellg();
#272        is2.seekg(0, std::ios::beg);
#273
#274
#275        textdata12->iodesc.total_txtlen = txtlen;
#276        textdata12->iodesc.log_on_update = CS_TRUE;
#277        retcode = ct_data_info(cmd, CS_SET, CS_UNUSED, &textdata12->iodesc);
#278        if (retcode != CS_SUCCEED)
#279        {
#280            ex_error("UpdateTextData: ct_data_info() failed");
#281            return retcode;
#282        }
#283        offset = 0;
#284        readsize = 0;
#285        while (!is2.eof())
#286        {
#287            readsize = offset + BLOCK_LEN < txtlen ? BLOCK_LEN : txtlen - offset;
#288            if (readsize <= 0) break;
#289
#290            is.read((char*)textdata12->textbuf, readsize);
#291            offset += readsize;
#292
#293            retcode = ct_send_data(cmd, textdata12->textbuf, readsize);
#294
#295            if (retcode != CS_SUCCEED)
#296            {
#297                ex_error("UpdateTextData: ct_send_data() failed");
#298                return retcode;
#299            }
#300        }
#301        is2.close();
#302
#303        if ((retcode = ct_send(cmd)) != CS_SUCCEED)
#304        {
#305            ex_error("UpdateTextData: ct_send() failed");
#306            return retcode;
#307        }
#308
#309        while ((retcode = ct_results(cmd, &res_type)) == CS_SUCCEED)
#310        {
#311            switch ((int)res_type)
#312            {
#313                case CS_PARAM_RESULT:
#314                    retcode = processTimestamp(cmd, textdata12);
#315                    if (retcode != CS_SUCCEED)
#316                    {
#317                        ex_error("UpdateTextData: ProcessTimestamp() failed");
#318                        ct_cancel(NULL, cmd, CS_CANCEL_ALL);
```

```
#319                return retcode;
#320            }
#321            break;
#322
#323            case CS_STATUS_RESULT:
#324            retcode = ct_cancel(NULL, cmd, CS_CANCEL_CURRENT);
#325            if (retcode != CS_SUCCEED)
#326            {
#327                ex_error("UpdateTextData: ct_cancel() failed");
#328                return retcode;
#329            }
#330            break;
#331
#332            case CS_CMD_SUCCEED:
#333            case CS_CMD_DONE:
#334            break;
#335
#336            case CS_CMD_FAIL:
#337            ex_error("UpdateTextData: ct_results() returned CS_CMD_FAIL");
#338            break;
#339
#340            default:
#341            ex_error("UpdateTextData: ct_results() returned unexpected result typ");
#342            ct_cancel(NULL, cmd, CS_CANCEL_ALL);
#343            break;
#344        }
#345    }
#346    switch ((int)retcode)
#347    {
#348        case CS_END_RESULTS:
#349            retcode = CS_SUCCEED;
#350            break;
#351        case CS_FAIL:
#352            ex_error("UpdateTextData: ct_results() failed");
#353            break;
#354            ex_error("UpdateTextData: ct_results() returned unexpected result");
#355            break;
#356    }
#357    return retcode;
#358 }
#359
#360 // 向表 multitype_t 中 ID 值为<id>的记录行里添加/更新 LOB 值,分别对应两个文件
#361 static int write_lob(char* col11_file, char* col12_file, int id)
#362 {
#363    char sql[512];
#364    sprintf(sql, "select col11, col12 from multitype_t    where id=%d", id);
#365    CS_RETCODE ret = CS_FAIL;
#366
```

```
#367        ret = ct_cmd_alloc(connection, &cmd);
#368        check_error(context, ret, "ct_cmd_alloc() failed");
#369
#370        ret = ct_command(cmd, CS_LANG_CMD, sql, CS_NULLTERM, CS_UNUSED);
#371
#372        if (ret != CS_SUCCEED)
#373        {
#374            std::cerr<<"ex_execute_cmd: ct_command() failed"<<std::endl;
#375            cmd_close(&cmd);
#376            return ret;
#377        }
#378
#379        ret = ct_send(cmd);
#380
#381        if (ret != CS_SUCCEED)
#382        {
#383            std::cerr<<"ex_execute_cmd: ct_send() failed"<<std::endl;
#384            cmd_close(&cmd);
#385            return ret;
#386        }
#387
#388        CS_INT res_type;
#389        CS_RETCODE retcode;
#390        TEXT_DATA textdata11;
#391        TEXT_DATA textdata12;
#392
#393        memset(&textdata11, 0, sizeof(textdata11));
#394        memset(&textdata12, 0, sizeof(textdata12));
#395
#396        while ((ret = ct_results(cmd, &res_type)) == CS_SUCCEED)
#397        {
#398            switch ((int)res_type)
#399            {
#400            case CS_ROW_RESULT:
#401                retcode = fetchResults(cmd, &textdata11, &textdata12);
#402                if (retcode != CS_SUCCEED)
#403                {
#404                    ex_error("RetrieveData: FetchResults() failed");
#405                    ct_cancel(NULL, cmd, CS_CANCEL_ALL);
#406                    return retcode;
#407                }
#408                break;
#409
#410            case CS_CMD_SUCCEED:
#411            case CS_CMD_DONE:
#412                break;
#413
#414            case CS_CMD_FAIL:
```

```
#415              ex_error("RetrieveData: ct_result() returned CS_CMD_FAIL");
#416              return retcode;
#417
#418          default:
#419              ex_error("RetrieveData: ct_results() returned unexpected result typ");
#420              ct_cancel(NULL, cmd, CS_CANCEL_ALL);
#421              break;
#422          }
#423      }
#424      switch ((int)ret)
#425      {
#426          case CS_END_RESULTS:
#427              retcode = CS_SUCCEED;
#428              break;
#429          case CS_FAIL:
#430              ex_error("RetrieveData: ct_results() failed");
#431              break;
#432          default:
#433              ex_error("RetrieveData: ct_results() returned unexpected result");
#434              break;
#435      }
#436      return update_image_text_data(&textdata11, col11_file, &textdata12, col12_file);
#437
#438 }
```

最终调用完"write_lob("C:\\jconn3.read.jar", "C:\\tmp.txt", 101);"以后，我们会发现 col11 和 col12 两列的长度与两个文件的长度一致：

```
1> select datalength(col11), datalength(col12) from multitype_t where id=101
2> go

 ----------- -----------
    553569     229622

(1 row affected)
```

14

嵌入式 SQL 编程

本章主要介绍如何在 Sybase ASE 下进行嵌入式 SQL-C（ESQLC）编程。

14.1 基本原理

嵌入式 SQL-C 预编译器（ESQLC）会将嵌入式 SQL 语句翻译成 C 的数据类型定义和相应的 C 语言调用。经过完整的预编译后，就可以将生成的源码当作普通的 C 程序，使用 C 语言编译器进行编译了。

预编译器会分两个阶段进行预处理。第一阶段，它会解析嵌入的 SQL 语句和变量声明，检查它们的语法，并显示任何可能错误信息。如果没发现严重错误，则会进入第二阶段：

- 为预编译变量添加声明，它们都以"_sql"开头，基于这个规定，我们在程序里不要定义"_sql"打头的变量。
- 将原始的 ESQL 语句转换成注释（因为 C 编译器不能识别）。
- 生成存储过程和存储过程的调用（如果在编译命令里设定了相应的选项）。
- 将 ESQL 语句转换成 Open Client 库的调用。ESQL 会使用 Open Client 库作为运行时库。
- 最多产生三个文件：一个目标文件、可选的清单文件、可选的 isql 脚本文件。

ESQL 可以有多个源文件进行预编译，它们有下述条件限制：

- 连接名在应用程序里必须是全局唯一。
- 一个连接里的游标名必须唯一。
- 一个连接里的动态或静态 SQL 语句必须唯一。
- 动态描述符在应用程序里全局唯一。

预编译生成的文件：

- 目标文件：与源输入文件类似，只不过所有的 ESQL（嵌入式 SQL）语句，会被翻译成 Open Client 运行时库的 API 函数调用。
- 清单文件：里边包含的是输入文件当中的源语句以及一些 Info、Warning、Error 消息。
- isql 脚本文件：主要包含了预编译生成的 Transact-SQL 存储过程。

除了包含宿主语言代码，ESQL 程序主要执行 5 项任务。每个 ESQL 程序必须将这些任务进行预编译、编译和执行。这些任务包括：

- 使用 SQLCA、SQLCODE 或者 SQLSTATE 建立 SQL 联系
- 创建 SQL 联系区（SQLCA、SQLCODE 或者 SQLSTATE），主要是为应用程序和 ASE 服务器之间提供联系的通道。这些结构体包含有 ASE 和 Open Client 库常用的错误、警告以及一般的消息。
- 变量声明。
- 用于标识 ESQL 语句要用到的宿主变量。
- 连接 ASE。
- 用于提供连接 ASE 的用户、密码、数据库名、服务名等相关信息。
- 发送 Transact-SQL 语句到 ASE。
- 主要是发送 SQL 语句到 ASE，用于具体的 SQL 查询或操作。
- 处理错误和返回值。
- 用于处理 Open Client 和 ASE 的返回值。

14.2 一个简单的示例

下面通过一个简单的示例来介绍如何进行 ESQLC 编程。下边的示例是从 iihero 数据库中的 student 表里输出 sno、sname、sgender 三列的值。

```
#001 #define ERREXIT      -1
#002 #define STDEXIT       0
#003
#004 #include <stdio.h>
#005
#006 /* Declare the SQLCA. */
#007 EXEC SQL INCLUDE SQLCA;
#008
#009 #define   EOLN      '\0'
#010
#011 void error_handler(void)
#012 {
#013      fprintf(stderr, "\n** SQLCODE=(%ld)", sqlca.sqlcode);
```

```
#014
#015        if (sqlca.sqlerrm.sqlerrml)
#016        {
#017            fprintf(stderr, "\n** ASE Error ");
#018            fprintf(stderr, "\n** %s", sqlca.sqlerrm.sqlerrmc);
#019        }
#020        fprintf(stderr, "\n\n");
#021        exit(ERREXIT);
#022 }
#023
#024 void warning_handler(void)
#025 {
#026
#027        if (sqlca.sqlwarn[1] == 'W')
#028        {
#029            fprintf(stderr,
#030                "\n** Data truncated.\n");
#031        }
#032        if (sqlca.sqlwarn[3] == 'W')
#033        {
#034            fprintf(stderr,
#035                "\n** Insufficient host variables to store results.\n");
#036        }
#037        return;
#038 }
#039
#040 int main()
#041 {
#042        EXEC SQL BEGIN DECLARE SECTION;
#043        /* storage for login name and password. */
#044        char    username[30];
#045        char    password[30];
#046        char    servername[30];
#047        EXEC SQL END DECLARE SECTION;
#048
#049        EXEC SQL BEGIN DECLARE SECTION;
#050        CS_INT sno;
#051        char    sname[33];
#052        char    sgender[2];
#053        CS_SMALLINT retcode;
#054        EXEC SQL END DECLARE SECTION;
#055
#056        EXEC SQL WHENEVER SQLERROR CALL error_handler();
#057        EXEC SQL WHENEVER SQLWARNING CALL warning_handler();
#058        EXEC SQL WHENEVER NOT FOUND CONTINUE;
#059
#060        strcpy(username, "spring");
#061        strcpy(password, "spring1");
```

```
#062         strcpy(servername, "SEANLAPTOP");
#063         EXEC SQL CONNECT :username IDENTIFIED BY :password USING :servername;
#064         printf("connect successfully!!!!\n\n");
#065         EXEC SQL USE iihero;
#066
#067         printf("Begin test for select sno, sname, sgender from student!\n");
#068
#069         /* Declare a cursor to select a list of student. */
#070         EXEC SQL DECLARE student CURSOR FOR
#071         SELECT DISTINCT sno, sname, sgender FROM student;
#072
#073         /* Open the cursor. */
#074         EXEC SQL OPEN student;
#075         printf("\n\nSelect a student:\n\n");
#076         for (;;)
#077         {
#078             EXEC SQL FETCH student INTO :sno, :sname, :sgender;
#079             if (sqlca.sqlcode == 100)
#080                 break;
#081             printf("\tsno=%ld, sname=%s, sgender=%s\n", sno, sname, sgender);
#082         }
#083         EXEC SQL CLOSE student;
#084         EXEC SQL DISCONNECT ALL;
#085
#086         printf("End test!\n");
#087
#088         return STDEXIT;
#089 }
```

该程序首先定义了连接数据库的相关信息，使用"EXEC SQL BEGIN DECLARE SECTION;"开始和"EXEC SQL END DECLARE SECTION;"结束，包括用户名、密码及 ASE 数据库服务名。这个服务名在 interfaces 文件中定义。

接着定义了查询 student 表，student 的表结构及各相关字段的关联变量如图 14-1 所示。

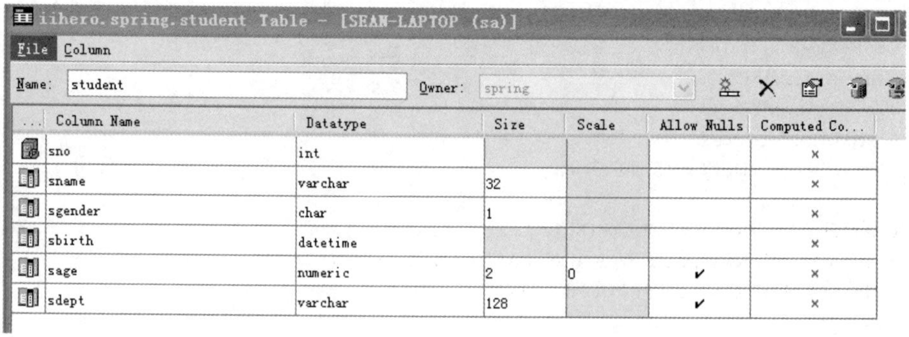

Column Name	Datatype	Size	Scale	Allow Nulls	Computed Co...
sno	int				×
sname	varchar	32			×
sgender	char	1			×
sbirth	datetime				×
sage	numeric	2	0	✓	×
sdept	varchar	128		✓	×

图 14-1 student 的表结构

sno 为整型；sname 对应的是字段 sname varchar(32)的映射，长度定义为 33，是预留一个字符串结束符；sgender 对应的是字段 sgender char(1)，长度定义为 2。

56～58 三行用于处理 ESQL 调用过程当可能出现的错误情况，一旦出错，则会触发这些调用。

- EXEC SQL WHENEVER SQLERROR CALL error_handler();: 一旦出现 SQLERROR，则会调用 error_handler()。
- EXEC SQL WHENEVER SQLWARNING CALL warning_handler(); : 一旦出现 SQLWARNING，则会调用 warning_handler()函数。
- EXEC SQL WHENEVER NOT FOUND CONTINUE;: 表示如果没有找到合适的行，则继续往下执行。

在定义完要用到的绑定变量和出错处理函数以后，就进入正常的处理逻辑。

1. 连接数据库

```
EXEC SQL CONNECT :username IDENTIFIED BY :password USING :servername;
EXEC SQL USE iihero;
```

使用用户名 username、密码 password 及服务名 servername 来连接数据库服务器，最终连接到的是数据库 iihero。实际上，这里我们也可以使用一个绑定变量来指定数据库名，如 EXEC SQL USE :dbname。

完整的连接数据库语法如下：

```
exec sql connect :user [identified by :password]
    [at :connection_name] [using :server]
```

connection_name 是为该连接指定的名字，如果不指定，则连接名为 using 指定的 server 变量值，如果 server 变量值也没有提供，则采用默认的连接名 DEFAULT。

2. 执行 SQL 查询，获取结果

```
EXEC SQL DECLARE student CURSOR FOR
SELECT DISTINCT sno, sname, sgender FROM student;
```

这里先定义一个游标，存储查询得到的结果集。然后在 for 循环里取游标中的每一行，输出相应列。

```
EXEC SQL FETCH student INTO :sno, :sname, :sgender;
```

一旦 sqlca.sqlcode == 100，意味着游标遍历结束，则跳出循环。循环结束以后，关闭游标。

在 ESQL 中调用 SQL 语句，可以在 SQL 之前显式地指定连接名，例如：

```
EXEC SQL at connetion1 update student set sgender = :sgender where sno = :sno;
```

不指定连接名时，则使用的是当前打开的数据库连接。

3. 断开数据库连接

EXEC SQL DISCONNECT ALL;

这些 ESQL 语法非常简明，节省了大量的代码空间。

断开数据库的 ESQL 语法如下：

exec sql disconnect {connection_name | current | DEFAULT | all}

其中：
- current：指的是当前连接，如果只有一个连接，可以使用这个标识符。
- DEFAULT：指的是默认连接名，如果全局只用一个连接，并且没有指定连接名或者指定 using 后面的服务名，使用它即可。
- all：则指的是当前使用或者打开的所有连接。

比较起来，disconnect 操作非常简便，节省了大量的 Open Client 客户端代码。在 disconnect 阶段会执行如下任务：
- 回滚没有提交的所有事务。
- 关闭数据库连接。
- 删除所有临时对象。
- 关闭所有游标。
- 释放当前事务用到的所有锁。
- 终止对服务器的访问。

需要注意的是，断开连接不会主动提交当前事务。在程序结束之前，我们务必要显式地调用 exec sql disconnect [all] 来释放所有打开的连接，如果没有此调用，这些连接将不会自动释放，从而造成服务器端的资源泄漏。

编译步骤如下：

在 Windows 下，进到 VC6 的 VC98\bin（使用 VC7、VS2008 均可）下，运行 VCVARS32.bat，然后再进到%SYBASE%目录下，运行 SYBASE.bat，这样把相应的环境变量全部导入，再进到当前目录。

然后执行（以 Windows 平台为例）：

E:\MyDocument\MYBOOKS\ASE\code\esqlc\example1>cpre -CMSVC -m -O student.c student.cp

Precompilation Successful. No Errors or Warnings found.
Statistical Report:
    Program name: cpre
    Options specified: /m
    Input file name: student.cp
    Listing file name:
    Target file name: student.c

```
ISQL file name:
Tag ID specified:
Compiler used: MSVC
Open Client version: CS_VERSION_150
Number of information messages: 11
Number of warning messages: 0
Number of error messages: 0
Number of SQL statements parsed: 14
Number of host variables declared: 8
Number of SQL cursors declared: 1
Number of dynamic SQL statements: 0
Number of stored Procedures generated: 0
Connection(s) information:
    User id:
    Server:
    Database:
```

```
E:\MyDocument\MYBOOKS\ASE\code\esqlc\example1>cl /DDEBUG=1 /D_DEBUG=1 /DNET_DEBU
G=1 /Od /Z7 /MD /nologo /DWIN32 -Id:\sybase\OCS-15_0\include d:\sybase\OCS-15_0\
include\sybesql.c student.c /c libsybct.lib libsybcs.lib libsybblk.lib advapi32.lib MSVCRT.lib kernel32.lib
Command line warning D4027 : source file 'libsybct.lib' ignored
Command line warning D4027 : source file 'libsybcs.lib' ignored
Command line warning D4027 : source file 'libsybblk.lib' ignored
Command line warning D4027 : source file 'advapi32.lib' ignored
Command line warning D4027 : source file 'MSVCRT.lib' ignored
Command line warning D4027 : source file 'kernel32.lib' ignored
sybesql.c
student.c
Generating Code...

E:\MyDocument\MYBOOKS\ASE\code\esqlc\example1>link  student.obj  sybesql.obj  /out:student.exe  libsybct.lib
libsybcs.lib MSVCRT.lib
Microsoft (R) Incremental Linker Version 6.00.8447
Copyright (C) Microsoft Corp 1992-1998. All rights reserved.
```

也可以把第二步的 cl 编译命令和第三步的 link 链接命令合为一步：

```
E:\MyDocument\MYBOOKS\ASE\code\esqlc\example2>cl /DDEBUG=1 /D_DEBUG=1 /DNET_DEBU
G=1 /Od /Z7 /MD /nologo /DWIN32 -Id:\sybase\OCS-15_0\include d:\sybase\OCS-15_0\
include\sybesql.c student.c libsybct.lib libsybcs.lib MSVCRT.lib /link /out:student.exe
sybesql.c
student.c
Generating Code...
```

Linux 下与其非常类似，同样要先行导入环境变量。然后经过上述三步：

```
iihero@seanlinux:~/notes/example1$ cpre -m student.cp

iihero@seanlinux:~/notes/example1$ cc -g -lsybct -lsybtcl -lsybcs -lsybcomn -lsybintl -lsybunic -rdynamic -ldl -lnsl -lm
```

```
student.c $SYBASE/$SYBASE_OCS/include/sybesql.c -o student.exe -I. -I$SYBASE/$SYBASE_OCS/include
-L$SYBASE/$SYBASE_OCS/lib
```

```
iihero@seanlinux:~/notes/example1$ ./student.exe
connect successfully!!!!

Begin test for select sno, sname, sgender from student!


Select a student:

        sno=2007002, sname=刘晨, sgender=F
        sno=2007003, sname=王敏, sgender=F
        sno=2007004, sname=张铁林, sgender=M
```

我们要留意一下 Linux 下 ASE Open Client 常用的几个动态链接库：

```
CC="cc ";
CFLAGS="-g";
CTLIB="-lsybct";
COMLIBS="-lsybtcl -lsybcs -lsybcomn -lsybintl -lsybunic";
SYSLIBS="-rdynamic -ldl -lnsl -lm" ;
```

分三步：预编译、编译、链接，最后生成可执行程序 student.exe。

这个示例的基本环境是：Server 端采用的是 cp936 字符集，客户端对应的 locale 项是 locale = default, chinese, cp936。不存在字符集转换问题。在实际应用当中，要谨慎选择字符集。

14.3　NULL 值及特殊字段类型的处理

前一节中介绍的例子里，student 表中的三个字段都是非空类型，因此不会在 cursor 里取这些字段值，不会出现错误。假如我们对 student 表作如下修改：

```
alter table spring.student modify sgender char(1) null
go
update spring.student set sgender = null where sname='王敏'
go
```

将姓名为"王敏"的性别 sgender 值设为 NULL，再运行前一节中的程序，则会在读取这条记录的时候报错：

```
Select a student:

        sno=2007002, sname=刘晨, sgender=F

** SQLCODE=(-16843097)
```

```
** ASE Error
** ct_fetch(): 用户 api 层: 外部错误: 列 3 的数据为空, 但无有效的指针
```

要避免出此问题,我们要对允许为空的列加入是否为空的处理。在 ESQL 里支持下边的语法:

```
:host_variable [[indicator] :indicator_variable]
```

宿主变量后紧跟着空指示符,建立两者之间的关联,这样在提取结果时,首先会提取空指示符的值,它是一个 short 类型(CS_SMALLINT)值,空指示符除了用于结果集提取以外,一样可以用于宿主变量的输入和输出(如存储过程的输出变量),它有三种可能值:

- -1,对应的数据库列值为 NULL。
- 0,宿主变量得到一个非空值。
- \>0,在宿主变量的数据转换过程中发生了溢出,得到一个截短的结果。这个指示符的值代表着宿主变量截短之前实际值的字节数。

因此,我们对 sgender 列添加一个指示符变量 sgender_ind,用以表明该列是否为空。对于某些类型的字段,如日期类型字段,虽然可以直接对其进行绑定,但是输出格式却是五花八门,要想得到一种比较标准的格式,我们可以在定义游标时直接显式地进行日期格式转换。

这里:

```
EXEC SQL DECLARE student CURSOR FOR
SELECT DISTINCT sno, sname, sgender, convert(char(20), sbirth, 23) FROM student;
```

使用 convert(char(20), sbirth, 23)将 sbirth 的日期型值转换为标准的第 23 个样式的字符串日期格式(YYYY-MM-DDTHH:mm:SS)。

```
1> use iihero
2> go
1> select convert(char(20), sbirth, 23) from spring.student
2> go

 --------------------
 1988-10-22T00:00:00
 1990-02-03T00:00:00
 1989-04-01T00:00:00

(3 rows affected)
1> select convert(char(20), getdate(), 23)
2> go

 --------------------
 2010-04-28T06:31:11
(1 row affected)
```

使用 convert 对日期等特殊字段类型进行转换的好处是,我们不用在程序中对字段格式进行特殊的处理。

在上一节中,我们留意到,对于字段 sname 值的输出,有大量的空格作为填充符,也就是说,

实际上输出的是 sname[33]，相当于 32 个字符的等宽字符串。这是因为我们采用的宿主变量使用的是 char 数组。要想 ASE 自动截短，我们必须使用 CS_CHAR 数组。

完整的例子如下所示：

```
#001 #define ERREXIT      -1
#002 #define STDEXIT      0
#003
#004 #include <stdio.h>
#005
#006 /* Declare the SQLCA. */
#007 EXEC SQL INCLUDE SQLCA;
#008
#009 #define    EOLN       '\0'
#010
#011 void error_handler(void)
#012 {
#013     fprintf(stderr, "\n** SQLCODE=(%ld)", sqlca.sqlcode);
#014
#015     if (sqlca.sqlerrm.sqlerrml)
#016     {
#017         fprintf(stderr, "\n** ASE Error ");
#018         fprintf(stderr, "\n** %s", sqlca.sqlerrm.sqlerrmc);
#019     }
#020     fprintf(stderr, "\n\n");
#021     exit(ERREXIT);
#022 }
#023
#024 void warning_handler(void)
#025 {
#026
#027     if (sqlca.sqlwarn[1] == 'W')
#028     {
#029         fprintf(stderr,
#030             "\n** Data truncated.\n");
#031     }
#032     if (sqlca.sqlwarn[3] == 'W')
#033     {
#034         fprintf(stderr,
#035             "\n** Insufficient host variables to store results.\n");
#036     }
#037     return;
#038 }
#039
#040 int main()
#041 {
#042     EXEC SQL BEGIN DECLARE SECTION;
#043     /* storage for login name and password. */
```

```
#044    char    username[30];
#045    char    password[30];
#046    char    servername[30];
#047    EXEC SQL END DECLARE SECTION;
#048
#049    EXEC SQL BEGIN DECLARE SECTION;
#050    CS_INT sno;
#051    CS_CHAR    sname[33];
#052    CS_CHAR    sgender[2];
#053    CS_CHAR    sbirth[21];
#054    CS_SMALLINT retcode;
#055    CS_SMALLINT sgender_ind;
#056    EXEC SQL END DECLARE SECTION;
#057
#058    EXEC SQL WHENEVER SQLERROR CALL error_handler();
#059    EXEC SQL WHENEVER SQLWARNING CALL warning_handler();
#060    EXEC SQL WHENEVER NOT FOUND CONTINUE;
#061
#062    strcpy(username, "spring");
#063    strcpy(password, "spring1");
#064    strcpy(servername, "SEANLAPTOP");
#065    EXEC SQL CONNECT :username IDENTIFIED BY :password USING :servername;
#066    printf("connect successfully!!!!\n\n");
#067    EXEC SQL USE iihero;
#068
#069    printf("Begin test for select sno, sname, sgender, sbirth from student!\n");
#070
#071    /* Declare a cursor to select a list of student. */
#072    EXEC SQL DECLARE student CURSOR FOR
#073    SELECT DISTINCT sno, sname, sgender, convert(char(20), sbirth, 23) FROM student;
#074
#075    /* Open the cursor. */
#076    EXEC SQL OPEN student;
#077    printf("\n\nSelect a student:\n\n");
#078    for (;;)
#079    {
#080        EXEC SQL FETCH student INTO :sno, :sname, :sgender indicator :sgender_ind, :sbirth;
#081        if (sqlca.sqlcode == 100)
#082            break;
#083        if (sgender_ind == -1)
#084        {
#085            printf("\tsno=%ld, sname=%s, sgender=NULL, sbirth=%s\n", sno, sname, sbirth);
#086        }
#087        else
#088        {
#089            printf("\tsno=%ld, sname=%s, sgender=%s, sbirth=%s\n", sno, sname, sgender, sbirth);
#090        }
#091    }
```

```
#092        EXEC SQL CLOSE student;
#093        EXEC SQL DISCONNECT ALL;
#094
#095        printf("End test!\n");
#096
#097        return STDEXIT;
#098 }
```

在第 83 行,我们对占位符 sgender_ind 进行判断,如果是-1,则意味着 sgender 列为 NULL。编译方式同前一节一样,最终输出结果如下:

```
Select a student:
        sno=2007002, sname=刘晨, sgender=F, sbirth=1988-10-22T00:00:00
        sno=2007003, sname=王敏, sgender=NULL, sbirth=1990-02-03T00:00:00
        sno=2007004, sname=张铁林, sgender=M, sbirth=1989-04-01T00:00:00
```

14.4 存储过程调用

本节将介绍如何在 ESQLC 中调用存储过程,在 ESQL 中调用存储过程就如同直接使用 SQL 调用存储过程一样,其语法如下:

EXEC SQL exec [:retcode =] <storeproc> [:arg1, :arg2, …… :argn];

其实每个参数都可以带可选的 output 后缀,表示它是一个输出参数。

下面来看一个完整的例子,我们先在 iihero 数据库里准备一个存储过程,我们以 sa 用户执行如下脚本(test_proc.sql):

```
use iihero
go

setuser 'spring'
go

if exists ( select 1 from sysobjects where name = 'test_proc' )
    drop proc test_proc
go

if exists ( select 1 from sysobjects where name = 't123' )
    drop table t123
go

create table t123(id int primary key, col2 varchar(32) not null)
insert into t123 values(1, 'iihero')
insert into t123 values(2, 'Sybase')
```

```
insert into t123 values(3, 'ASE')
go

create proc test_proc (@id_min int, @num_t123 int output) with recompile
as
select @num_t123 = count( a.id ) from iihero.spring.t123 a where a.id > = @id_min
go

setuser
go

declare    @num_t123 int
exec spring.test_proc 1, @num_t123 output
-- select @num_123 = count(a.id)    from iihero.spring.t123 a where a.id>=1
select @num_t123
go
```

存储过程 test_proc 就是返回表 t123 当中，ID 值不小于输入值的记录个数。

调用此存储过程 test_proc 的 ESQLC 程序片段如下：

```
#001 #define ERREXIT        -1
#002 #define STDEXIT        0
#003
#004 #include <stdio.h>
#005
#006 /* Declare the SQLCA. */
#007 EXEC SQL INCLUDE SQLCA;
#008
#009 #define    EOLN       '\0'
#010
#011 void error_handler(void)
#012 {
#013      fprintf(stderr, "\n** SQLCODE=(%ld)", sqlca.sqlcode);
#014
#015      if (sqlca.sqlerrm.sqlerrml)
#016      {
#017           fprintf(stderr, "\n** ASE Error ");
#018           fprintf(stderr, "\n** %s", sqlca.sqlerrm.sqlerrmc);
#019      }
#020
#021      fprintf(stderr, "\n\n");
#022
#023      exit(ERREXIT);
#024 }
#025
#026 void warning_handler(void)
#027 {
#028
```

```
#029        if (sqlca.sqlwarn[1] == 'W')
#030        {
#031            fprintf(stderr,
#032                "\n** Data truncated.\n");
#033        }
#034
#035        if (sqlca.sqlwarn[3] == 'W')
#036        {
#037            fprintf(stderr,
#038                "\n** Insufficient host variables to store results.\n");
#039        }
#040        return;
#041 }
#042
#043 int main()
#044 {
#045        EXEC SQL BEGIN DECLARE SECTION;
#046        /* storage for login name and password. */
#047        char    username[30];
#048        char    password[30];
#049        char    servername[30];
#050        char    dbname[30];
#051        EXEC SQL END DECLARE SECTION;
#052
#053        EXEC SQL BEGIN DECLARE SECTION;
#054        CS_INT id_min;
#055        CS_INT num_t123;
#056        CS_SMALLINT retcode;
#057        EXEC SQL END DECLARE SECTION;
#058
#059        EXEC SQL WHENEVER SQLERROR CALL error_handler();
#060        EXEC SQL WHENEVER SQLWARNING CALL warning_handler();
#061        EXEC SQL WHENEVER NOT FOUND CONTINUE;
#062
#063        strcpy(username, "spring");
#064        strcpy(password, "spring1");
#065        strcpy(servername, "SEANLAPTOP");
#066        strcpy(dbname, "iihero");
#067        EXEC SQL CONNECT :username IDENTIFIED BY :password USING :servername;
#068        EXEC SQL USE :dbname;
#069
#070        printf("Begin test!\n");
#071
#072        id_min = 0;
#073        exec sql exec :retcode = test_proc :id_min, :num_t123 output;
#074
#075        printf("return code=%ld, num_t123=%ld\n", retcode, num_t123);
#076
```

```
#077        EXEC SQL DISCONNECT ALL;
#078
#079        printf("End test!\n");
#080
#081        return STDEXIT;
#082 }
```

最后执行结果：

```
E:\MyDocument\MYBOOKS\ASE\code\esqlc\example3>testproc
Begin test!
return code=0, num_t123=3
End test!
```

14.5 插入/更新数据

ESQLC 程序里往表中插入或更新数据，通常有两种方式，一种是直接使用 Insert 或 Update 来操作；另一种是使用游标更新操作。

14.5.1 直接 Insert/Update

首先我们来看第一种情况——直接 Insert 或 Update。它是通过 ESQL 直接调用 Insert 或 Update 语句执行插入更新操作。

我们来看下边的示例（test_cu.cp, example4）：

```
#001 #define ERREXIT      -1
#002 #define STDEXIT       0
#003
#004 #include <stdio.h>
#005
#006 /* Declare the SQLCA. */
#007 EXEC SQL INCLUDE SQLCA;
#008
#009 #define   EOLN        '\0'
#010
#011 void error_handler(void)
#012 {
#013      fprintf(stderr, "\n** SQLCODE=(%ld)", sqlca.sqlcode);
#014
#015      if (sqlca.sqlerrm.sqlerrml)
#016      {
#017          fprintf(stderr, "\n** ASE Error ");
#018          fprintf(stderr, "\n** %s", sqlca.sqlerrm.sqlerrmc);
#019      }
#020
```

```
#021        fprintf(stderr, "\n\n");
#022
#023        exit(ERREXIT);
#024 }
#025
#026 void warning_handler(void)
#027 {
#028
#029        if (sqlca.sqlwarn[1] == 'W')
#030        {
#031            fprintf(stderr,
#032                "\n** Data truncated.\n");
#033        }
#034
#035        if (sqlca.sqlwarn[3] == 'W')
#036        {
#037            fprintf(stderr,
#038                "\n** Insufficient host variables to store results.\n");
#039        }
#040        return;
#041 }
#042
#043 int main()
#044 {
#045        int i = 0;
#046        EXEC SQL BEGIN DECLARE SECTION;
#047        /* storage for login name and password. */
#048        char    username[30];
#049        char    password[30];
#050        char    servername[30];
#051        char    dbname[30];
#052        EXEC SQL END DECLARE SECTION;
#053
#054        EXEC SQL BEGIN DECLARE SECTION;
#055        CS_INT id;
#056        CS_CHAR col2[33];
#057        CS_SMALLINT retcode;
#058        EXEC SQL END DECLARE SECTION;
#059
#060        EXEC SQL WHENEVER SQLERROR CALL error_handler();
#061        EXEC SQL WHENEVER SQLWARNING CALL warning_handler();
#062        EXEC SQL WHENEVER NOT FOUND CONTINUE;
#063
#064        strcpy(username, "spring");
#065        strcpy(password, "spring1");
#066        strcpy(servername, "SEANLAPTOP");
#067        strcpy(dbname, "iihero");
#068        EXEC SQL CONNECT :username IDENTIFIED BY :password USING :servername;
```

```
#069        EXEC SQL USE :dbname;
#070
#071        printf("Begin test!\n");
#072
#073        /* create a demo table first */
#074        EXEC SQL create table foo(id int primary key, col2 varchar(32) null);
#075        printf("table foo created. \n");
#076
#077        /* test insert and update data in table foo */
#078        for (i = 0; i < 5; i++)
#079        {
#080            id = i + 1;
#081            sprintf((char*)col2, "foo_%ld", id);
#082            EXEC SQL insert into foo (id, col2) values( :id, :col2 );
#083        }
#084
#085        strcpy(col2, "foo_4_updated");
#086        id = 4;
#087        EXEC SQL update foo set col2 = :col2 where id = :id;
#088        printf("the 4th row updated, col2 = 'foo_4_updated'. \n");
#089
#090        /* show all the rows of table foo */
#091        EXEC SQL DECLARE foo CURSOR FOR
#092        SELECT id, col2 FROM foo order by id;
#093
#094        EXEC SQL OPEN foo;
#095        printf("The rows of foo: \n");
#096        for (;;)
#097        {
#098            EXEC SQL FETCH foo INTO :id, :col2;
#099            if (sqlca.sqlcode == 100)
#100                break;
#101            printf("id=%ld, col2=%s\n", id, col2);
#102        }
#103        EXEC SQL CLOSE foo;
#104
#105        /* drop the demo table: foo */
#106        EXEC SQL drop table foo;
#107        printf("table foo dropped. \n");
#108
#109        EXEC SQL DISCONNECT ALL;
#110
#111        printf("End test!\n");
#112
#113        return STDEXIT;
#114 }
```

在本例中，首先创建示例表 foo(id int primary key, col2 varchar(32) null)。然后通过一个 5 次循

环，插入 5 条记录，在每次调用 insert 之前，都对 id 和 col2 绑定的变量进行赋值。第 86 至 88 行则对 id 为 4 的那一行进行更新操作。

第 94 至 103 行则打开一个查询 foo 表的游标，遍历所有记录行并输出结果。

最后删除该表。

该例程编译和执行过程如下：

```
E:\MyDocument\MYBOOKS\ASE\code\esqlc\example4>cpre -CMSVC -p -b -m -O test_cu.c test_cu.cp
Precompilation Successful. No Errors or Warnings found.
Statistical Report:
            Program name: cpre
            Options specified: /m
            Input file name: test_cu.cp
            Listing file name:
            Target file name: test_cu.c
            ISQL file name:
            Tag ID specified:
            Compiler used: MSVC
            Open Client version: CS_VERSION_150
            Number of information messages: 11
            Number of warning messages: 0
            Number of error messages: 0
            Number of SQL statements parsed: 17
            Number of host variables declared: 8
            Number of SQL cursors declared: 1
            Number of dynamic SQL statements: 0
            Number of stored Procedures generated: 0
            Connection(s) information:
                    User id:
                    Server:
                    Database:

E:\MyDocument\MYBOOKS\ASE\code\esqlc\example4>cl /DDEBUG=1 /D_DEBUG=1 /DNET_DEBU
G=1 /Od /Z7 /MD /nologo /DWIN32 -Id:\sybase\OCS-15_0\include d:\sybase\OCS-15_0\
include\sybesql.c test_cu.c libsybct.lib libsybcs.lib MSVCRT.lib /link /out:test_cu.exe
sybesql.c
test_cu.c
Generating Code...

E:\MyDocument\MYBOOKS\ASE\code\esqlc\example4>test_cu
Begin test!
table foo created.
the 4th row updated, col2 = 'foo_4_updated'.
The rows of foo:
id=1, col2=foo_1
id=2, col2=foo_2
id=3, col2=foo_3
```

```
id=4, col2=foo_4_updated
id=5, col2=foo_5
table foo dropped.
End test!
```

这里为预编译命令特意加入了-p 和-b 选项，是出于性能的考虑，我们注意到第 79 至 83 行，每次都执行一次绑定，但是这个动态的 SQL 语句是一样的，除了变量值不同以外，加入-p 选项主要就是可以实现 SQL 语句绑定共享，只要绑定的输入变量值可变即可。-b 是针对输出变量的。

14.5.2　通过游标来更新数据

除了通过 update 语句直接进行更新以外，我们还可以通过打开一个 select 游标，对符合条件的记录行进行更新。如下例所示：

```
#001 #define ERREXIT        -1
#002 #define STDEXIT         0
#003
#004 #include <stdio.h>
#005
#006 /* Declare the SQLCA. */
#007 EXEC SQL INCLUDE SQLCA;
#008
#009 #define   EOLN         '\0'
#010
#011 void error_handler(void)
#012 {
#013    fprintf(stderr, "\n** SQLCODE=(%ld)", sqlca.sqlcode);
#014
#015    if (sqlca.sqlerrm.sqlerrml)
#016    {
#017        fprintf(stderr, "\n** ASE Error ");
#018        fprintf(stderr, "\n** %s", sqlca.sqlerrm.sqlerrmc);
#019    }
#020
#021    fprintf(stderr, "\n\n");
#022
#023    exit(ERREXIT);
#024 }
#025
#026 void warning_handler(void)
#027 {
#028
#029    if (sqlca.sqlwarn[1] == 'W')
#030    {
#031        fprintf(stderr,
#032            "\n** Data truncated.\n");
#033    }
```

```
#034
#035           if (sqlca.sqlwarn[3] == 'W')
#036           {
#037               fprintf(stderr,
#038                   "\n** Insufficient host variables to store results.\n");
#039           }
#040           return;
#041 }
#042
#043 int main()
#044 {
#045       int i = 0;
#046       EXEC SQL BEGIN DECLARE SECTION;
#047       /* storage for login name and password. */
#048       char    username[30];
#049       char    password[30];
#050       char    servername[30];
#051       char    dbname[30];
#052       EXEC SQL END DECLARE SECTION;
#053
#054       EXEC SQL BEGIN DECLARE SECTION;
#055       CS_INT id;
#056       CS_CHAR col2[33];
#057       CS_SMALLINT retcode;
#058       EXEC SQL END DECLARE SECTION;
#059
#060       EXEC SQL WHENEVER SQLERROR CALL error_handler();
#061       EXEC SQL WHENEVER SQLWARNING CALL warning_handler();
#062       EXEC SQL WHENEVER NOT FOUND CONTINUE;
#063
#064       strcpy(username, "spring");
#065       strcpy(password, "spring1");
#066       strcpy(servername, "SEANLAPTOP");
#067       strcpy(dbname, "iihero");
#068       EXEC SQL CONNECT :username IDENTIFIED BY :password USING :servername;
#069       EXEC SQL USE :dbname;
#070
#071       printf("Begin test!\n");
#072
#073       /* create a demo table first */
#074       EXEC SQL create table foo(id int primary key, col2 varchar(32) null);
#075       printf("table foo created. \n");
#076
#077       /* test insert and update data in table foo */
#078       for (i = 0; i < 5; i++)
#079       {
#080           id = i + 1;
#081           sprintf((char*)col2, "foo_%ld", id);
```

```
#082                EXEC SQL insert into foo (id, col2) values( :id, :col2 );
#083            }
#084
#085        /* update the 4th row of table foo */
#086        EXEC SQL DECLARE foo CURSOR FOR
#087        SELECT id, col2 FROM foo order by id;
#088
#089        EXEC SQL OPEN foo;
#090        printf("The rows of foo: \n");
#091        for (;;)
#092        {
#093            EXEC SQL FETCH foo INTO :id, :col2;
#094            if (sqlca.sqlcode == 100)
#095                break;
#096            printf("id=%ld, col2=%s\n", id, col2);
#097            if ( id == 4)
#098            {
#099                strcpy(col2, "foo_4_updated");
#100                EXEC SQL update foo set col2 = :col2 WHERE CURRENT OF foo;
#101                printf("*** the 4th row updated, col2 = 'foo_4_updated'. \n");
#102                break;
#103            }
#104        }
#105        EXEC SQL CLOSE foo;
#106
#107        /* drop the demo table: foo */
#108        EXEC SQL drop table foo;
#109        printf("table foo dropped. \n");
#110
#111        EXEC SQL DISCONNECT ALL;
#112
#113        printf("End test!\n");
#114
#115        return STDEXIT;
#116 }
```

第100行"EXEC SQL update foo set col2 = :col2 WHERE CURRENT OF foo;"直接对游标中的当前记录行进行更新。这种方式比较直观，是在打开符合条件的记录的基础上进行操作。

14.6 BLOB/CLOB 数据处理

BLOB/CLOB 类型对应于 ASE 数据库，就是 IMAGE/TEXT 类型，它的读写与其他普通字段可能不太一样。

首先，如果直接使用动态 insert 或者 update，绑定的 IMAGE/TEXT 值有长度限制，不能超过

64KB。如果要处理超过此长度的 IMAGE/TEXT，需要借助于-y 预编译选项，用于生成支持长度超过 64KB 的 IMAGE/TEXT。此功能在 ASE 11.0 以后版本开始得到支持。

下边的示例展示的是向表 foo 中插入一行带有 IMAGE/TEXT 类型的记录（example6，test_lob.pc）：

```
#001 #define ERREXIT        -1
#002 #define STDEXIT         0
#003
#004 #include <stdio.h>
#005
#006 /* Declare the SQLCA. */
#007 EXEC SQL INCLUDE SQLCA;
#008
#009 #define    EOLN         '\0'
#010
#011 void error_handler(void)
#012 {
#013       fprintf(stderr, "\n** SQLCODE=(%ld)", sqlca.sqlcode);
#014       if (sqlca.sqlerrm.sqlerrml)
#015       {
#016            fprintf(stderr, "\n** ASE Error ");
#017            fprintf(stderr, "\n** %s", sqlca.sqlerrm.sqlerrmc);
#018       }
#019       fprintf(stderr, "\n\n");
#020       exit(ERREXIT);
#021 }
#022
#023 void warning_handler(void)
#024 {
#025
#026       if (sqlca.sqlwarn[1] == 'W')
#027       {
#028            fprintf(stderr,
#029                 "\n** Data truncated.\n");
#030       }
#031
#032       if (sqlca.sqlwarn[3] == 'W')
#033       {
#034            fprintf(stderr,
#035                 "\n** Insufficient host variables to store results.\n");
#036       }
#037       return;
#038 }
#039
#040 int main()
#041 {
#042      int i = 0;
```

```
#043    EXEC SQL BEGIN DECLARE SECTION;
#044    /* storage for login name and password. */
#045    char    username[30];
#046    char    password[30];
#047    char    servername[30];
#048    char    dbname[30];
#049    EXEC SQL END DECLARE SECTION;
#050
#051    EXEC SQL BEGIN DECLARE SECTION;
#052    CS_INT id;
#053    CS_IMAGE col2_var[10000];
#054    CS_TEXT   col3_var[10000];
#055    CS_SMALLINT retcode;
#056    CS_INT col2_len;
#057    CS_INT col3_len;
#058    EXEC SQL END DECLARE SECTION;
#059
#060    EXEC SQL WHENEVER SQLERROR CALL error_handler();
#061    EXEC SQL WHENEVER SQLWARNING CALL warning_handler();
#062    EXEC SQL WHENEVER NOT FOUND CONTINUE;
#063
#064    strcpy(username, "spring");
#065    strcpy(password, "spring1");
#066    strcpy(servername, "SEANLAPTOP");
#067    strcpy(dbname, "iihero");
#068    EXEC SQL CONNECT :username IDENTIFIED BY :password USING :servername;
#069    EXEC SQL USE :dbname;
#070
#071    printf("Begin test!\n");
#072
#073    /* set max text size 1000KB */
#074    EXEC SQL set textsize 1024000;
#075
#076    /* create a demo table first */
#077    EXEC SQL create table foo(id int primary key, col2 image null, col3 text null);
#078    printf("table foo created. \n");
#079
#080    /* test insert and update data in table foo */
#081    id = 1;
#082    strcpy(col2_var, "abcd");
#083    strcpy(col3_var, "中文");
#084    for (i = 4; i < 10000; i++)
#085    {
#086        col2_var[i] = 'f';
#087        col3_var[i] = '@';
#088    }
#089    EXEC SQL insert into foo (id, col2, col3) values( :id, :col2_var, :col3_var );
#090    if ( sqlca.sqlcode == 0 )
```

```
#091            {
#092                    printf("Row successfully inserted! \n");
#093                    EXEC SQL COMMIT WORK;
#094            }
#095            /* reset col2 and col3 */
#096            strcpy(col2_var, "");
#097            strcpy(col3_var, "");
#098
#099            /* fetch the image/text rows */
#100            EXEC SQL DECLARE foo CURSOR FOR
#101            SELECT id, col2, col3, datalength(col2), datalength(col3) FROM foo order by id;
#102
#103            EXEC SQL OPEN foo;
#104            printf("The rows of foo: \n");
#105            for (;;)
#106            {
#107                    EXEC SQL FETCH foo INTO :id, :col2_var, :col3_var, :col2_len, :col3_len;
#108                    if (sqlca.sqlcode == 100)
#109                            break;
#110                    printf("id=%ld, col2 len=%ld, col3 len=%ld\n", id, col2_len, col3_len);
#111            }
#112            EXEC SQL CLOSE foo;
#113
#114            /* drop the demo table: foo */
#115            EXEC SQL drop table foo;
#116            printf("table foo dropped. \n");
#117
#118            EXEC SQL DISCONNECT ALL;
#119
#120            printf("End test!\n");
#121
#122            return STDEXIT;
#123 }
```

在表 foo(id int primary key, col2 image null, col3 text null)中，我们对 col2 使用 CS_IMAGE 数组进行绑定，对 col3 使用 CS_TEXT 数组进行绑定。

编译命令：

```
E:\MyDocument\MYBOOKS\ASE\code\esqlc\example6>cpre -CMSVC -y -m -O test_lob.c te
st_lob.cp
Precompilation Successful. No Errors or Warnings found.
Statistical Report:
            Program name: cpre
            Options specified: /m
            Input file name: test_lob.cp
            Listing file name:
            Target file name: test_lob.c
```

```
            ISQL file name:
            Tag ID specified:
            Compiler used: MSVC
            Open Client version: CS_VERSION_150
            Number of information messages: 11
            Number of warning messages: 0
            Number of error messages: 0
            Number of SQL statements parsed: 18
            Number of host variables declared: 11
            Number of SQL cursors declared: 1
            Number of dynamic SQL statements: 0
            Number of stored Procedures generated: 0
            Connection(s) information:
                    User id:
                    Server:
                    Database:
E:\MyDocument\MYBOOKS\ASE\code\esqlc\example6>cl /DDEBUG=1 /D_DEBUG=1 /DNET_DEBU
G=1 /Od /Z7 /MD /nologo /DWIN32 -Id:\sybase\OCS-15_0\include d:\sybase\OCS-15_0\
include\sybesql.c test_lob.c libsybct.lib libsybcs.lib MSVCRT.lib /link /out:test_lob.exe
sybesql.c
test_lob.c
Generating Code...
```

如果我们仔细检查一下生成的代码 test_lob.c，会发现在加了"-y"选项以后，为 col2 和 col3 绑定的变量值确实有不同的处理方式：

```
                            _sql->retcode = ct_command(_sql->conn.command,
# line 89 "test_lob.cp"
                                CS_LANG_CMD, "insert into foo (id, col2, col3) value"
"s( @sql0_id , ", 52, CS_MORE);
# line 89 "test_lob.cp"
                            _sql->retcode = ct_command(_sql->conn.command,
# line 89 "test_lob.cp"
                                CS_LANG_CMD, "0x", 2, CS_MORE);
# line 89 "test_lob.cp"
                            _sql->retcode = _sqlProcessParam(_sql, CT_COMMAND,
# line 89 "test_lob.cp"
                                (CS_CHAR *)NULL, 0, (CS_VOID *)col2_var, 10000,
# line 89 "test_lob.cp"
                                CS_IMAGE_TYPE);
# line 89 "test_lob.cp"
                            _sql->retcode = ct_command(_sql->conn.command,
# line 89 "test_lob.cp"
                                CS_LANG_CMD, " , ", 3, CS_MORE);
# line 89 "test_lob.cp"
                            _sql->retcode = ct_command(_sql->conn.command,
# line 89 "test_lob.cp"
                                CS_LANG_CMD, "'", 1, CS_MORE);
```

```
# line 89 "test_lob.cp"
                    _sql->retcode = _sqlProcessParam(_sql, CT_COMMAND,
# line 89 "test_lob.cp"
                        (CS_CHAR *)NULL, 0, (CS_VOID *)col3_var, 10000,
# line 89 "test_lob.cp"
                        CS_TEXT_TYPE);
# line 89 "test_lob.cp"
                    _sql->retcode = ct_command(_sql->conn.command,
# line 89 "test_lob.cp"
                        CS_LANG_CMD, "'", 1, CS_MORE);
# line 89 "test_lob.cp"
                    _sql->retcode = ct_command(_sql->conn.command,
# line 89 "test_lob.cp"
                        CS_LANG_CMD, " )", 2, CS_END);
```

我们可以从上边生成的代码片段中看出,"-y"选项把 IMAGE/TEXT 字段的绑定生成多条 ct_command 命令调用,也就是把一个长的 IMAGE/TEXT 值分成多段进行发送。

对 IMAGE 字段,它会先发送"0x"前缀,然后发送 IMAGE 绑定变量的二进制值;针对 TEXT 字段,会依次发送"'"、文本串、"'"。

15

使用 ODBC 开发 ASE 应用

ODBC 是数据库最基本的标准访问接口之一，ASE 提供了对该接口的完备支持。本章主要介绍如何使用 ODBC 接口访问 ASE 数据库。

15.1 ODBC 简要介绍

15.1.1 ODBC 介绍

ODBC（Open Database Connection，开放数据库互连）是一种完全开放的数据库访问接口，是由微软公司提出的一个用于访问数据库的统一接口标准，随着 C/S 体系在各行业领域广泛应用，多种数据库之间的互连访问成为一个突出的问题，而 ODBC 成为目前一个强有力的解决方案。ODBC 之所以能够操作众多的数据库，是由于当前绝大部分数据库全部或部分地遵从关系数据库概念，并且有了统一的查询语言。支持众多的数据库并不意味着 ODBC 会变得复杂，因为 ODBC 是基于结构化查询语言（SQL），使用 SQL 可大大简化其应用程序设计接口（API），由于 ODBC 思想上的先进性，而且没有同类标准或产品与之竞争，因而越来越受到众多厂家和用户的青睐。目前，几乎所有商业数据库都实现了 ODBC 接口。

在 1994 年，ODBC 有了第一个版本，这种名为 Open Database Connection（开放式数据库互连）的技术很快通过了标准化，并且得到各个数据库厂商的支持。ODBC 在当时解决了两个问题，一个是在 Windows 平台下的数据库开发，另一个是建立一个统一的标准。只要数据厂商提供的开发包支持这个标准，那么开发人员通过 ODBC 开发的程序可以在不同的数据库之间自由转换，这对开发人员来说的确值得庆贺。

ODBC 参照了 X/Open Data Management: SQL Call-Level Interface 和 ISO/ICE1995 Call-Level Interface 标准（这些都由 IBM 提出，至今 DB2 数据库的 CLI 和 ODBC 都非常的相似），在 ODBC

版本 3.X 中已经完全实现了这两个标准的所有要求。所以本书所有内容都基于 ODBC 3.0 以上版本。

最开始支持 ODBC 的数据库只有 SQL Server、Access、FoxPro，这些都是微软的产品，所以它们能够支持 ODBC 一点也不奇怪，但是那时候 Windows 的图形界面已经成为了客户端软件最理想的载体，所以各大数据厂商也在不久后发布了针对 ODBC 的驱动程序。

在 Windows 平台下，到了 Windows 98 及以后的版本，当你安装完操作系统后，ODBC 不需要另行安装，因为它已经成为了操作系统的一部分。这对于开发人员来说，简化了大量的安装配置工作。

ODBC 支持数据库的一些重要接口，但并不要求数据库完全实现 ODBC 所提出的所有接口。如果 ODBC 只提出所有数据库的公共接口（即接口的交集），那么它的使用会受到很大的限制；相反，当今众多数据库共存的原因在于它们都有各自的特性，当然，如果 ODBC 将所有数据库的所有特性都提炼出来，那么特定的 ODBC 驱动将难以实现。

15.1.2　ODBC 体系结构

ODBC 的基本结构有四个组成部分，如图 15-1 所示。

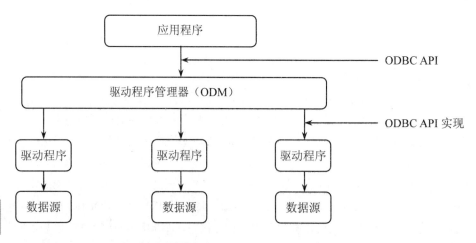

图 15-1　ODBC 的基本体系结构

- 应用程序：不与数据库直接打交道，主要负责处理并调用 ODBC 函数，发送对数据库的 SQL 请求及取得结果。
- 驱动程序管理器（ODBC Driver Manager，ODM）：驱动程序管理器是一个带有输入程序的动态链接库（DLL），主要目的是加载驱动程序，处理 ODBC 调用的初始化调用，提供 ODBC 调用的参数有效性和序列有效性。典型的 ODM 在 Windows 平台下有微软提供的 ODM，通过 odbcad32 就可以打开这个 ODM 管理器界面。而在 Linux/UNIX 下，有开源的 unixODBC 可供使用，功能也非常强大。

- 驱动程序（Driver）：驱动程序是一个或多个完成 ODBC 函数调用并与数据库相互影响的 DLL，这些驱动程序可以处理特定的数据库访问请求。对于应用驱动程序管理器送来的命令，驱动程序再进行解释形成自己的数据库所能理解的命令。驱动程序将处理所有的数据库访问请求，对于应用程序来讲不需要关注所使用的是本地数据库还是远程数据库。

15.2　ASE 中的 ODBC 环境

从 ASE 12.5.4 开始，ASE 数据库的 ODBC 驱动名称从 SYBASE ASE ODBC Driver 改为 Adaptive Server Enterprise。前者是由 DataDirect 为 Sybase 开发的 ODBC 驱动，后者是由 Sybase 自己实现的 ODBC 驱动。并且，DataDirect 实现的 ODBC 驱动从 12.5.4 版本开始，不再出现在 ASE 的安装版本当中。Sybase 对它的支持到 2008 年 12 月份全面终止，全部切换到自己实现的 Adaptive Server Enterprise ODBC 驱动。

旧版本 DataDirect 的实现和新版本 Sybase 自己的实现有如下不同：

- DataDirect ODBC 驱动库位于%SYBASE%\ODBC 目录下，该目录下共有 11 个 DLL 文件，其中 ODBC API 的实现 DLL 是 syodase.dll。
- Sybase ODBC 驱动库位于%SYBASE%\DataAccess\ODBC\dll 目录下，总共有 6 个 DLL 文件，其中 sybdrvodb.dll 是 ODBC 的实现库。

无论新旧版本，ASE 的 ODBC 驱动都有一个好处，它们都是完全独立于 ASE 运行时的环境的，不依赖于 Open Client 库。也就是说，作为应用程序开发者，他们完全可以把 ODBC 驱动连同自己的应用程序发布到单独的一台机器上，而不用在这台机器上安装 ASE 的 Open Client 客户端。

在 Windows 平台下，手动注册旧版本 DataDirect ODBC 驱动，可以采用如下脚本 (install_aseodbc_datadirect.reg)更新注册表：

```
Windows Registry Editor Version 5.00
[HKEY_LOCAL_MACHINE\SOFTWARE\ODBC]

[HKEY_LOCAL_MACHINE\SOFTWARE\ODBC\ODBCINST.INI\ODBC Drivers]
"SYBASE ASE ODBC Driver"="Installed"

[HKEY_LOCAL_MACHINE\SOFTWARE\ODBC\ODBCINST.INI\SYBASE ASE ODBC Driver]
"Driver"="D:\\ASE150\\ODBC\\dll\\syodase.dll"
"Setup"="D:\\ASE150\\ODBC\\dll\\syodases.dll"
```

手动注册新版本 Adaptive Server Enterprise ODBC 驱动，可以使用如下脚本更新注册表：

```
Windows Registry Editor Version 5.00
[HKEY_LOCAL_MACHINE\SOFTWARE\ODBC]

[HKEY_LOCAL_MACHINE\SOFTWARE\ODBC\ODBCINST.INI\ODBC Drivers]
"Adaptive Server Enterprise"="Installed"
```

```
[HKEY_LOCAL_MACHINE\SOFTWARE\ODBC\ODBCINST.INI\Adaptive Server Enterprise]
"Driver"="D:\\ASE150\\DataAccess\\ODBC\\dll\\sybdrvodb.dll"
"Setup"="D:\\ASE150\\DataAccess\\ODBC\\dll\\sybdrvodb.dll"
```

只要将里边的 D:\\ASE150 换成真正的 SYBASE ASE 根目录即可。还有一种更简单的办法是直接运行命令：regsvr32 D:\ASE150\DataAccess\ODBC\dll\sybdrvodb.dll，它会生成上述注册表内容。两者效果完全一样。SYBASE ASE ODBC 与它自身提供的 Open Client 库相比，一个明显的好处是它比较小巧独立，不依赖于 Open Client 库，与稍显庞大的 Open Client 库相比，发布应用更加方便。同时，它内部也是通过 TDS 协议与 ASE 服务器进行交互，运行效率不会有太大差别。同时，由于 ODBC 接口是标准化的接口，使用 ODBC 进行客户端编程比 Open Client 接口编程要简单许多。

在 Linux 下，如果要开发 ODBC 应用程序来访问 ASE 数据库，必须预先安装和配置 unixODBC。一个是 unixODBC 应用程序，另一个是 unixODBC-dev（开发库）。unixODBC 目前已经发展到 2.3.0 版本，我们可以从官网http://www.unixodbc.org/下载源码包自行编译，也可以直接从 Linux 发行版直接安装，目前最常用的还是 unixODBC-2.2.11。

15.3　连接 ASE

在介绍 ODBC 如何连接 ASE 之前，我们要简单看一下几个 ODBC API 中常用数据类型与 C 语言中使用的数据类型的对应关系，如表 15-1 所示。

表 15-1　ODBC 中类型与 C 数据类型对应关系

类型标识符	ODBC 数据类型	C 数据类型
SQL_C_CHAR	SQLCHAR *	unsigned char *
SQL_C_SSHORT	SQLSMALLINT	short int
SQL_C_USHORT	SQLUSMALLINT	unsigned short int
SQL_C_SLONG	SQLINTEGER	long int
SQL_C_FLOAT	SQLREAL	float
SQL_C_DOUBLE	SQLDOUBLE, SQLFLOAT	double
SQL_C_BINARY	SQLCHAR *	unsigned char *
SQL_C_TYPE_DATE	SQL_DATE_STRUCT	struct tagDATE_STRUCT {SQLSMALLINT year; SQLUSMALLINT month; SQLUSMALLINT day; } DATE_STRUCT;
SQL_C_TYPE_TIME	SQL_TIME_STRUCT	struct tagTIME_STRUCT {SQLSMALLINT hour; SQLUSMALLINT minute; SQLUSMALLINT second; } TIME_STRUCT;

这些对应关系主要用于程序中输入/输出变量与关系表中字段的绑定。

15.3.1 连接 ASE 的过程

下面将通过一个简单的例子来介绍一下如何访问 ASE 数据库。使用的数据源名称（DSN）为 iihero_ase，用户名为 spring，密码为 spring1。这个数据源需要提前通过 ODM 建好。运行 odbcad32（Windows 下）：

选择用户 DSN，然后添加一个数据源，名为 iihero_ase，驱动选择新版的 Adaptive Server Enterprise，然后填入表单，如图 15-2 所示。

可以通过 Test Connection 来验证创建的数据源是否有效。如果你选择的是低版本的 Sybase ODBC Driver 驱动，则按如图 15-3 所示的格式填入相关信息。

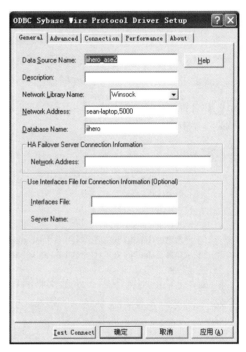

图 15-2　ASE ODBC 连接　　　　　　图 15-3　低版本的 ASE ODBC 连接

图 15-3 中的网络地址（Network Address）必须使用格式：<主机名或 IP 地址>,端口号

下边有一个可选项，是用于指定 interfaces 文件及 server name 的，实际上就是指定 Open Client 的 interface 文件的位置及相应的 server name 项，这样可以通过这个配置信息间接生成到 ASE 的连接信息。不过，它在运行时并不依赖于 Open Client 库，稍后我们可以证明这一点。

通常连接一个数据库的基本步骤如下：

Step 1　创建环境句柄并设置相关参数。

首先，声明一个 ODBC 环境句柄（SQLHENV），它可以用来获取相关的 ODBC 环境信息，通过调用 API SQLAllocHandle(SQL_HANDLE_ENV, SQL_NULL_HANDLE, &env)初始化这个句柄，分配好句柄之后，需要设定你要使用的 ODBC 版本，它可以调用 SQLSetEnvAttr(env, SQL_ATTR_ODBC_VERSION, (void*)SQL_OV_ODBC3, 0)，SQL_ATTR_ODBC_VERSION 是存放你定义的 ODBC 版本号的变量，SQL_OV_ODBC3 则说明你的程序使用的是 ODBC 3.0。

Step 2 创建连接句柄并设置超时参数。

然后需要通过环境句柄 env，创建一个连接句柄（SQLHDBC）dbc，用来存储数据库连接信息，调用 SQLAllocHandle(SQL_HANDLE_DBC, env, &dbc)，接着通过调用 SQLSetConnectOption(dbc, SQL_LOGIN_TIMEOUT, 30)设置连接超时间为 30 秒。

Step 3 连接数据库。

在连接句柄分配成功以后，调用 SQLConnect(dbc,(SQLCHAR*) "iihero_ase", SQL_NTS, (SQLCHAR*)"spring", SQL_NTS, (SQLCHAR*)"spring1", SQL_NTS)，这里需要的就是前边提到的三个参数：数据源名、用户名、密码。

Step 4 分配 SQL 语句句柄并执行相关处理。

后边将有介绍。

Step 5 断开数据库连接并释放所有相关句柄。

在程序的末尾，要断开数据库连接并释放所有相关句柄，避免资源泄露。通过调用以下命令实现。

```
SQLDisconnect(dbc);
SQLFreeHandle(SQL_HANDLE_DBC,dbc);
SQLFreeHandle(SQL_HANDLE_ENV, env);
```

下边是完整的示例清单（详见代码清单 odbc\odbc_demo\odbc_conn.cpp）：

```
#001 #if defined (WIN32)
#002     #include <windows.h>
#003 #endif
#004     #include <sql.h>
#005     #include <sqlext.h>
#006     #include <sqltypes.h>
#007
#008 #include <stdlib.h>
#009 #include <stddef.h>
#010 #include <string.h>
#011 #include <stdio.h>
#012
#013
#014
#015 #include <iostream>
```

```
#016
#017 #define IS_SQL_ERR    !IS_SQL_OK
#018 #define IS_SQL_OK(res) (res==SQL_SUCCESS_WITH_INFO || res==SQL_SUCCESS)
#019
#020 static    bool ODBCError(SQLSMALLINT hType, SQLHANDLE hHandle)
#021 {
#022     SQLCHAR       sqlstate[64];
#023     SQLCHAR       errmsg[SQL_MAX_MESSAGE_LENGTH];
#024     SQLCHAR       msgbuf[SQL_MAX_MESSAGE_LENGTH*2];
#025     SQLINTEGER    msgnumber;
#026     SQLSMALLINT actualmsglen;
#027     int           connected = 1;
#028     int           rc;
#029     int i=1;
#030     rc = SQLGetDiagRec(hType, hHandle, (SQLSMALLINT)i++, (SQLCHAR *)sqlstate, &msgnumber, (SQLCHAR *)errmsg,
#031             (SQLSMALLINT)(SQL_MAX_MESSAGE_LENGTH - 1), &actualmsglen);
#032     if (rc == SQL_SUCCESS)
#033     {
#034         sprintf((char *)msgbuf, "ODBCError: SQLError: %d, SQLState: %s, %s.\n", msgnumber, sqlstate, errmsg);
#035         std::cout<<msgbuf<<std::endl;
#036         std::cout.flush();
#037         connected = !( ((sqlstate[0] == '0') && (sqlstate[1] == '8'))
#038                     || (strcmp((const char *)sqlstate, "00000")) );
#039     }
#040     else
#041     {
#042         return false;
#043     }
#044     return true;
#045 }
#046
#047 static const char* DSN_ASE = "iihero_ase";
#048 static const char* USER = "spring";
#049 static const char* PASSWD = "spring1";
#050
#051 static const char* HOST = "sean-laptop";
#052
#053
#054 void quitODBC(SQLHENV henv, SQLHDBC hdbc)
#055 {
#056     SQLDisconnect(hdbc);
#057     SQLFreeHandle(SQL_HANDLE_DBC,hdbc);
#058     SQLFreeHandle(SQL_HANDLE_ENV, henv);
#059 }
#060
#061 int main()
#062 {
```

```
#063
#064        SQLHENV        env;
#065        SQLHDBC        dbc;
#066        SQLHSTMT       stmt;
#067        SQLRETURN      retcode;
#068        SQLCHAR        selectstmt[] = "select sno, sname from student ";
#069        long sno;
#070        char sname[33];
#071
#072        // 初始化环境句柄
#073        retcode = SQLAllocHandle( SQL_HANDLE_ENV, SQL_NULL_HANDLE, &env );
#074        if (IS_SQL_ERR(retcode))
#075        {
#076            std::cerr<<"Env create failed"<<std::endl;
#077            exit(-1);
#078        }
#079        // 设置 ODBC 为 ODBC 3.0 兼容
#080        retcode = SQLSetEnvAttr( env, SQL_ATTR_ODBC_VERSION, (void*)SQL_OV_ODBC3, 0);
#081        if ( IS_SQL_ERR(retcode) )
#082        {
#083            ODBCError(SQL_HANDLE_ENV, env);
#084            SQLFreeHandle(SQL_HANDLE_ENV, env);
#085            exit(-1);
#086        }
#087
#088
#089        // 初始化连接句柄
#090        if ( IS_SQL_ERR(SQLAllocHandle( SQL_HANDLE_DBC, env, &dbc )) )
#091        {
#092            ODBCError(SQL_HANDLE_ENV, env);
#093            SQLFreeHandle(SQL_HANDLE_ENV, env);
#094            exit(-1);
#095        }
#096
#097        // 设置连接等待时间
#098        SQLSetConnectOption(dbc, SQL_LOGIN_TIMEOUT, 30);
#099
#100        // 连接数据库
#101        retcode = SQLConnect( dbc,(SQLCHAR*) DSN_ASE, SQL_NTS,
#102                              (SQLCHAR*)USER, SQL_NTS, (SQLCHAR*)PASSWD, SQL_NTS );
#103        if ( IS_SQL_ERR(retcode ) )
#104        {
#105            ODBCError(SQL_HANDLE_DBC, dbc);
#106            SQLFreeHandle(SQL_HANDLE_DBC, dbc);
#107            SQLFreeHandle(SQL_HANDLE_ENV, env);
#108            exit(-1);
#109        }
#110
```

```
#111        std::cout<<"Connect to DSN: "<<DSN_ASE<<" successfully!"<<std::endl;
#112        retcode = SQLAllocHandle(SQL_HANDLE_STMT, dbc, &stmt);
#113        if ( IS_SQL_ERR(retcode) )
#114        {
#115            ODBCError(SQL_HANDLE_DBC, dbc);
#116            quitODBC(env, dbc);
#117            exit(-1);
#118        }
#119
#120        // 对 student 表: sno(int), sname varchar(32)进行绑定
#121        long cbVal;
#122        SQLBindCol(stmt, 1, SQL_C_LONG, &sno, sizeof(sno), &cbVal);
#123        SQLBindCol(stmt, 2, SQL_C_CHAR, &sname, sizeof(sname), &cbVal);
#124        retcode = SQLExecDirect(stmt,   selectstmt, SQL_NTS);
#125        if ( IS_SQL_ERR(retcode) )
#126        {
#127            ODBCError(SQL_HANDLE_DBC, dbc);
#128            SQLFreeHandle(SQL_HANDLE_STMT, stmt);
#129            quitODBC(env, dbc);
#130            exit(-1);
#131        }
#132        // 获取数据
#133        while (SQL_NO_DATA != SQLFetch(stmt))
#134        {
#135            std::cout<<"sno: "<<sno<<", "<<"sname: "<<sname<<std::endl;
#136        }
#137
#138        SQLFreeHandle(SQL_HANDLE_STMT, stmt);
#139
#140        // 断开连接
#141        quitODBC(env, dbc);
#142        std::cout<<"disconnect from the database"<<std::endl;
#143
#144        return 0;
#145    }
```

第 112～137 行描述的是 SQL 查询获取结果集的过程，首先通过连接句柄 DBC 来创建语句句柄 stmt，通过调用 SQLBindCol 对列 sno 和 sname 分别进行绑定，sno 为整型，sname 为 varchar(32)，需要 char[33]对其进行绑定，多余的一个字符为字符串结束符。接着调用 SQLExecDirect 来执行 select 查询：select sno, name from student。通过 SQLFetch 的返回值来判断结果集中是否还有记录行，从而获取所有的记录行。

15.3.2 配置及编译运行

如何让 odbc_conn.cpp 编译并运行起来呢？我们分两个平台分别进行介绍：

在 Windows 下，直接编译示例中的 odbc_demo.dsp 即可，或者在导入 Sybase.bat 及 vcvars32.bat 中的环境变量以后运行：

cl /DDEBUG=1 /D_DEBUG=1 /DNET_DEBUG=1 /Od /Z7 /MD /GX /nologo /DWIN32 odbc32.lib odbccp32.lib msvcrt.lib odbc_conn.cpp /link /out:odbc_conn.exe

即可得到可执行程序 odbc_conn.exe。这里的编译器你可以采用 VC6、VC7、VC8 任一版本均可。

在 Linux 下，以 Ubuntu 9.0.4 为例（其他发行版 Linux 与此类似）。

1. 安装 unixODBC 及 unixODBC-dev

```
sudo apt-get install unixodbc
sudo apt-get install unixodbc-dev
```

unixODBC-dev 主要提供开发 ODBC 程序要用到的头文件以及共享库。

我们甚至可以安装 unixODBC 的图形管理界面：

```
sudo apt-get install unixodbc-bin
```

2. 在 unixODBC 下配置 ODBC 数据源

假设你的 Sybase ASE 数据库对应的操作系统用户为 iihero，环境变量文件 SYBASE.sh 已经导入到该用户下，要使 ASE 的 ODBC 库结合 unixODBC 使用，需要 iihero 用户对/etc/odbc.ini 及 /etc/odbcinst.ini 都有读写权限。

设置好权限以后，将$SYBASE/DataAccess/ODBC/samples/drivertemplate.txt 里边的内容复制到 /etc/odbcinst.ini，并修改成合适的 ODBC 驱动库登记信息：

```
[Adaptive Server Enterprise]
Description     = Sybase ODBC Driver
Driver          = /home/iihero/ase15/DataAccess/ODBC/lib/libsybdrvodb.so
FileUsage       = -1
UsageCount      = 1
```

[Adaptive Server Enterprise]是 ODBC 驱动名称，这是 ASE 15.0 当中 ODBC 驱动的标准名。Driver 项的值是 ASE ODBC 的实现库的全路径。

在配置好驱动以后，要配置好目标数据源，即要到 odbc.ini 中添加适当的数据源信息：

```
[iihero_ase]
Description     = Sybase ODBC Data Source
UserID          = spring
Password        = spring1
Driver          = Adaptive Server Enterprise
Server          = 192.168.1.3
Port            = 5000
Database        = iihero
UseCursor       = 1
~
```

其中[iihero_ase]里登记的是数据源名，UserID 为用户名，Password 为密码，Driver 为 ODBC 驱动名，而不是驱动库的全路径。Server 为要连接的数据库服务器主机名或者 IP 地址，Port 为服务器上数据库服务的端口号，Database 为目标数据库名，UseCursor 标识为是否允许使用游标。

上述两个文件的生成也可以通过运行 ODBCConfig 命令打开图形界面来进行配置，如图 15-4 所示。

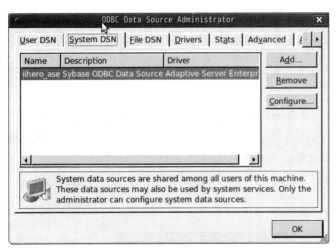

图 15-4　unixODBC 中的 ASE 连接配置

验证 ODBC 数据源是否配置成功，可以运行/usr/bin/isql 命令（注意与 ASE 提供的 isql 命令区分开）：

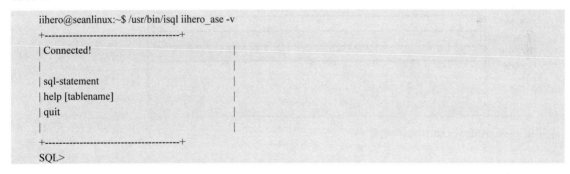

3. 编译可执行程序

在 Linux 的 Sybase 用户下边，在已经获取到 Sybase 数据库相关环境变量的前提下运行：

g++ -g odbc_conn.cpp -lodbc -lodbcinst -o odbc_conn

即可得到可执行文件。如果没有将 SYBASE.sh 导入 iihero 用户下，可以在命令行下边运行 source SYBASE.sh，然后再执行上述编译动作，在后边执行程序时，依然要导入上述环境变量。

4. 执行结果

unixODBC 是怎么找到/etc/odbc.ini 的呢？默认情况下，并不是查找这个路径下边的 odbc.ini。当没有指定 ODBCINI 环境变量时，unixODBC 将从当前路径下边查找 odbc.ini。

因此，在 iihero 的用户环境变量里，可以将 ODBCINI 设成/etc/odbc.ini。

```
iihero@seanlinux:~/notes/odbc_demo$ ./odbc_conn
Connect to DSN: iihero_ase successfully!
sno: 2007002, sname: 刘晨
sno: 2007003, sname: 王敏
sno: 2007004, sname: 张铁林
```

由于 ODBC 库本身的独立性，它自身完全可以脱离于 ASE 的环境运行，为了证明这个结论，我们可以做如下实验：

1. 将 Sybase ASE 的 ODBC 驱动库从/home/iihero/ase15/DataAccess/ODBC 复制到/home/iihero/ODBC，并将/home/iihero/ase15/DataAccess/ODBC 重命名为/home/iihero/ase15/DataAccess/ODBC.orig，这样确保以前的 ODBC 无法找到 ODBC.orig 中的运行时库。

2. 重新配置一个新的 ODBC 驱动，名为 ASE，在 iihero 用户下边运行 ODBC Config，如图 15-5 所示。

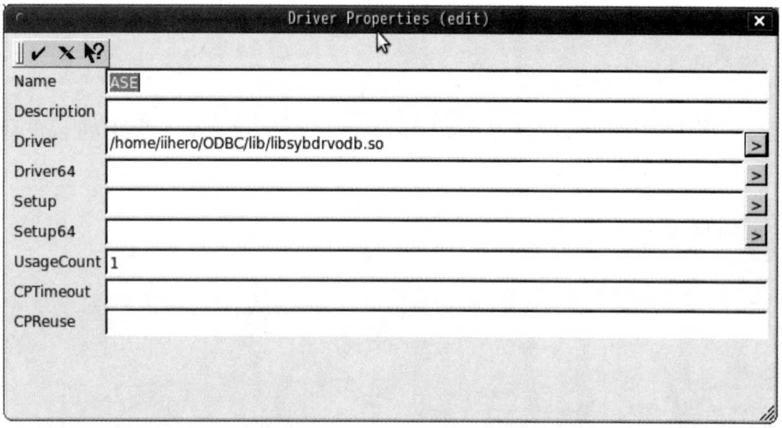

图 15-5　ASE 的 ODBC 驱动配置

然后保存。

3. 配置基于驱动 ASE 的数据源 iihero_ase。

直接编辑/etc/odbc.ini：

```
[iihero_ase2]
Description     = Sybase ODBC Data Source
UserID          = spring
Password        = spring1
```

```
Driver          = ASE
Server          = 192.168.1.3
Port            = 5000
Database        = iihero
UseCursor       = 1~
~
```

4. 去掉环境变量 Sybase 的影响，运行/usr/bin/isql 来验证新的数据源。

```
iihero@seanlinux:~/notes/odbc_demo$ export SYBASE=
iihero@seanlinux:~/notes/odbc_demo$ /usr/bin/isql iihero_ase2 -v
+---------------------------------------+
| Connected!                            |
|                                       |
| sql-statement                         |
| help [tablename]                      |
| quit                                  |
|                                       |
+---------------------------------------+
SQL>
```

在 Windows 下验证这个结论就更简单了，直接将%SYBASE%下的 Open Client 库目录重命名，而程序依然可以独立运行。ODBC 整个实现库的大小总共才几兆，比 Open Client 库要小得多。同时由于都使用 TDS 网络协议，我们有理由相信，两者效率上相差无几。

在这点上，ASE 比 Oracle 要来得轻便。使用过 Oracle ODBC 的人都知道，至少需要装一个客户端，哪怕是后来的 Oracle Instance Client，也是一个庞然大物。

如果要发布 ASE ODBC 应用，Windows 下直接将 ODBC 实现库复制出单独一个目录，然后使用前一节介绍的注册表注册方法注册 ODBC 驱动，连同应用一起，可以发布为独立的应用。在 Linux 平台下，可以将应用程序连同 ASE 的 ODBC 库目录连同/etc/odbc.ini, /etc/odbcinst.ini 一起发布。它不需要 ASE 当中其他任何文件就可以独立运行。

15.3.3 一种增强的连接方式

我们注意到，前一节中连接数据库时调用的是 API：

SQLConnect(dbc,(SQLCHAR*) DSN_ASE, SQL_NTS, (SQLCHAR*)USER, SQL_NTS, (SQLCHAR*)PASSWD, SQL_NTS)

这里需要间接指定一个 ODBC 数据源名，也就是说，需要预先使用 ODM 创建好一个数据源。我们完全可以调用另一个增强的 API：

```
SQLRETURN SQLDriverConnect(
    SQLHDBC         connectionHandle,
    SQLHWND         windowHandle,
    SQLCHAR         * connectStrIn,
```

```
SQLCHAR        * connectStrOut,
SQLSMALLINT    connectStrOutMax,
SQLSMALLINT *  requiredBytes,
SQLUSMALLINT   driverCompletion )
```

第一个参数是连接句柄，第二个参数就有些特殊了，它是 SQLHWND 类型，意味着传进来的是一个 Windows 窗口句柄，可以提示用户输入更多的连接参数，尤其是在带图形界面的用户程序当中。第三个参数是输入的连接字符串，第四个参数为其容量，接下来的三个参数（connectStrOut、connectStrOutMax 和 requiredBytes）则用于返回一个完整的连接串给客户端程序，如果你成功地连接到了数据库，则该函数调用会生成以 null 结尾的完整的连接串信息。

如果输出缓冲区的长度不够，连接串将被截短，并且返回实际需要的长度给*requiredBytes。最后一个参数在你提供的连接串参数不完整时，会作一些额外的补救动作，它有如下有效值：

1. SQL_DRIVER_PROMPT：用户将看到一个连接对话框，即使实际上并不需要。
2. SQL_DRIVER_COMPLETE：如果连接串不完整，将看到连接对话框，提示用户输入所有必需的和可选的连接属性。
3. SQL_DRIVER_COMPLETE_REQUIRED：如果连接串不完整，将看到连接对话框，提示用户输入所有必需的连接属性。
4. SQL_DRIVER_NOPROMPT：如果连接串不完整，将返回 SQL_ERROR。

需要说明的是，如果第二个参数 WindowHandle 为 NULL 时，因为没有合适的窗口句柄，将不会弹出连接对话框。

比如在使用了 SQLDriverConnect 调用以后，将生成如下连接字符串：

SQLDriverConnect: DRIVER={Adaptive Server Enterprise};UID=spring;PWD=spring1;server=sean-laptop;port=5000;db=iihero;

相应修改的代码片段如下：

```
char connstr_out[1024];
    sprintf(connstr, "Driver={Adaptive Server Enterprise};server=%s;port=%d;uid=%s;pwd=%s;db=%s",
        HOST, 5000, USER, PASSWD, "iihero");
    short result;
    retcode = SQLDriverConnect(dbc, NULL, (SQLCHAR*)connstr, sizeof(connstr), (SQLCHAR*)connstr_out, sizeof(connstr_out), &result, SQL_DRIVER_NOPROMPT);
    if ( IS_SQL_ERR(retcode ) )
    {
        ODBCError(SQL_HANDLE_DBC, dbc);
        SQLFreeHandle(SQL_HANDLE_DBC, dbc);
        SQLFreeHandle(SQL_HANDLE_ENV, env);
        exit(-1);
    }
    std::cout<<"SQLDriverConnect: "<<connstr_out<<std::endl;
```

驱动 Adaptive Server Enterprise 的 ODBC 连接中最常用的是以下 5 个参数（参数名不区分大小写）：

- UID 或 UserID，用于表示连接 ASE 的用户名。
- PWD 或 Password，连接 ASE 的密码（当然，如果该用户没有密码，则可以不提供此参数）。
- Server，ASE 服务器的主机名或者 IP 地址。
- Port，ASE 服务器的端口号。
- Database，想要连接的数据库名。

我们同时也要注意以下几个有用的连接参数：

- UseCursor，默认值为 0，表示是否使用游标（是为 1，否为 0）。
- Charset，连接将要使用的字符集，当然 ASE 服务器必须安装了此字符集。
- Language，用于指定 ASE 返回错误消息使用的语言，默认为英文。
- DynamicPrepare，是否支持动态绑定，当设为 1 时，调用 SQLPrepare 则会为相同的查询提升性能。
- LoginTimeOut，连接返回之前等待的时间，单位为秒，默认值为 10。这个值也可以通过连接句柄直接设置连接属性。在我们的示例中设定的连接超时为 30 秒。
- TextSize，TEXT/IMAGE 字段最大传输长度，默认值为 32KB，所以要想传超过这个大小的 TEXT/IMAGE 字段值，必须将此参数值设为稍大一点的值，比如，如果要存储的图片最大为 1.2MB，你可以将其设为 1024 × 1024 × 2 = 2097152。

15.4 错误处理

在使用 ODBC API 时，一个很重要的地方就是错误处理。针对每个 API 调用都有一个返回值，成功的返回值通常是 SQL_SUCCESS_WITH_INFO 或者 SQL_SUCCESS。

因此，我们可以初步定义如下宏：

```
#define IS_SQL_ERR   !IS_SQL_OK
#define IS_SQL_OK(res) (res==SQL_SUCCESS_WITH_INFO || res==SQL_SUCCESS)
```

ODBC API 3.0 里提供了标准的错误诊断 API：

```
SQLRETURN SQLGetDiagRec(
    SQLSMALLINT       HandleType,
    SQLHANDLE         Handle,
    SQLSMALLINT       RecNumber,
    SQLCHAR *         Sqlstate,
    SQLINTEGER *      NativeErrorPtr,
    SQLCHAR *         MessageText,
    SQLSMALLINT       BufferLength,
    SQLSMALLINT *     TextLengthPtr);
```

参数说明：

- HandleType，要诊断的句柄类型，通常有如下类型：SQL_HANDLE_ENV、SQL_HANDLE_DBC、SQL_HANDLE_STMT、SQL_HANDLE_DESC，它们分别是环境句柄、连接句柄、语句句柄以及描述句柄。
- Handle，实际的句柄结构，与第一个参数相对应。
- Sqlstate，状态记录，记录号从 1 计起，它是输出变量。
- NativeError，返回原始错误码的缓冲，也是输出变量，指向一个整数。
- MessageText，诊断出的错误消息的缓冲，是输出变量。
- BufferLength，用户指定的*MessageText 缓冲区长度。
- TextLengthPtr，实际长度的缓冲，指向一个整数，返回*MessageText 缓冲里实际的消息长度，不包括结束符 "\0"，如果返回的实际长度比 BufferLength 要长，则返回的消息会被截短。

返回值为 SQL_SUCCESS,SQL_SUCCESS_WITH_INFO,SQL_ERROR,SQL_INVALID_HANDLE。

下面是使用 SQLGetDiagRec 进行错误处理的示例：

```
#define IS_SQL_ERR    !IS_SQL_OK
#define IS_SQL_OK(res) (res==SQL_SUCCESS_WITH_INFO || res==SQL_SUCCESS)

static    bool ODBCError(SQLSMALLINT hType, SQLHANDLE hHandle)
{
    SQLCHAR       sqlstate[64];
    SQLCHAR       errmsg[SQL_MAX_MESSAGE_LENGTH];
    SQLCHAR       msgbuf[SQL_MAX_MESSAGE_LENGTH*2];
    SQLINTEGER    msgnumber;
    SQLSMALLINT   actualmsglen;
    int           connected = 1;
    int           rc;
    int i=1;
    do
    {
        rc = SQLGetDiagRec(hType, hHandle, (SQLSMALLINT)i++, (SQLCHAR *)sqlstate, &msgnumber, (SQLCHAR *)errmsg,
            (SQLSMALLINT)(SQL_MAX_MESSAGE_LENGTH - 1), &actualmsglen);
        if (rc == SQL_SUCCESS)
        {
            sprintf((char *)msgbuf, "ODBCError: SQLError: %d, SQLState: %s, %s.\n", msgnumber, sqlstate, errmsg);
            std::cout<<msgbuf<<std::endl;
            std::cout.flush();
            connected = !( (((sqlstate[0] == '0') && (sqlstate[1] == '8'))
                        || (strcmp((const char *)sqlstate, "00000")) );
        }
        else
```

```
            {
                return false;
            }
    } while ( (rc != SQL_INVALID_HANDLE && rc != SQL_ERROR) && (connected == 1) );
    return true;
}
```

只要此 API 的返回值不是 SQL_INVALID_HANDLE 和 SQL_ERROR，则不断循环提取错误消息状态码和详细的错误消息。这里加了一层判断，即它是在处于连接的状态下进行循环的，如果不是连接状态，则只调用一次。

这样，针对环境句柄，我们可以调用：

```
if(IS_SQL_ERR(SQLAllocHandle( SQL_HANDLE_ENV, SQL_NULL_HANDLE, &env ))
{
    ODBCError(SQL_HANDLE_ENV, env);
    SQLFreeHandle(SQL_HANDLE_ENV, env);
    exit(-1);
}
```

针对连接成功以后的连接句柄，可以调用：

```
if(IS_SQL_ERR (SQLConnect( dbc,(SQLCHAR*) DSN_ASE, SQL_NTS,
                (SQLCHAR*)USER, SQL_NTS, (SQLCHAR*)PASSWD, SQL_NTS ))
{
    ODBCError(SQL_HANDLE_DBC, dbc);
    SQLFreeHandle(SQL_HANDLE_DBC, dbc);
    SQLFreeHandle(SQL_HANDLE_ENV, env);
    exit(-1);
}
```

针对语句句柄，可以调用：

```
if(IS_SQL_ERR (SQLExecDirect(stmt,  selectstmt, SQL_NTS)))
{
    ODBCError(SQL_HANDLE_DBC, dbc);
    SQLFreeHandle(SQL_HANDLE_STMT, stmt);
    SQLFreeHandle(SQL_HANDLE_DBC, dbc);
    SQLFreeHandle(SQL_HANDLE_ENV, env);
    exit(-1)
}
```

当然还有一种省略的形式，我们也可以使用，即函数 SQLError：

SQLError(henv, hdbc, hstmt, szSqlState, pfNativeError, szErrorMsg, cbErrorMsgMax, pcbErrorMsg)

它完全等效于：

SQLGetDiagRec(HandleType, Handle, RecNumber, szSqlstate, pfNativeErrorPtr, szErrorMsg, cbErrorMsgMax, pcbErrorMsg)

其中，HandleType 值分别是 SQL_HANDLE_ENV、SQL_HANDLE_DBC 或者 SQL_HANDLE_STMT，Handle 值分别设为这三种类型的句柄，RecNumber 则由 ODBC 的实现决定，这种方式似乎更精简。我们可以将 ODBCError 的实现改为如下形式：

```
static   bool ODBCError(SQLHANDLE env, SQLHANDLE dbc, SQLHANDLE stmt)
{
    SQLCHAR         sqlstate[64];
    SQLCHAR         errmsg[SQL_MAX_MESSAGE_LENGTH];
    SQLCHAR         msgbuf[SQL_MAX_MESSAGE_LENGTH*2];
    SQLINTEGER      msgnumber;
    SQLSMALLINT actualmsglen;
    int             connected = 1;
    int             rc;
    int i=1;
    do
    {
        rc = SQLError(env, dbc, stmt, (SQLCHAR *)sqlstate, &msgnumber, (SQLCHAR *)errmsg,
            (SQLSMALLINT)(SQL_MAX_MESSAGE_LENGTH - 1), &actualmsglen);
        if (rc == SQL_SUCCESS)
        {
            sprintf((char *)msgbuf, "ODBCError: SQLError: %d, SQLState: %s, %s.\n", msgnumber, sqlstate, errmsg);
            std::cout<<msgbuf<<std::endl;
            std::cout.flush();
            connected = !( ((sqlstate[0] == '0') && (sqlstate[1] == '8'))
                        || (strcmp((const char *)sqlstate, "00000")) );
        }
        else
        {
            return false;
        }
    } while ( ( rc != SQL_INVALID_HANDLE && rc != SQL_ERROR) && (connected == 1) );
    return true;

}
```

完整的代码清单见 odbc/odbc_error：

```
#001 #if defined (WIN32)
#002     #include <windows.h>
#003 #endif
#004     #include <sql.h>
#005     #include <sqlext.h>
#006     #include <sqltypes.h>
#007
#008 #include <stdlib.h>
#009 #include <stddef.h>
#010 #include <string.h>
```

```
#011 #include <stdio.h>
#012
#013
#014
#015 #include <iostream>
#016
#017 #define IS_SQL_ERR !IS_SQL_OK
#018 #define IS_SQL_OK(res) (res==SQL_SUCCESS_WITH_INFO || res==SQL_SUCCESS)
#019
#020 static    bool ODBCError(SQLHANDLE env, SQLHANDLE dbc, SQLHANDLE stmt)
#021 {
#022      SQLCHAR      sqlstate[64];
#023      SQLCHAR      errmsg[SQL_MAX_MESSAGE_LENGTH];
#024      SQLCHAR      msgbuf[SQL_MAX_MESSAGE_LENGTH*2];
#025      SQLINTEGER   msgnumber;
#026      SQLSMALLINT actualmsglen;
#027      int          connected = 1;
#028      int          rc;
#029      int i=1;
#030      do
#031      {
#032          rc = SQLError(env, dbc, stmt, (SQLCHAR *)sqlstate, &msgnumber, (SQLCHAR *)errmsg,
#033              (SQLSMALLINT)(SQL_MAX_MESSAGE_LENGTH - 1), &actualmsglen);
#034          if (rc == SQL_SUCCESS)
#035          {
#036              sprintf((char *)msgbuf, "ODBCError: SQLError: %d, SQLState: %s, %s.\n", msgnumber, sqlstate, errmsg);
#037              std::cout<<msgbuf<<std::endl;
#038              std::cout.flush();
#039              connected = !( (((sqlstate[0] == '0') && (sqlstate[1] == '8'))
#040                  || (strcmp((const char *)sqlstate, "00000")) );
#041          }
#042          else
#043          {
#044              return false;
#045          }
#046      } while ( (rc != SQL_INVALID_HANDLE && rc != SQL_ERROR) && (connected == 1) );
#047      return true;
#048
#049 }
#050
#051 static const char* DSN_ASE = "iihero_ase";
#052 static const char* USER = "spring";
#053 static const char* PASSWD = "spring1";
#054
#055 static const char* HOST = "sean-laptop";
#056
#057
#058 void quitODBC(SQLHENV henv, SQLHDBC hdbc)
```

```
#059 {
#060     SQLDisconnect(hdbc);
#061     SQLFreeHandle(SQL_HANDLE_DBC,hdbc);
#062     SQLFreeHandle(SQL_HANDLE_ENV, henv);
#063 }
#064
#065 int main()
#066 {
#067
#068     SQLHENV    env;
#069     SQLHDBC    dbc;
#070     SQLHSTMT   stmt;
#071     SQLRETURN      retcode;
#072     SQLCHAR    selectstmt[] = "select sno, sname from student ";
#073     long sno;
#074     char sname[33];
#075
#076     // 初始化环境句柄
#077     retcode = SQLAllocHandle( SQL_HANDLE_ENV, SQL_NULL_HANDLE, &env );
#078     if (IS_SQL_ERR(retcode))
#079     {
#080         std::cerr<<"Env create failed"<<std::endl;
#081         exit(-1);
#082     }
#083     // 设置 ODBC 为 ODBC 3.0 兼容
#084     retcode = SQLSetEnvAttr( env, SQL_ATTR_ODBC_VERSION, (void*)SQL_OV_ODBC3, 0);
#085     if ( IS_SQL_ERR(retcode) )
#086     {
#087         ODBCError(env, NULL, NULL);
#088         SQLFreeHandle(SQL_HANDLE_ENV, env);
#089         exit(-1);
#090     }
#091
#092
#093     // 初始化连接句柄
#094     if ( IS_SQL_ERR(SQLAllocHandle( SQL_HANDLE_DBC, env, &dbc )) )
#095     {
#096         ODBCError(env, NULL, NULL);
#097         SQLFreeHandle(SQL_HANDLE_ENV, env);
#098         exit(-1);
#099     }
#100
#101     // 设置连接等待时间
#102     SQLSetConnectOption(dbc, SQL_LOGIN_TIMEOUT, 30);
#103
#104     // 连接数据库
#105     //retcode = SQLConnect( dbc,(SQLCHAR*) DSN_ASE, SQL_NTS,
#106     //  (SQLCHAR*)USER, SQL_NTS, (SQLCHAR*)PASSWD, SQL_NTS );
```

```
#107        // or
#108        char connstr[1024];
#109        char connstr_out[1024];
#110        sprintf(connstr, "Driver={Adaptive Server Enterprise};server=%s;port=%d;uid=%s;pwd=%s;db=%s",
#111            HOST, 5000, USER, PASSWD,  "iihero");
#112        short result;
#113        retcode = SQLDriverConnect( dbc, NULL, (SQLCHAR*)connstr, sizeof(connstr), (SQLCHAR*)connstr_out, sizeof(connstr_out), &result, SQL_DRIVER_NOPROMPT);
#114        if ( IS_SQL_ERR(retcode) )
#115        {
#116            ODBCError(env, dbc, NULL);
#117            SQLFreeHandle(SQL_HANDLE_DBC, dbc);
#118            SQLFreeHandle(SQL_HANDLE_ENV, env);
#119            exit(-1);
#120        }
#121        std::cout<<"SQLDriverConnect: "<<connstr_out<<std::endl;
#122
#123        std::cout<<"Connect to DSN: "<<DSN_ASE<<" successfully!"<<std::endl;
#124        retcode = SQLAllocHandle(SQL_HANDLE_STMT, dbc, &stmt);
#125        if ( IS_SQL_ERR(retcode) )
#126        {
#127            ODBCError(env, dbc, NULL);
#128            quitODBC(env, dbc);
#129            exit(-1);
#130        }
#131
#132        // 对 student 表: sno(int), sname varchar(32)进行绑定
#133        long cbVal;
#134        SQLBindCol(stmt, 1, SQL_C_LONG, &sno, sizeof(sno), &cbVal);
#135        SQLBindCol(stmt, 2, SQL_C_CHAR, &sname, sizeof(sname), &cbVal);
#136        retcode = SQLExecDirect(stmt,   selectstmt, SQL_NTS);
#137        if ( IS_SQL_ERR(retcode) )
#138        {
#139            ODBCError(env, dbc, stmt);
#140            SQLFreeHandle(SQL_HANDLE_STMT, stmt);
#141            quitODBC(env, dbc);
#142            exit(-1);
#143        }
#144        // 获取数据
#145        while (SQL_NO_DATA != SQLFetch(stmt))
#146        {
#147            std::cout<<"sno: "<<sno<<", "<<"sname: "<<sname<<std::endl;
#148        }
#149
#150        SQLFreeHandle(SQL_HANDLE_STMT, stmt);
#151
#152        // 断开连接
#153        quitODBC(env, dbc);
```

```
#154        std::cout<<"disconnect from the database"<<std::endl;
#155
#156        return 0;
#157 }
```

第二种形式的 ODBCError 的好处就是不用记住句柄类型,只要按顺序传入环境句柄、连接句柄和语句句柄即可,如果没有使用该类型的句柄,传入 NULL 即可。

15.5　一个 CRUD 的综合示例

本节通过一个综合性的示例来介绍如何使用 ODBC 访问 ASE 中的表。

```
#001 #if defined (WIN32)
#002        #include <windows.h>
#003 #endif
#004        #include <sql.h>
#005        #include <sqlext.h>
#006        #include <sqltypes.h>
#007
#008 #include <stdlib.h>
#009 #include <stddef.h>
#010 #include <string.h>
#011 #include <stdio.h>
#012
#013 #include <iostream>
#014
#015 #define IS_SQL_ERR   !IS_SQL_OK
#016 #define IS_SQL_OK(res) (res==SQL_SUCCESS_WITH_INFO || res==SQL_SUCCESS)
#017
#018
#019
#020 static const char* DSN_ASE = "iihero_ase";
#021 static const char* USER = "spring";
#022 static const char* PASSWD = "spring1";
#023
#024 static const char* HOST = "192.168.56.1";
#025
#026 static SQLHENV henv = NULL;
#027 static SQLHDBC hdbc = NULL;
#028
#029 static   bool ODBCError(SQLHANDLE env, SQLHANDLE dbc, SQLHANDLE stmt)
#030 {
#031        SQLCHAR      sqlstate[64];
#032        SQLCHAR      errmsg[SQL_MAX_MESSAGE_LENGTH];
#033        SQLCHAR      msgbuf[SQL_MAX_MESSAGE_LENGTH*2];
#034        SQLINTEGER   msgnumber;
```

```
#035        SQLSMALLINT actualmsglen;
#036        int         connected = 1;
#037        int         rc;
#038        int i=1;
#039        do
#040        {
#041            rc = SQLError(env, dbc, stmt, (SQLCHAR *)sqlstate, &msgnumber, (SQLCHAR *)errmsg,
#042                (SQLSMALLINT)(SQL_MAX_MESSAGE_LENGTH - 1), &actualmsglen);
#043            if (rc == SQL_SUCCESS)
#044            {
#045                sprintf((char *)msgbuf, "ODBCError: SQLError: %d, SQLState: %s, %s.\n", msgnumber, sqlstate, errmsg);
#046                std::cout<<msgbuf<<std::endl;
#047                std::cout.flush();
#048                connected = !( (((sqlstate[0] == '0') && (sqlstate[1] == '8'))
#049                        || (strcmp((const char *)sqlstate, "00000")) );
#050            }
#051            else
#052            {
#053                return false;
#054            }
#055        } while ( (rc != SQL_INVALID_HANDLE && rc != SQL_ERROR) && (connected == 1) );
#056        return true;
#057
#058 }
#059
#060 static void quitODBC(SQLHENV env, SQLHDBC dbc)
#061 {
#062        SQLDisconnect(dbc);
#063        SQLFreeHandle(SQL_HANDLE_DBC,dbc);
#064        SQLFreeHandle(SQL_HANDLE_ENV, env);
#065        std::cout<<"quit ODBC connection ....."<<std::endl;
#066 }
#067
#068 // env and dbc handler 初始化
#069 static void connect(SQLHENV& env, SQLHDBC& dbc)
#070 {
#071        SQLRETURN      retcode;
#072
#073        // 初始化环境句柄
#074        retcode = SQLAllocHandle( SQL_HANDLE_ENV, SQL_NULL_HANDLE, &env );
#075        if (IS_SQL_ERR(retcode))
#076        {
#077            std::cerr<<"Env create failed"<<std::endl;
#078            exit(-1);
#079        }
#080        // 设置 ODBC 为 ODBC 3.0 兼容
#081        retcode = SQLSetEnvAttr( env, SQL_ATTR_ODBC_VERSION, (void*)SQL_OV_ODBC3, 0);
```

```
#082        if ( IS_SQL_ERR(retcode) )
#083        {
#084            ODBCError(env, NULL, NULL);
#085            SQLFreeHandle(SQL_HANDLE_ENV, env);
#086            exit(-1);
#087        }
#088
#089
#090        // 初始化连接句柄
#091        if ( IS_SQL_ERR(SQLAllocHandle( SQL_HANDLE_DBC, env, &dbc )) )
#092        {
#093            ODBCError(env, NULL, NULL);
#094            SQLFreeHandle(SQL_HANDLE_ENV, env);
#095            exit(-1);
#096        }
#097
#098        // 设置连接等待时间
#099        SQLSetConnectOption(dbc, SQL_LOGIN_TIMEOUT, 30);
#100
#101        // 连接数据库
#102        //retcode = SQLConnect( dbc,(SQLCHAR*) DSN_ASE, SQL_NTS,
#103        //                      (SQLCHAR*)USER, SQL_NTS, (SQLCHAR*)PASSWD, SQL_NTS );
#104        // or
#105        char connstr[1024];
#106        char connstr_out[1024];
#107        sprintf(connstr, "Driver={Adaptive Server Enterprise};server=%s;port=%d;uid=%s;pwd=%s;db=%s",
#108                HOST, 5000, USER, PASSWD,   "iihero");
#109        short result;
#110        retcode = SQLDriverConnect( dbc, NULL, (SQLCHAR*)connstr, sizeof(connstr), (SQLCHAR*)connstr_out, sizeof(connstr_out),  &result, SQL_DRIVER_NOPROMPT);
#111        if ( IS_SQL_ERR(retcode ) )
#112        {
#113            ODBCError(env, dbc, NULL);
#114            SQLFreeHandle(SQL_HANDLE_DBC, dbc);
#115            SQLFreeHandle(SQL_HANDLE_ENV, env);
#116            exit(-1);
#117        }
#118        std::cout<<"SQLDriverConnect: "<<connstr_out<<std::endl;
#119        std::cout<<"Connect to DSN: "<<DSN_ASE<<" successfully!"<<std::endl;
#120 }
#121
#122 struct Multitype
#123 {
#124        int id;
#125        char col2[33];
#126        char col3[33];
#127        char col4[80];   // mapped into numeric
#128        double col5;
```

```
#129        double col6;
#130        float col7;         // mapped into money
#131        TIMESTAMP_STRUCT col8;
#132        unsigned char col9[4];
#133        int col10;
#134        unsigned char* col11;
#135        int col11_len;
#136        char* col12;
#137        int col12_len;
#138 public:
#139        Multitype() : col11(NULL), col12(NULL)
#140        {
#141        }
#142        ~Multitype()
#143        {
#144            if (col12 != NULL) delete[] col12;
#145            if (col11 != NULL) delete[] col11;
#146        }
#147 };
#148
#149 static void checkError(SQLRETURN retcode, SQLHANDLE env, SQLHANDLE dbc, SQLHANDLE stmt)
#150 {
#151        if ( IS_SQL_ERR(retcode) )
#152        {
#153            ODBCError(env, dbc, stmt);
#154            quitODBC(env, dbc);
#155            exit(-1);
#156        }
#157 }
#158
#159 static int createMultitype(SQLHENV env, SQLHANDLE dbc, const Multitype& t)
#160 {
#161        SQLRETURN       retcode;
#162        SQLHSTMT    stmt;
#163        retcode = SQLAllocHandle(SQL_HANDLE_STMT, dbc, &stmt);
#164        if ( IS_SQL_ERR(retcode) )
#165        {
#166            ODBCError(env, dbc, NULL);
#167            quitODBC(env, dbc);
#168            exit(-1);
#169        }
#170        long indVal;
#171        short nCols;
#172        SQLCHAR     insert[] = "insert into multitype_t values(?,?,?,?,?,?,?,?,?,?,?,?) ";
#173        // prepare statement
#174        retcode = SQLPrepare(stmt, insert, SQL_NTS);
#175        checkError(retcode, env, dbc, stmt);
#176
```

```
#177        retcode = SQLNumParams(stmt, &nCols);
#178        checkError(retcode, env, dbc, stmt);
#179        std::cout<<"Binding parameters count = "<<nCols<<std::endl;
#180
#181        // bind parameters (except the blob/clob/binary column)
#182        checkError(SQLBindParameter(stmt, 1, SQL_PARAM_INPUT, SQL_C_LONG, SQL_INTEGER, 0, 0, (void*)
&t.id, 0, SQL_NULL_HANDLE), env, dbc, stmt);
#183        checkError(SQLBindParameter(stmt, 2, SQL_PARAM_INPUT, SQL_C_CHAR, SQL_CHAR,      strlen(t.col2),
0, (void*)&t.col2, sizeof(t.col2), SQL_NULL_HANDLE), env, dbc, stmt);
#184        checkError(SQLBindParameter(stmt, 3, SQL_PARAM_INPUT, SQL_C_CHAR, SQL_VARCHAR, strlen(t.col3),
0, (void*)&t.col3, sizeof(t.col3), SQL_NULL_HANDLE), env, dbc, stmt);
#185            checkError(SQLBindParameter(stmt,    4,  SQL_PARAM_INPUT,   SQL_C_CHAR,  SQL_NUMERIC,
8/*strlen(t.col4)*/, 3, (void*)&t.col4, sizeof(t.col4), SQL_NULL_HANDLE), env, dbc, stmt);
#186        checkError(SQLBindParameter(stmt, 5, SQL_PARAM_INPUT, SQL_C_FLOAT, SQL_FLOAT, 0, 0, (void*)
&t.col5, 0, SQL_NULL_HANDLE), env, dbc, stmt);
#187
#188        checkError(SQLBindParameter(stmt, 6, SQL_PARAM_INPUT, SQL_C_DOUBLE, SQL_DOUBLE, 0, 0, (void*)
&t.col6, 0, SQL_NULL_HANDLE), env, dbc, stmt);
#189        checkError(SQLBindParameter(stmt, 7, SQL_PARAM_INPUT, SQL_C_FLOAT, SQL_FLOAT, 0, 0, (void*)
&t.col7, 0, SQL_NULL_HANDLE), env, dbc, stmt);
#190        checkError(SQLBindParameter(stmt, 8, SQL_PARAM_INPUT, SQL_C_TYPE_TIMESTAMP, SQL_TYPE_
TIMESTAMP, 0, 0, (void*)&t.col8, sizeof(t.col8), SQL_NULL_HANDLE), env, dbc, stmt);
#191
#192        int col9_len = sizeof(t.col9);
#193        SQLINTEGER cb9 = SQL_LEN_DATA_AT_EXEC(col9_len);
#194        checkError(SQLBindParameter(stmt, 9, SQL_PARAM_INPUT, SQL_C_BINARY, SQL_BINARY, 0, 0, (void*)
&t.col9, 0, /*sizeof(t.col9),*/ &cb9), env, dbc, stmt);
#195        checkError(SQLBindParameter(stmt, 10, SQL_PARAM_INPUT, SQL_C_BIT, SQL_INTEGER, 0, 0, (void*)
&t.col10, 0, SQL_NULL_HANDLE), env, dbc, stmt);
#196
#197        SQLINTEGER cb11 = SQL_LEN_DATA_AT_EXEC(t.col11_len);
#198        checkError(SQLBindParameter(stmt, 11, SQL_PARAM_INPUT, SQL_C_BINARY, SQL_LONGVARBINARY, 0,
0, NULL, 0, &cb11), env, dbc, stmt);
#199
#200        SQLINTEGER cb = SQL_LEN_DATA_AT_EXEC(t.col12_len);
#201        checkError(SQLBindParameter(stmt, 12, SQL_PARAM_INPUT, SQL_C_CHAR, SQL_LONGVARCHAR, 0, 0,
NULL, 0, &cb), env, dbc, stmt);
#202
#203        retcode = SQLExecute(stmt);
#204        SQLPOINTER pToken;
#205        int BLOCKSIZE = 6;
#206        if (retcode != SQL_NEED_DATA)
#207        {
#208            checkError(retcode, env, dbc, stmt);
#209        }
#210        else
#211        {
#212            // col9
```

```
#213        if (SQL_NEED_DATA == SQLParamData(stmt, &pToken))
#214        {
#215            SQLPutData(stmt, (void*)t.col9, sizeof(t.col9));
#216        }
#217
#218        // col11
#219        if (SQL_NEED_DATA == SQLParamData(stmt, &pToken))
#220        {
#221            for (int j=0; j<t.col11_len/BLOCKSIZE; j++)
#222            {
#223                SQLPutData(stmt, t.col11 + j*BLOCKSIZE, BLOCKSIZE);
#224            }
#225            int nTmp = t.col11_len % BLOCKSIZE;
#226            if (nTmp > 0)
#227            {
#228                SQLPutData(stmt, t.col11 + j*BLOCKSIZE, nTmp);
#229            }
#230        }
#231        // col12
#232        if ( SQL_NEED_DATA == SQLParamData(stmt, &pToken) )
#233        {
#234            for (int j=0; j<t.col12_len/BLOCKSIZE; j++)
#235            {
#236                SQLPutData(stmt, t.col12 + j*BLOCKSIZE, BLOCKSIZE);
#237            }
#238            int nTmp = t.col12_len % BLOCKSIZE;
#239            if (nTmp > 0)
#240            {
#241                SQLPutData(stmt, t.col12 + j*BLOCKSIZE, nTmp);
#242            }
#243        }
#244        while (SQL_NEED_DATA == SQLParamData(stmt, &pToken))
#245        {
#246        }
#247    }
#248    // checkError(SQLEndTran(SQL_HANDLE_DBC,   dbc , SQL_COMMIT), env, dbc, stmt);
#249    long nRows = 0;
#250    checkError(SQLRowCount(stmt,&nRows), env, dbc, stmt);
#251    std::cout<<"affected row count="<<nRows<<std::endl;
#252
#253    SQLFreeHandle(SQL_HANDLE_STMT, stmt);
#254    return 0;
#255 }
#256
#257 static int updateMultitype(SQLHENV env, SQLHANDLE dbc, const Multitype& t)
#258 {
#259    SQLRETURN    retcode;
#260    SQLHSTMT     stmt;
```

```
#261        retcode = SQLAllocHandle(SQL_HANDLE_STMT, dbc, &stmt);
#262        if ( IS_SQL_ERR(retcode) )
#263        {
#264            ODBCError(env, dbc, NULL);
#265            quitODBC(env, dbc);
#266            exit(-1);
#267        }
#268        long indVal;
#269        short nCols;
#270        SQLCHAR      update_sql[] = "update multitype_t set col2=?,col3=?,col4=?,col5=?,col6=?,col7=?,col8=?,col9=?,col10=?,col11=?,col12=? where id=?";
#271        // prepare statement
#272        retcode = SQLPrepare(stmt, update_sql, SQL_NTS);
#273        checkError(retcode, env, dbc, stmt);
#274
#275        retcode = SQLNumParams(stmt, &nCols);
#276        checkError(retcode, env, dbc, stmt);
#277        std::cout<<"Binding parameters count = "<<nCols<<std::endl;
#278
#279        // bind parameters (except the blob/clob/binary column)
#280
#281        checkError(SQLBindParameter(stmt, 1, SQL_PARAM_INPUT, SQL_C_CHAR, SQL_CHAR,    strlen(t.col2), 0, (void*)&t.col2, sizeof(t.col2), SQL_NULL_HANDLE), env, dbc, stmt);
#282        checkError(SQLBindParameter(stmt, 2, SQL_PARAM_INPUT, SQL_C_CHAR, SQL_VARCHAR, strlen(t.col3), 0, (void*)&t.col3, sizeof(t.col3), SQL_NULL_HANDLE), env, dbc, stmt);
#283        checkError(SQLBindParameter(stmt, 3, SQL_PARAM_INPUT, SQL_C_CHAR, SQL_NUMERIC, 8/*strlen(t.col4)*/, 3, (void*)&t.col4, sizeof(t.col4), SQL_NULL_HANDLE), env, dbc, stmt);
#284        checkError(SQLBindParameter(stmt, 4, SQL_PARAM_INPUT, SQL_C_FLOAT, SQL_FLOAT, 0, 0, (void*)&t.col5, 0, SQL_NULL_HANDLE), env, dbc, stmt);
#285
#286        checkError(SQLBindParameter(stmt, 5, SQL_PARAM_INPUT, SQL_C_DOUBLE, SQL_DOUBLE, 0, 0, (void*)&t.col6, 0, SQL_NULL_HANDLE), env, dbc, stmt);
#287        checkError(SQLBindParameter(stmt, 6, SQL_PARAM_INPUT, SQL_C_FLOAT, SQL_FLOAT, 0, 0, (void*)&t.col7, 0, SQL_NULL_HANDLE), env, dbc, stmt);
#288        checkError(SQLBindParameter(stmt, 7, SQL_PARAM_INPUT, SQL_C_TYPE_TIMESTAMP, SQL_TYPE_TIMESTAMP, 0, 0, (void*)&t.col8, sizeof(t.col8), SQL_NULL_HANDLE), env, dbc, stmt);
#289
#290        int col9_len = sizeof(t.col9);
#291        SQLINTEGER cb9 = SQL_LEN_DATA_AT_EXEC(col9_len);
#292        checkError(SQLBindParameter(stmt, 8, SQL_PARAM_INPUT, SQL_C_BINARY, SQL_BINARY, 0, 0, (void*)&t.col9, 0, /*sizeof(t.col9),*/ &cb9), env, dbc, stmt);
#293        checkError(SQLBindParameter(stmt, 9, SQL_PARAM_INPUT, SQL_C_BIT, SQL_INTEGER, 0, 0, (void*)&t.col10, 0, SQL_NULL_HANDLE), env, dbc, stmt);
#294
#295        SQLINTEGER cb11 = SQL_LEN_DATA_AT_EXEC(t.col11_len);
#296        checkError(SQLBindParameter(stmt, 10, SQL_PARAM_INPUT, SQL_C_BINARY, SQL_LONGVARBINARY, 0, 0, NULL, 0, &cb11), env, dbc, stmt);
#297
```

```
#298        SQLINTEGER cb = SQL_LEN_DATA_AT_EXEC(t.col12_len);
#299        checkError(SQLBindParameter(stmt, 11, SQL_PARAM_INPUT, SQL_C_CHAR, SQL_LONGVARCHAR, 0, 0,
NULL, 0, &cb), env, dbc, stmt);
#300        // the last parameter
#301        checkError(SQLBindParameter(stmt, 12, SQL_PARAM_INPUT, SQL_C_LONG, SQL_INTEGER, 0, 0, (void*)
&t.id, 0, SQL_NULL_HANDLE), env, dbc, stmt);
#302
#303        retcode = SQLExecute(stmt);
#304        SQLPOINTER pToken;
#305        int BLOCKSIZE = 6;
#306        if (retcode != SQL_NEED_DATA)
#307        {
#308            checkError(retcode, env, dbc, stmt);
#309        }
#310        else
#311        {
#312            // col9
#313            if (SQL_NEED_DATA == SQLParamData(stmt, &pToken))
#314            {
#315                SQLPutData(stmt, (void*)t.col9, sizeof(t.col9));
#316            }
#317
#318            // col11
#319            if (SQL_NEED_DATA == SQLParamData(stmt, &pToken))
#320            {
#321                for (int j=0; j<t.col11_len/BLOCKSIZE; j++)
#322                {
#323                    SQLPutData(stmt, t.col11 + j*BLOCKSIZE, BLOCKSIZE);
#324                }
#325                int nTmp = t.col11_len % BLOCKSIZE;
#326                if (nTmp > 0)
#327                {
#328                    SQLPutData(stmt, t.col11 + j*BLOCKSIZE, nTmp);
#329                }
#330            }
#331            // col12
#332            if ( SQL_NEED_DATA == SQLParamData(stmt, &pToken) )
#333            {
#334                for (int j=0; j<t.col12_len/BLOCKSIZE; j++)
#335                {
#336                    SQLPutData(stmt, t.col12 + j*BLOCKSIZE, BLOCKSIZE);
#337                }
#338                int nTmp = t.col12_len % BLOCKSIZE;
#339                if (nTmp > 0)
#340                {
#341                    SQLPutData(stmt, t.col12 + j*BLOCKSIZE, nTmp);
#342                }
#343            }
```

```
#344            while (SQL_NEED_DATA == SQLParamData(stmt, &pToken))
#345            {
#346            }
#347        }
#348        long nRows = 0;
#349        checkError(SQLRowCount(stmt,&nRows), env, dbc, stmt);
#350        std::cout<<"After update, affected row count="<<nRows<<std::endl;
#351
#352        SQLFreeHandle(SQL_HANDLE_STMT, stmt);
#353        return 0;
#354 }
#355
#356 static int deleteMultitype(SQLHENV env, SQLHANDLE dbc, int id)
#357 {
#358        SQLRETURN     retcode;
#359        SQLHSTMT       stmt;
#360        retcode = SQLAllocHandle(SQL_HANDLE_STMT, dbc, &stmt);
#361        if ( IS_SQL_ERR(retcode) )
#362        {
#363            ODBCError(env, dbc, NULL);
#364            quitODBC(env, dbc);
#365            exit(-1);
#366        }
#367        long indVal;
#368        short nCols;
#369        SQLCHAR      sql_delete[] = "delete from multitype_t where id=? ";
#370        // prepare statement
#371        retcode = SQLPrepare(stmt, sql_delete, SQL_NTS);
#372        checkError(retcode, env, dbc, stmt);
#373
#374        retcode = SQLNumParams(stmt, &nCols);
#375        checkError(retcode, env, dbc, stmt);
#376        std::cout<<"Binding parameters count = "<<nCols<<std::endl;
#377
#378        // bind parameters (except the blob/clob/binary column)
#379        checkError(SQLBindParameter(stmt, 1, SQL_PARAM_INPUT, SQL_C_LONG, SQL_INTEGER, 0, 0, (void*)
&id, 0, SQL_NULL_HANDLE), env, dbc, stmt);
#380        checkError(SQLExecute(stmt), env, dbc, stmt);
#381        long nRows = 0;
#382        checkError(SQLRowCount(stmt,&nRows), env, dbc, stmt);
#383        std::cout<<"affected row count="<<nRows<<std::endl;
#384
#385        SQLFreeHandle(SQL_HANDLE_STMT, stmt);
#386
#387        return 0;
#388 }
#389
#390 static int queryMultitype(SQLHENV env, SQLHANDLE dbc, int id)
```

```
#391 {
#392     SQLRETURN     retcode;
#393     SQLHSTMT      stmt;
#394     retcode = SQLAllocHandle(SQL_HANDLE_STMT, dbc, &stmt);
#395     if ( IS_SQL_ERR(retcode) )
#396     {
#397         ODBCError(env, dbc, NULL);
#398         quitODBC(env, dbc);
#399         exit(-1);
#400     }
#401
#402     SQLCHAR       query_sql[] = "select * from multitype_t where id=?";
#403     // prepare statement
#404     retcode = SQLPrepare(stmt, query_sql, SQL_NTS);
#405     checkError(retcode, env, dbc, stmt);
#406     short nCols;
#407     retcode = SQLNumParams(stmt, &nCols);
#408     checkError(retcode, env, dbc, stmt);
#409     std::cout<<"Binding parameters count = "<<nCols<<std::endl;
#410
#411     retcode = SQLNumResultCols(stmt, &nCols);
#412     checkError(retcode, env, dbc, stmt);
#413     std::cout<<"Result column count = "<<nCols<<std::endl;
#414
#415     // bind parameter
#416     checkError(SQLBindParameter(stmt, 1, SQL_PARAM_INPUT, SQL_C_LONG, SQL_INTEGER, 0, 0, (void*)&id, 0, SQL_NULL_HANDLE), env, dbc, stmt);
#417
#418     Multitype t;
#419
#420     t.col11 = new unsigned char[64];
#421     t.col11_len = 64;
#422     memset(t.col11, 0, 64);
#423     t.col12 = new char[64];
#424     t.col12_len = 64;
#425
#426     // bind column
#427     long ind[12] = {0, 0, 0, 0, 0, 0, 0, 0, 0, 0, 0, 0}; int i=0;
#428     checkError(SQLBindCol(stmt, 1, SQL_INTEGER, &t.id, sizeof(t.id), &ind[i++]), env, dbc, stmt);
#429     checkError(SQLBindCol(stmt, 2, SQL_CHAR, &t.col2, sizeof(t.col2), &ind[i++]), env, dbc, stmt);
#430     checkError(SQLBindCol(stmt, 3, SQL_CHAR, &t.col3, sizeof(t.col3), &ind[i++]), env, dbc, stmt);
#431     // use char to bind numeric
#432     checkError(SQLBindCol(stmt, 4, SQL_CHAR, &t.col4, sizeof(t.col4), &ind[i++]), env, dbc, stmt);
#433
#434     checkError(SQLBindCol(stmt, 5, SQL_DOUBLE, &t.col5, sizeof(t.col5), &ind[i++]), env, dbc, stmt);
#435     checkError(SQLBindCol(stmt, 6, SQL_DOUBLE, &t.col6, sizeof(t.col6), &ind[i++]), env, dbc, stmt);
#436     checkError(SQLBindCol(stmt, 7, SQL_DOUBLE, &t.col7, sizeof(t.col7), &ind[i++]), env, dbc, stmt);
#437     checkError(SQLBindCol(stmt, 8, SQL_TYPE_TIMESTAMP, &t.col8, sizeof(t.col8), &ind[i++]), env, dbc, stmt);
```

```
#438        checkError(SQLBindCol(stmt, 9, SQL_BINARY, &t.col9, sizeof(t.col9), &ind[i++]), env, dbc, stmt);
#439        checkError(SQLBindCol(stmt, 10, SQL_INTEGER, &t.col10, sizeof(t.col10), &ind[i++]), env, dbc, stmt);
#440        checkError(SQLBindCol(stmt, 11, SQL_BINARY, t.col11, t.col11_len, &ind[i++]), env, dbc, stmt);
#441        checkError(SQLBindCol(stmt, 12, SQL_CHAR, t.col12, t.col12_len, &ind[i++]), env, dbc, stmt);
#442
#443
#444        retcode = SQLExecute(stmt);
#445        checkError(retcode, env, dbc, stmt);
#446
#447        retcode = SQLFetch(stmt);
#448        while (retcode == SQL_SUCCESS_WITH_INFO || retcode == SQL_SUCCESS)
#449        {
#450            // dump resultset
#451            std::cout<<"id: "<<t.id<<std::endl;
#452            std::cout<<"col2: "<<t.col2<<std::endl;
#453            std::cout<<"col3: "<<t.col3<<std::endl;
#454            std::cout<<"col4: "<<t.col4<<std::endl;
#455
#456            std::cout<<"col5: "<<t.col5<<std::endl;
#457            std::cout<<"col6: "<<t.col6<<std::endl;
#458            std::cout<<"col7: "<<t.col7<<std::endl;
#459            std::cout<<"col8: "<<t.col8.year<<'-'<<t.col8.month<<'-'<<t.col8.day<<std::endl;
#460            std::cout<<"col9: "<<t.col9<<std::endl;
#461            std::cout<<"col10: "<<t.col10<<std::endl;
#462            std::cout<<"col11: "<<t.col11<<std::endl;
#463            std::cout<<"col12: "<<t.col12<<std::endl;
#464
#465            retcode = SQLFetch(stmt);
#466        }
#467
#468        SQLFreeHandle(SQL_HANDLE_STMT, stmt);
#469
#470
#471        return 0;
#472 }
#473
#474
#475 int main()
#476 {
#477        Multitype t;
#478        t.id = 100;
#479        strcpy(t.col2, "abc");
#480        strcpy(t.col3, "abcd");
#481        strcpy(t.col4, "123");
#482        t.col5 = 123.00f;
#483        t.col6 = 124.00f;
#484        t.col7 = 25.02f;
#485        TIMESTAMP_STRUCT N = {2012, 2, 15, 14, 21, 55, 123};
```

```
#486        memcpy(&t.col8, &N, sizeof(N));
#487        strcpy((char*)t.col9, "abc");
#488        t.col10 = 1;
#489
#490        t.col11 = new unsigned char[32];
#491        strcpy((char*)t.col11, "abcdefghij0123456789");
#492        t.col11_len = strlen((char*)t.col11);
#493
#494        t.col12 = new char[32];
#495        strcpy(t.col12, "abcdefghij0123456789");
#496        t.col12_len = strlen(t.col12);
#497
#498        connect(henv, hdbc);
#499        createMultitype(henv, hdbc, t);
#500
#501        strcpy(t.col2, "abc_new");
#502        strcpy(t.col3, "abcd_updated");
#503        strcpy((char*)t.col11, "abcdefghij0123456789_new");
#504        t.col11_len = strlen((char*)t.col11);
#505        strcpy(t.col12, "abcdefghij0123456789_update");
#506        t.col12_len = strlen(t.col12);
#507
#508        updateMultitype(henv, hdbc, t);
#509        queryMultitype(henv,hdbc,100);
#510        deleteMultitype(henv, hdbc, t.id);
#511        quitODBC(henv, hdbc);
#512        std::cout<<"finished"<<std::endl;
#513        return 0;
#514 }
```

在本例中,我们针对表 multitype_t,使用 ODBC 进行典型的 CRUD 操作,并获取相应结果。表 multitype_t 的定义如下:

```
create table multitype_t
(
        id int not null primary key,
        col2 char(32) not null,
        col3 varchar(32) null,
        col4 numeric(8,2) null,
        col5 float null,
        col6 double precision null,
        col7 money null,
        col8 datetime null,
        col9 binary(4) null,
        col10 bit not null, /* bit 不允许为空 */
        col11 image null,
```

```
        col12 text null
)
go
```

该表基本上涵盖了 ASE 数据库的各种数据类型。访问数据库表的基本流程如下：

Step 1 提供连接信息，连接数据库。

Step 2 创建对应的 SQLHSTMT 句柄，初始化 SQL 语句。

Step 3 绑定相应的输入/输出变量。

Step 4 执行 SQL 语句。

Step 5 获取输出结果。

Step 6 断开连接，关闭释放创建的句柄资源。

下面分别介绍如何进行各种类型的操作。

15.5.1 Insert/Update 操作

在执行 Insert 操作前，必须获得有效的数据库连接，即 SQLHDBC 句柄。整个操作通过函数 static int createMultitype(SQLHENV env, SQLHANDLE dbc, const Multitype& t)实现。

1. 通过 env, dbc 创建有效的语句句柄，第 163 行为 SQLAllocHandle(SQL_HANDLE_STMT, dbc, &stmt)，进行相应的错误处理。
2. 准备 SQL 语句，这里采用动态绑定的方式，相应的 SQL 语句为 insert into multitype_t values(?,?,?,?,?,?,?,?,?,?,?,?)，需要绑定 12 个输入变量。第 174 行调用 SQLPrepare 来准备这个 SQL 语句。
3. 对输入变量调用 SQLBindParameter 进行绑定，这里有 12 个字段，需要进行 12 次绑定。

我们看看 SQLBindParameter 函数的用法说明：

```
SQLRETURN SQLBindParameter(
    SQLHSTMT            StatementHandle,
    SQLUSMALLINT        ParameterNumber,
    SQLSMALLINT         InputOutputType,
    SQLSMALLINT         ValueType,
    SQLSMALLINT         ParameterType,
    SQLUINTEGER         ColumnSize,
    SQLSMALLINT         DecimalDigits,
    SQLPOINTER          ParameterValuePtr,
    SQLINTEGER          BufferLength,
    SQLINTEGER *        StrLen_or_IndPtr);
```

参数说明如下：

- StatementHandle，语句句柄。

- ParameterNumber，参数的序号，按照递增的顺序，从 1 开始计起。
- InputOutputType，参数的输入/输出类型，根据参数的输入/输出类型来决定。一般在 SQL 语句当中，其值都是输入类型 SQL_PARAM_INPUT。输出类型 SQL_PARAM_OUTPUT 常用于存储过程的调用，例如调用存储过程{? = call GetEmpById(?)}，通过用户的 ID 值得到用户的姓名，第一个参数就是输出类型 SQL_PARAM_OUTPUT，用于返回值；第二个参数是输入类型 SQL_PARAM_INPUT，用于指示输入用户 ID 值。
- ValueType，参数对应的 C 语言数据类型。
- ParameterType，参数对应的 SQL 数据类型。
- ColumnSize，列大小。
- DecimalDigits，针对数字类型的列或表达式的小数位数。
- ParameterValuePtr，指向参数值缓冲区的指针。
- BufferLength，参数值缓冲区的长度。
- StrLen_or_IndPtr，指向参数值缓冲区长度值的指针。

我们再来看看代码 182~201 行，有几个需要注意的地方：

1. 第 4 个参数和第 5 个参数的对应关系。

    大部分都是直接对应的，如 C 类型的 SQL_C_LONG 对应 SQL_INTEGER 用于描述整数，SQL_C_CHAR 对应 SQL_CHAR 用于描述字符串，浮点类型用 SQL_C_FLOAT 对应 SQL_FLOAT，数据库类型的 SQL_NUMERIC 需要转换成 C 里的字符串类型 SQL_C_CHAR。

2. 最终的参数值的表示，严格按照 C 类型的定义进行。

3. 对于 TEXT/IMAGE（即 BLOB/CLOB）字段，考虑到字段长度可能很长，可能需要多次写入，这里采用了回调函数的方法进行处理。这是一种通用的处理方法。通过调用 SQLParamData(stmt, &pToken)，只要返回值为 SQL_NEED_DATA，意味着还需要往参数里放入值，每次顺序放入一个列，放完为止。直到该函数返回值不为 SQL_NEED_DATA。

4. 在 BindParameter 操作结束以后，执行 SQLExecute 操作用以执行 SQL 语句，对其返回值进行判断，如果返回值为 SQL_NEED_DATA，则需要往对应的参数里多次放入值，比如这里对 col9、col11、col12 列的绑定，见第 213~246 行。主要用于大的 BLOB/CLOB 字段的处理，每次可以按块（32KB 大小）放入，以提高效率。

    以 col11（IMAGE 类型，即 BLOB 类型）为例，首先我们通过调用 "SQLINTEGER cb11 = SQL_LEN_DATA_AT_EXEC(t.col11_len);" 这是一个宏调用，cb11 这个值将传递给后边的 BindParameter 参数——SQLBindParameter(stmt, 11, SQL_PARAM_INPUT, SQL_C_BINARY, SQL_LONGVARBINARY, 0, 0, NULL, 0, &cb11)。这里的缓冲区指针为 NULL，长度为 0，通过 cb11 来表示该参数的值，在执行的时候再分一次或多次传入。

再看 219～229 行：

```
#219        if (SQL_NEED_DATA == SQLParamData(stmt, &pToken))
#220        {
#221            for (int j=0; j<t.col11_len/BLOCKSIZE; j++)
#222            {
#223                SQLPutData(stmt, t.col11 + j*BLOCKSIZE, BLOCKSIZE);
#224            }
#225            int nTmp = t.col11_len % BLOCKSIZE;
#226            if (nTmp > 0)
#227            {
#228                SQLPutData(stmt, t.col11 + j*BLOCKSIZE, nTmp);
#229            }
#230        }
```

处理完第 9 列，就是第 11 列需要后期传值了，这里为了实验的目的，只将 BLOCKSIZE 设为 6。在实际应用当中，可以将其值设为操作系统页大小的整数倍，比如 16KB、32KB 等。通过多次 SQLPutData 调用，将长度为 t.col11_len（实际缓冲长度）的缓冲区内容全部放入绑定的变量。每次 PutData 都会与服务器交互一次。

5．释放相应的语句句柄。

Update 操作除了语句有些不同以外，绑定过程及执行过程与 Insert 基本类似。

Delete 操作相对简单，因为它的绑定变量肯定是作为查询条件，查询条件中的字段不可能是 BLOB/CLOB 字段。所以没有复杂的绑定过程，详见代码第 356～388 行。

15.5.2　Select 查询操作

Select 操作的基本过程如下：

Step 1 生成 SQL_HANDLE_STMT 类型语句句柄。

Step 2 准备 SQL 查询语句。

Step 3 调用 SQLBindParameter 绑定查询变量值。

Step 4 调用 SQLBindCol，绑定输出结果列到具体的列值变量。

详见第 427～441 行。SQLBindCol 的使用与 SQLBindParameter 有些类似，同样涉及到列类型、绑定列值的缓冲区及其大小。其函数说明如下：

```
SQLRETURN SQLBindCol(
    SQLHSTMT          StatementHandle,
    SQLUSMALLINT      ColumnNumber,
    SQLSMALLINT       TargetType,
    SQLPOINTER        TargetValuePtr,
    SQLINTEGER        BufferLength,
    SQLLEN *          StrLen_or_Ind);
```

参数说明如下：
- StatementHandle，输入参数，语句句柄。
- ColumnNumber，列序号，从 1 计起。
- TargetType，最终的输出类型，数据库驱动会将数据库的实际类型转换到这个目标类型。
- TargetValuePtr，输出值缓冲区指针。
- BufferLengh，缓冲区长度。
- StrLen_or_IndPtr，指示符，最终结果实际长度的指针可能会返回 SQL_NO_TOTAL，SQL_NULL_DATA（该列值为 NULL 时）。

Step 5 调用 SQLExecute，执行 SQL 语句。

Step 6 调用 SQLFetch，取结果集，每次取一行，直至没有结果为止。

Step 7 从绑定的缓冲区读取实际结果。

Step 8 释放语句句柄。

另外，针对 Select 操作，也可以不用事先绑定列值，可以在 SQLExecute 操作之后调用 SQLGetData 函数，直接取各列的值，对于取大的 BLOB/CLOB 值尤其有效，取一个大的 BLOB/CLOB 列，其逻辑如下：

```
while ( (errcode = SQLGetData(stmt, 11, SQL_C_BINARY, buf, BLOCKSIZE, &cbCol11)) != SQL_NO_DATA )
{
}
```

每次取固定大小 BLOCKSIZE 的 buf 值，直到列 col11 中的值读取完为止。

16

使用 Java 访问 ASE

在众多访问数据库的接口当中，Java 的 JDBC 接口无疑是使用最多、最广泛的。因为 Java 语言的跨平台特性以及 J2EE 的广泛应用（企业级应用、互联网应用），致使 JDBC 接口得到普遍的使用。

JDBC API 是 Java 语言用于访问数据库的一套标准 API，比较有意思的是，JDBC 实际上是一个商标名，而不是一个缩写词（参考http://en.wikipedia.org/wiki/Java_Database_Connectivity。但是，大家都已经默认它指的就是 Java Database Connectivity）。目前，JDBC 规范已经发展到了 4.1 版本。Sybase ASE 数据库的 JDBC 最新驱动 Jconnect 7.0 也实现了 JDBC 4.0 规范，但由于 Jconnect 7.0 并未得到广泛应用，而且是在 ASE 15.5 以后才推出，因此，本章将以 Jconnect 6.x（支持 JDBC 3.0）为基础进行系统介绍。

16.1 环境和工具

16.1.1 DBISQL

使用 JDBC 访问 ASE 数据库，Sybase 提供了相应的管理工具 Sybase Central。这个工具在访问数据库表的时候会启动一个叫做 DBISQL 的子工具，它就是基于 JDBC 来访问 ASE 数据库的。它有一个内部名称，叫做 Interactive SQL。在安装完 Sybase ASE 以后，执行菜单命令：Sybase→Adaptive Enterprise Server→Interactive SQL，即可打开它的图形界面，如图 16-1 所示。

填入正确的用户名、密码及服务名，可以单击 Settings 按钮进行服务名的详细设置或者更正，如图 16-2 所示。

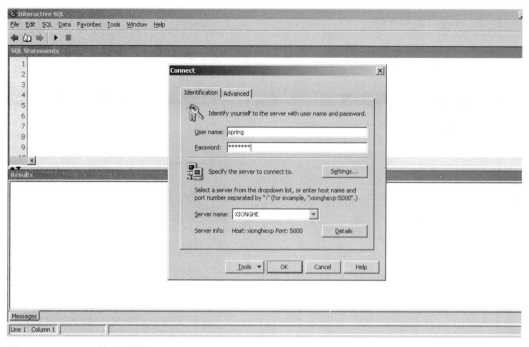

图 16-1　DBISQL 连接界面

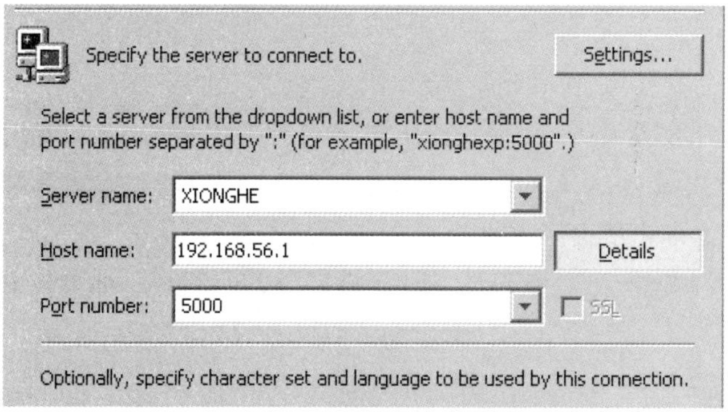

图 16-2　连接信息

单击 OK 按钮以后，即可打开 DBISQL 的主界面，如图 16-3 所示。

用户可以在编辑框输入多条查询语句，按 F5 键以后，可以在结果栏里得到多个结果集，分多个 tab 页显示（见图 16-3），有三个结果集，分别在 Result Set 1, 2, 3 里头展示。也可以选中某一行或几行，只执行被选中的部分（或按快捷键 F9）。还有一个重要的功能，按快捷键 F7 可以列出当前数据库下所有的表，这对于构造有效的 SQL 语句非常有用。

DBISQL 另一种打开方式是直接进入目录%SYBASE%\DBISQL\bin，直接打开 dbisql.com，同样可以打开该应用程序。

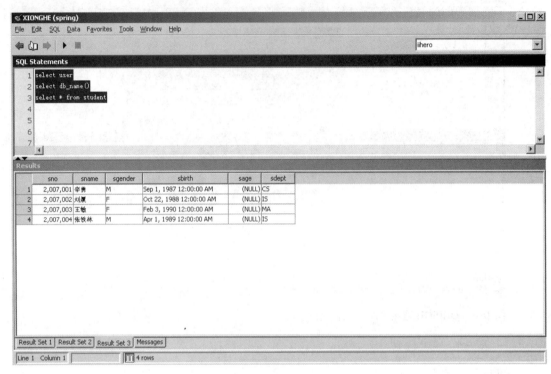

图 16-3　在 DBISQL 中使用查询

这里需要提醒的是，DBISQL 是 Sybase 一个公共的数据库查询管理工具框架，如果在同一台机器上安装了 Sybase 的其他数据库，比如 ASA（SQL Anywhere）或者 IQ，那么它们也会带有一份 DBISQL 工具。多个 DBISQL 相互之间可能会产生冲突。

其实，dbisql.com 的内部实现就是动态产生一个 bat，该 bat 再调用 Java 程序。如果使用下边的命令行：

```
D:\SybaseASE\DBISQL\bin>dbisql.com /batch
```

你会看到它会产生一个批处理文件：

```
D:\SybaseASE\DBISQL\bin>more dbisql.bat
setlocal
set path=D:\SybaseASE\Shared\win32;%path%
set classpath=D:\SybaseASE\DBISQL\lib\isql.jar;D:\SybaseASE\DBISQL\lib\jlogon.jar;D:\SybaseASE\Shared\java\SCEditor600.jar;D:\SybaseASE\Shared\java\JComponents1100.jar;D:\SybaseASE\Shared\java\jsyblib600.jar;D:\SybaseASE\Shared\java\JavaHelp-2_0\jh.jar;;
```

"D:\SybaseASE\Shared\JRE-6_0_6_32BIT\bin\java.exe" -Xmx500m -Xms50m -Djava.security.policy="D:\SybaseASE\DBISQL\lib\java.policy" -Disql.helpFolder="D:\SybaseASE\DBISQL\help" -Dsybase.native.executable="D:\SybaseASE\DBISQL\bin\dbisql.com" -Dsun.java2d.noddraw=true -Dsun.java2d.d3d=false -ea sybase.isql.isql
endlocal

实际上，这就是 dbisql 真正要执行的内容。多个 dbisql 发生冲突，主要原因在于一个配置文件 %ALLUSERSPROFILE%\DBISQL 11.0.x\dbisql.rep。请看 ASE 15.0.3 下这个文件的详细内容：

[Adaptive Server Enterprise]
classLoaderName=aseisqlplugin11
mainclass=com.sybase.aseisqlplugin.ASEISQLPlugin
classpath=D:\SybaseASE\Shared\lib\jconn3.jar;D:\SybaseASE\DBISQL\lib\aseisqlplugin.jar;D:\SybaseASE\DBISQL\lib\jodbc.jar;D:\SybaseASE\DBISQL\lib\xml4j.jar;D:\SybaseASE\DBISQL\lib\planviewer.jar;D:\SybaseASE\DBISQL\lib\dsparser.jar;D:\SybaseASE\DBISQL\lib\asa.jar;D:\SybaseASE\DBISQL\lib\uaf-client.jar;D:\SybaseASE\DBISQL\lib\jini-core.jar;D:\SybaseASE\DBISQL\lib\jini-ext.jar;D:\SybaseASE\DBISQL\lib\jmxremote.jar;D:\SybaseASE\DBISQL\lib\jmxri.jar;D:\SybaseASE\DBISQL\lib\commons-logging.jar;D:\SybaseASE\DBISQL\lib\log4j-1.2.6.jar

要想不出问题，只要把这个文件备份并还原即可，把 D:\SybaseASE 目录换成你的 Sybase ASE 安装目录即可。这是一个重要的配置文件，用于加载 dbisql 插件。如果没有这个文件，打开的对话框可能会变成图 16-4 这个样子。

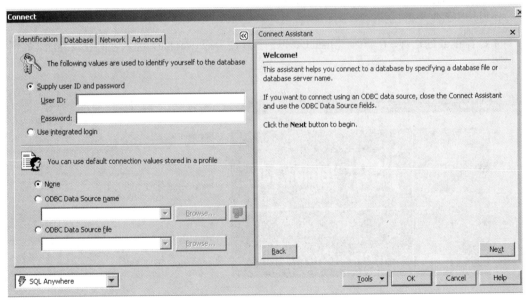

图 16-4 错误的 DBISQL 登录对话框

这已经变成了 ASA（SQL Anywhere）的连接对话框了，无法完成对 ASE 数据库的连接。

总之，Sybase Central 的插件 DBISQL 是图形界面程序中访问 ASE 数据库的首选。

16.1.2　JUtils

JUtils 工具并未包含在 Sybase ASE 的默认安装当中，只有当你选择安装它的时候，才会被包括进来。而且，它在 Sybase 的官方文档中也未提及其用法。但是，它确实是一个非常轻量级的 jdbc 工具包。

JUtils 位于%SYBASE%/jutils-2_0 子目录下，它的运行仅依赖于 JDK 及 ASE 的 JDBC 驱动。JUtils 工具包其实包含了两个小工具：jisql 和 Ribo。前者用于 JDBC 数据库连接查询，后者用于程序诊断。

要运行 jisql，必须修改它的运行主程序：%SYBASE%/jutils-2_0/jisql/jisql.bat，在该文件的开始@echo off 后添加如下两项内容：

```
set JAVA_HOME=D:\SybaseASE\Shared\JRE-6_0_6_32BIT
set JDBC_HOME=D:\SybaseASE\jConnect-6_0
```

这里 D:\SybaseASE 是 Sybase ASE 安装的根目录，即%SYBASE%指向的位置。

对于小工具 Ribo 而言，也需要修改它的运行主程序：%SYBASE%/jutils-2_0/ribo/ribo.bat，在该文件的开始处@echo off 后添加如下内容：

```
set JAVA_HOME=D:\SybaseASE\Shared\JRE-6_0_6_32BIT
set RIBO_HOME=D:\SybaseASE\jutils-2_0\ribo
```

先看看 jisql 的运行，在修改完 jisql.bat 以后，直接运行 jisql.bat，进入它的登录界面，如图 16-5 所示。

图 16-5　jisql 登录界面

在输入用户名、密码、主机名、端口，并选择合适的语言之后，单击 Connect 按钮，即可进入该工具的主界面，如图 16-6 所示。

使用 Java 访问 ASE　第 16 章

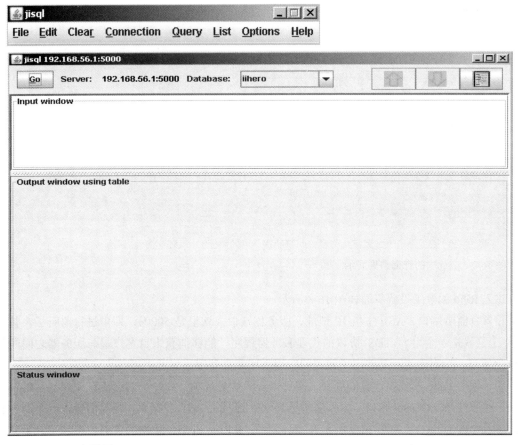

图 16-6　jisql 主界面

我们只要在输入窗口输入任意 SQL 语句，然后单击 Go 按钮，即可得到结果。一次还可以输入多条 SQL 语句。不过需要分行隔开，并且不能用分号。

例如，我们执行三条 SQL 查询：

```
select user
select db_name()
select * from spring.student
```

将这些内容输入输入窗口，然后单击 Go 按钮，最后会得到三个结果集，分别在 ResultSet 1/2/3 中显示，如图 16-7 所示。

由于 jisql 工具只有 1MB 左右大小，相对于 Sybase Central 庞大的体积而言，其占用空间非常小，因此，DBA 甚至可以用它作日常的 SQL 命令行管理工具，用以替代 isql 命令行。

Ribo 的启动则需要带有命令行参数，进入目录%SYBASE%\jutils-2_0\ribo，运行命令：

```
Ribo.bat -gui
```

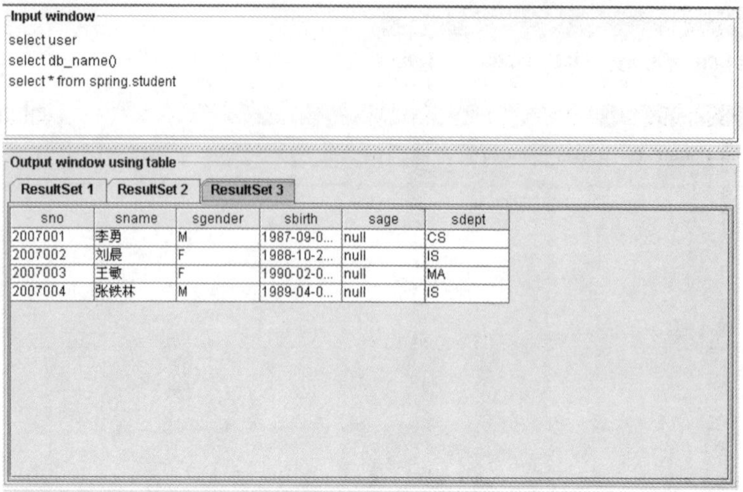

图 16-7 jisql 中的查询结果展示

进入 Ribo 的登录设置界面如图 16-8 所示。

设置好监听端口、ASE 主机 IP 地址、服务器端口（这里是 5000），即可进行监听了。其实，Ribo 工具就是一个用于 TDS 协议的代理服务器程序，能够捕获来自客户端与服务器之间的所有 TDS 网络包，这对于诊断某些出了问题的应用程序非常有用。

为了说明 Ribo 的使用，我们选择"文件"→"优先选项"命令，勾选"在窗口显示转换"复选项。然后打开一个 jisql 窗口，填入端口号 9999，连接上 ASE 数据库，当我们输入一个简单的查询：select db_name()（如图 16-9 所示），我们会发现 Ribo 开启一个新的窗口，里边记录了完整的转换 TDS 包以后的内容，如图 16-10 所示。

图 16-8 Ribo 登录设置界面

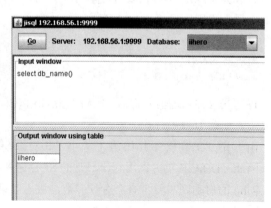

图 16-9 jisql 中的查询结果

使用 Java 访问 ASE　第 16 章

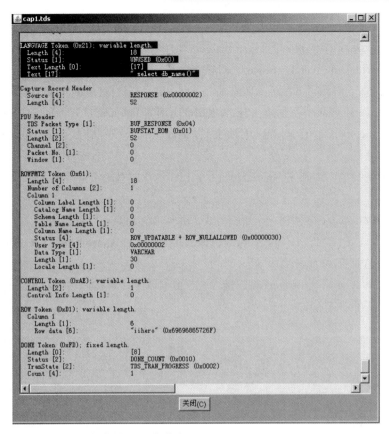

图 16-10　Ribo 捕获的 TDS 包信息

这个工具对于当应用程序（无论是客户端的，还是服务器端的）碰到非常复杂的问题，怀疑是 ASE 内部的问题的时候，非常有用，只要将相应的 TDS 转换包及其对应的文件打包，发给 Sybase 的工程师，他们针对包里的内容就可以诊断问题的根源了。

16.1.3　DBeaver

这里还要介绍一个第三方的自由软件工具：DBeaver，它完全基于 Java 开发，并采用 JDBC 来连接主流数据库，其中也包括 Sybase ASE/ASA 等。推荐使用它的原因如下：

1. 该工具完全免费，并且还有一个好消息是，根据官网的最新消息，从 1.6 版本开始，将实行开源（详见官网上的说明：http://dbeaver.jkiss.org/）。目前最新版本是 1.5.6，你可以用它做任何事情，没有任何限制。相比于 DBVisualizer（该工具分社区版和专业版，社区版很多功能受到限制）而言，无疑是一个很大的优势。
   虽然有类似的软件，如 SQuirrel，但是 DBeaver 比它更节省内存。
2. 可配置性高，界面友好，很容易支持对几乎所有具备 JDBC 驱动的数据库的访问。

下面来看看它的使用方法：

1. 从官网http://dbeaver.jkiss.org/download/直接下载压缩版，如1.5.6-x86-32位版本：
   http://dbeaver.jkiss.org/files/dbeaver-1.5.6-win32.win32.x86.zip
2. 将压缩版 dbeaver*x86.zip 解压至本地机器的目录上，设为 D:\，即有目录 D:\dbeaver，创建子目录 D:\dbeaver\drivers\sybase，然后将 ASE 的 JDBC 驱动库 Jconnect 6 下的 jconn3.jar 复制到 D:\dbeaver\drivers\sybase 目录。该驱动文件 jconn3.jar 位于目录 D:\SybaseASE\jConnect-6_0\classes 下边。
3. 下载一个 jdk1.6 的安装包，装到 C:\shared\jdk1.6.0_18，本章所有的示例都可以使用这个 jdk1.6。
4. 在 D:\dbeaver 下，为 dbeaver.exe 创建一个快捷方式，并修改它的启动应用程序：
   由 D:\dbeaver\dbeaver.exe 改为 D:\dbeaver\dbeaver.exe -vm C:\shared\jdk1.6.0_18\bin**javaw**，即为它指定 JVM 的位置。

完成上述配置以后，就可以使用 DBeaver 来连接 ASE 数据库了。直接通过快捷方式打开 DBeaver，出现如图 16-11 所示的主界面。

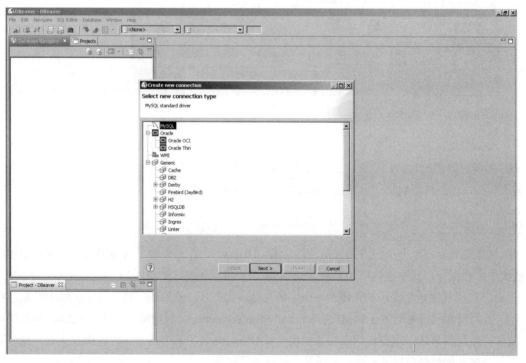

图 16-11　DBeaver 的主界面

我们先把 Sybase 的 ASE JDBC 驱动注册上去，先单击 Cancel 按钮，然后选择 Database→Driver

Manager→New 命令，填入下面的信息，如图 16-12 所示。

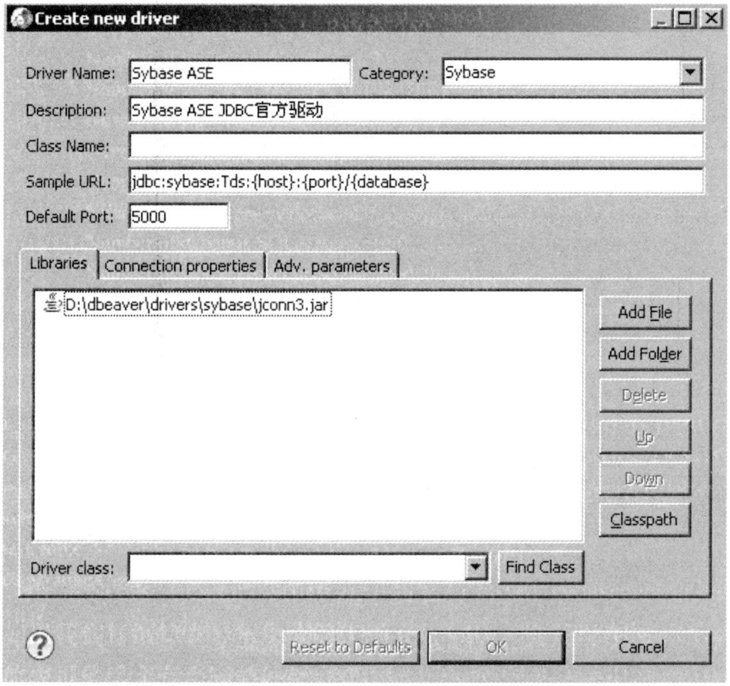

图 16-12 填入信息

在把 jconn3.jar 填入之后，单击 Find Class 按钮，会出现多个可能的 Driver class，如图 16-13 所示。

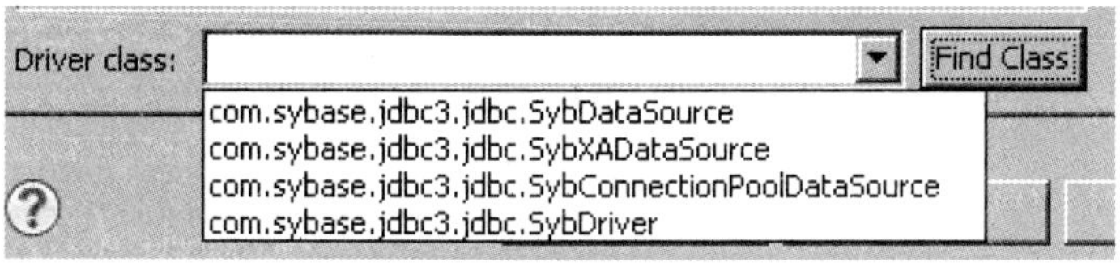

图 16-13 可能的 ASE JDBC 驱动类列表

我们选择 com.sybase.jdbc3.jdbc.SybDriver 选项即可，然后单击 OK 按钮完成注册。在这之后，我们可以通过 Database→New Connection，选择 Sybase→Sybase ASE→Next 命令，进入图 16-14 所示的连接信息设置界面。

只要填入正确的主机 IP 地址（或主机名）、端口号、数据库名以及用户名和密码，然后单击 Test Connection 按钮，可以验证是否能够与目标 ASE 数据库服务器相连，这样就可以成功地创建一个 ASE 数据库连接了，完成之后即进入新建连接 Sybase - iihero 的主界面，如图 16-15 所示。

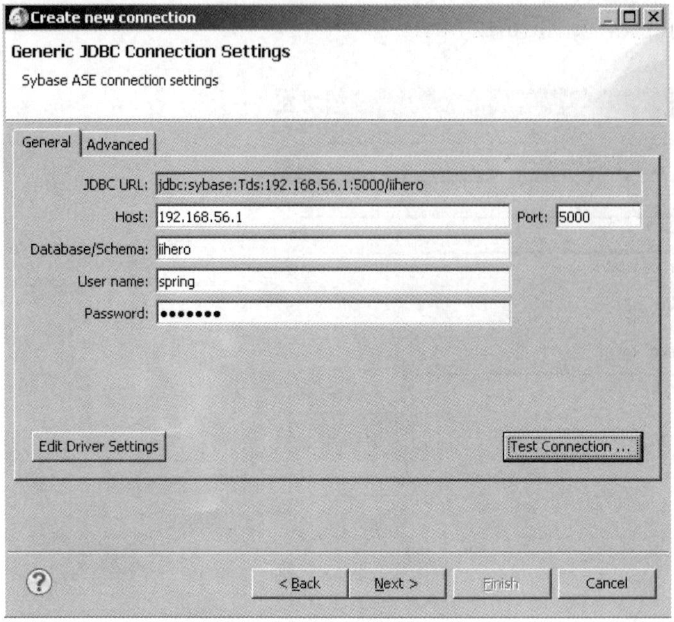

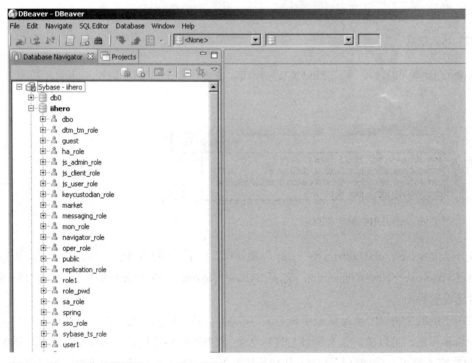

图 16-14　ASE 在 DBeaver 中的连接信息设置

图 16-15　ASE 中的数据库 iihero 相关信息

选择你要管理的 iihero 数据库——用户 spring，接着可以看到用户 spring 所拥有的表资源列表。浏览表数据，执行 SQL 查询非常便利，如图 16-6 所示就是浏览表 course 的结果。

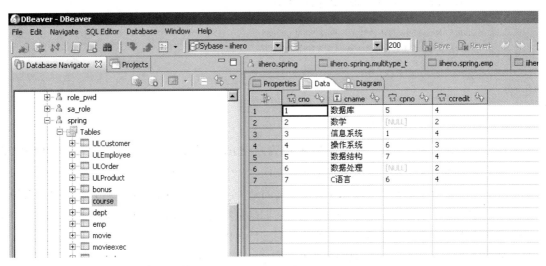

图 16-6　iihero 数据库中的库表浏览

事实上，为了提供对多种数据库的支持，你可以直接下载由 DBeaver 整理的驱动包合集：http://dbeaver.jkiss.org/files/driver-pack-1.5.6.zip，但是此包里并未包含 Sybase ASE 的官方 JDBC 驱动，而只为 Sybase 数据库提供了开源的 JTDS 驱动，此驱动相对于官方驱动而言是有缺陷的，在实际的生产环境中，最好使用由 Sybase 官方提供的 JDBC 驱动来实现自己的应用。

DBeaver 将用户注册的 JDBC 驱动信息写入到% USERPROFILE%\ .dbeaver\ drivers.xml 文件当中，以 xml 的形式来描述：

例如：

```
<provider id="generic">
    <driver id="B8005B73-FB13-21D5-6BA8-E8A75313ED84"
        category="Sybase" custom="true"
        name="Sybase ASE"
        class="com.sybase.jdbc3.jdbc.SybDriver" url="jdbc:sybase:Tds:{host}:{port}/{database}"
        port="5000" description="Sybase ASE JDBC 官方驱动">
        <library path="D:\dbeaver\drivers\sybase\jconn3.jar"/>
    </driver>
</provider>
```

添加的数据库连接信息放到% USERPROFILE%\.dbeaver\DBeaver\data-sources.xml 配置文件当中，如我们前边添加的连接 Sybase - iihero，其内容如下：

```
<data-sources>
    <data-source id="B8005B73-FB13-21D5-6BA8-E8A75313ED84-1338634021123--507606854" provider="generic" driver=
```

```
"B8005B73-FB13-21D5-6BA8-E8A75313ED84"
    name="Sybase - iihero" create-date="1338634021123"
    save-password="true" show-system-objects="true">
        <connection host="192.168.56.1" port="5000"
        server="" database="iihero"
        url="jdbc:sybase:Tds:192.168.56.1:5000/iihero"
        user="spring" password="ABQUKU9EFXbk"/>
    </data-source>
</data-sources>
```

16.1.4 JDBC 驱动 Jconnect 6.0.5 简介

目前，ASE 的 JDBC 驱动库使用最为广泛的还是 Jconnect 6.x，它位于%SYBASE%\jConnect-6_0\classes 下边，主要有两个 jar 文件，一个是 jdbc 驱动的实现：jconn3.jar，另一个是 jTDS3.jar，主要用于 TDS 包的处理。值得一提的是，Sybase 的 ASE 数据库采用的是 TDS 5.0 协议，微软的 SQL Server 采用的是 TDS 8.0 协议，它们虽然都用 TDS 协议，但是版本不同。开源的 jTDS jdbc 驱动就是根据这两个协议来实现的，从而支持两种数据库的访问。

要想得到 JConnect 驱动的具体版本号，可以直接运行 jconn3.jar，如：

```
D:\SybaseASE\jConnect-6_0\classes>java -jar jconn3.jar
jConnect (TM) for JDBC(TM)/6.05(Build 26312)/P/EBF15807/JDK14/Tue Jun 10   1:06:20 2008

Confidential property of Sybase, Inc.
Copyright 1997, 2004
Sybase, Inc.    All rights reserved.
Unpublished rights reserved under U.S. copyright laws.
This software contains confidential and trade secret information of Sybase,
Inc.   Use, duplication or disclosure of the software and documentation by
the U.S. Government is subject to restrictions set forth in a license
agreement between the Government and Sybase, Inc. or other written
agreement specifying the Government's rights to use the software and any
applicable FAR provisions, for example, FAR 52.227-19.

Sybase, Inc. One Sybase Drive, Dublin, CA 94568
```

从中可以得到 Build 号和 EBF 版本号。在 JConnect 包中有一个子目录 sp，其中有对应的各种版本的 ASE/ASA 数据库的 JDBC 执行脚本。

```
D:\SybaseASE\jConnect-6_0\classes>cd ..\sp
D:\SybaseASE\jConnect-6_0\sp>dir
 Directory of D:\SybaseASE\jConnect-6_0\sp
2012-04-23   07:01    <DIR>          .
2012-04-23   07:01    <DIR>          ..
2008-05-29   05:10            98,066 sql_asa.sql
2008-05-29   05:10            97,730 sql_asa10.sql
2008-05-29   05:10            98,137 sql_asa11.sql
```

```
2008-06-10  16:11         304,316 sql_server.sql
2008-06-10  16:11         315,474 sql_server12.5.sql
2008-06-10  16:11         309,821 sql_server12.sql
2008-06-10  16:11         312,176 sql_server15.0.sql
             7 File(s)  1,535,720 bytes
```

对应于 ASE 15.0.x 版本的就是 sql_server15.0.sql（历史上，ASE 数据库刚开始就叫做 Sybase SQL Server，为了与微软的 MS SQL Server 相区分，才叫现在的 ASE（Adaptive Enterprise Server））。它是 ASE 的 JDBC 实现在 Server 端需要安装的存储过程脚本。一般情况下，如果不升级，这是因为 ASE 安装程序已经为你安装好了。但是如果你想打新的 JDBC EBF 补丁，就需要手动安装 JConnect 包自带的这个存储过程脚本。

16.2 通过 JDBC 连接 ASE 数据库

在 ASE 数据库服务器已经启动以后，我们可以通过常规的 JDBC 调用连接 ASE 数据库。本节将介绍如何通过 DriverManager 来创建到数据库的连接。

使用 DriverManager 创建连接，分两步完成：

1. 加载 JDBC 驱动的 Java 类

对于 ASE 数据库而言，要加载的驱动类就是 com.sybase.jdbc3.jdbc.SybDriver，通过下述调用完成：

```
Class.forName("com.sybase.jdbc3.jdbc.SybDriver");
```

调用此方法之后，会自动完成驱动的注册。

2. 创建 JDBC 连接对象

通过下面的代码片段可以完成对 ASE 数据库的连接：

```
Connection conn = DriverManager.getConnection(url, "login", "password");
```

只需要三个参数、连接的 URL 字符串、数据库登录用户名和密码。对照前一节中的 DBeaver 连接信息，我们可以做一个连接：

```
Connection conn = DriverManager.getConnection("jdbc:sybase:Tds:192.168.56.1:5000/iihero", "spring", " spring1");
```

这是最简单的连接构造方式了。

如果将其写成完整的 Java 应用程序（本章所有 Java 代码及工程都使用 Eclipse 开发环境，编译前，注意添加依赖的 JDBC 库 jconn3.jar）。其代码如下所示：

```java
package com.iihero;

import java.sql.*;

public class ASEConnect
```

```java
{
    final static String ASEDriver = "com.sybase.jdbc3.jdbc.SybDriver";
    private Connection _conn;

    private    void connect(String username, String password, String url)
    {
        try
        {
            Class.forName(ASEDriver);
            _conn = DriverManager.getConnection(url, username, password);
            System.out.println("成功连接到： " + url);
        }
        catch (Exception ex)
        {
            System.out.println("连接到： " + url + "   失败. 原因为: " + ex.toString());
        }
    }

    private void close()
    {
        if (_conn != null)
        {
            try
            {
                _conn.close();
                System.out.println("断开连接.");
            }
            catch (SQLException ex)
            {
                System.out.println("断开连接有异常...");
                System.out.println(ex.toString());
            }
        }
    }

    public static void main(String[] args)
    {
        ASEConnect ac = new ASEConnect();
        ac.connect("spring", "spring1", "jdbc:sybase:Tds:192.168.56.1:5000/iihero");
        ac.close();
    }
}
```

以上代码是针对 Sybase ASE 数据库所写，其实稍加改动，我们可以将其写为一个诊断程序，适用于所有数据库，只要有目标数据库的 JDBC 驱动即可。

基本思路：针对某 DBMS，它的 JDBC 驱动的类名是固定的，连接串 URL 的格式也是固定的，变化的部分是主机名、端口号、数据库名，而用户名和密码可以作为另外两个参数。

我们为每种已知的数据库指定好它的 Driver 类名以及 URL 格式串（可以带模板化参数），这样，在连接的时候直接替换 URL 串中的相应模板即可。这里列出主流的数据库 Driver 类名及 URL 连接串。

1. Sybase ASE 15.0.x 或以上版本。

    Driver：com.sybase.jdbc3.jdbc.SybDriver

    URL：jdbc:sybase:Tds:{host}:{port}/{database}

    port 默认端口号为 5000，Driver 来自官方提供的 JDBC 库 jconn3.jar。

    另有非官方驱动：

    Driver：net.sourceforge.jtds.jdbc.Driver

    URL：jdbc:jtds:sybase://{host}:{port}/{database}

    port 默认端口号为 5000，Driver 来自 sourceforge 上的开源驱动类库：jtds-1.2.5.jar（http://jtds.sourceforge.net）。

2. Sybase ASA 9.x 或以上版本。

    Driver：com.sybase.jdbc3.jdbc.SybDriver

    URL：jdbc:sybase:Tds:{host}:{port}/{database}

    port 默认端口号为 2638，Driver 来自库 jconn3.jar。

3. Sybase Advantage Database 9.0 或以上版本。

    Driver：com.extendedsystems.jdbc.advantage.ADSDriver

    URL：jdbc:extendedsystems:advantage://{host}:{port};catalog={database}

    port 默认端口号为 6262，Driver 来自类库 adsjdbc.jar。

4. Oracle 9i 或以上版本。

    Driver: oracle.jdbc.driver.OracleDriver

    URL：jdbc:oracle:thin:@{host}:{port}:{database}

    port 默认端口号为 1521，Driver 来自 9i 或以上版本中的 JDBC 类库：jdbc14.jar。

5. PostgreSQL 8.x 或以上版本。

    Driver：org.postgresql.Driver

    URL：jdbc:postgresql://{host}:{port}/{database}

    port 默认端口号为 5432，Driver 来自 8.x 或以上版本中的 JDBC 类库：postgresql-8.4-701.jdbc4.jar 或者 postgresql-9.1-901.jdbc4.jar。

6. My SQL 5.0 或以上版本。

    Driver：com.mysql.jdbc.Driver

    URL：jdbc:mysql://{host}:{port}/{database}

    port 默认端口号为 3306，Driver 来自 My SQL 5.0 或以上版本所带的 JDBC 类库

MySQL-Connect/J：mysql-connector-java-5.1.13-bin.jar。

7. DB2 V8.1 或以上版本。

   Driver：com.ibm.db2.jcc.DB2Driver

   URL：jdbc:db2://{host}:{port}/{database}

   port 默认端口号为 50000，Driver 来自 DB2 V8.1 或以上版本所带的 JDBC 类库：db2jcc.jar 及 db2jcc_license_cu.jar。

8. MS SQL Server 2005 或以上版本。

   Driver：com.microsoft.sqlserver.jdbc.SQLServerDriver

   URL：jdbc:sqlserver://{host}:{port};DatabaseName={database}，

   port 默认端口号为 1433，Driver 来自 SQL Server 官方提供的 JDBC 类库：sqljdbc4.jar。

   另有非官方开源驱动 jtds：

   Driver：net.sourceforge.jtds.jdbc.Driver

   URL：jdbc:jtds:sqlserver://{host}:{port}/{database}

   Driver 来自 sourceforge 上的开源驱动类库：jtds-1.2.5.jar（http://jtds.sourceforge.net）。

9. SQLite3.x。

   Driver：org.sqlite.JDBC

   URL：jdbc:sqlite:{database}

   这里的 database 指的是 sqlite3 数据库文件的全路径，Driver 来自 www.zentus.com/ sqlitejdbc/ 提供的开源类库：sqlitejdbc-v056.jar。

虽然也有开源的 jtds 类库 jtds-1.2.5.jar 为 SQL Server 及 Sybase ASE 提供驱动，但是它与 MS 及 Sybase 官方提供的 JDBC 驱动相比，还是有差距的，可以用于实验环境，可以用于测试，但是不推荐用于生产环境。

上述列表对于多种数据库中的 Java 开发来说非常重要。我们再来看看通用的 JDBC 连接诊断程序的实现：

```
#001 package com.iihero;
#002
#003 import java.sql.Connection;
#004 import java.sql.DriverManager;
#005 import java.util.HashMap;
#006 import java.util.Map;
#007
#008 /**
#009  * @author xionghe
#010  */
#011 class DriverInfo
#012 {
#013     String driver;
```

```
#014       String url;
#015       public DriverInfo(String driver, String url)
#016       {
#017           this.driver = driver;
#018           this.url = url;
#019       }
#020       // 已知可注册的 jdbc 驱动列表及 URL 串信息
#021       final static DriverInfo ASE = new DriverInfo("com.sybase.jdbc3.jdbc.SybDriver", "jdbc:sybase:Tds:{host}:{port}/{database}");
#022       final static DriverInfo ASA = new DriverInfo("com.sybase.jdbc3.jdbc.SybDriver", "jdbc:sybase:Tds:{host}:{port}/{database}");
#023       final static DriverInfo ASE_JTDS = new DriverInfo("net.sourceforge.jtds.jdbc.Driver", "jdbc:jtds:sybase://{host}:{port}/{database}");
#024       final static DriverInfo ADS = new DriverInfo("com.extendedsystems.jdbc.advantage.ADSDriver", "jdbc:extendedsystems: advantage://{host}:{port};catalog={database}");
#025       final static DriverInfo ORACLE = new DriverInfo("oracle.jdbc.driver.OracleDriver", "jdbc:oracle:thin:@{host}:{port}:{database}");
#026       final static DriverInfo POSTGRESQL = new DriverInfo("org.postgresql.Driver", "jdbc:postgresql://{host}:{port}/{database}");
#027       final static DriverInfo MYSQL = new DriverInfo("com.mysql.jdbc.Driver", "jdbc:mysql://{host}:{port}/{database}");
#028       final static DriverInfo DB2 = new DriverInfo("com.ibm.db2.jcc.DB2Driver", "jdbc:db2://{host}:{port}/{database}");
#029       final static DriverInfo MSSQL = new DriverInfo("com.microsoft.sqlserver.jdbc.SQLServerDriver", "jdbc:sqlserver://{host}:{port};DatabaseName={database}");
#030       final static DriverInfo MSSQL_JTDS = new DriverInfo("net.sourceforge.jtds.jdbc.Driver", "jdbc:jtds:sqlserver://{host}:{port}/{database}");
#031       final static DriverInfo SQLITE = new DriverInfo("org.sqlite.JDBC", "jdbc:sqlite:{database}");
#032
#033       final static Map<String, DriverInfo> DRIVERS = new HashMap<String, DriverInfo>();
#034       static
#035       {
#036           DRIVERS.put("ase", ASE);
#037           DRIVERS.put("asa", ASA);
#038           DRIVERS.put("ase_jtds", ASE_JTDS);
#039           DRIVERS.put("ads", ADS);
#040           DRIVERS.put("oracle", ORACLE);
#041           DRIVERS.put("pgsql", POSTGRESQL);
#042           DRIVERS.put("mysql", MYSQL);
#043           DRIVERS.put("db2", DB2);
#044           DRIVERS.put("mssql", MSSQL);
#045           DRIVERS.put("mssql_jtds", MSSQL_JTDS);
#046           DRIVERS.put("sqlite", SQLITE);
#047       }
#048 }
#049
#050 public class GenericConnection
#051 {
```

```
#052        public static void verify(String dbtype, String host, String port, String database, String username, String password)
#053        {
#054            DriverInfo driverInfo = DriverInfo.DRIVERS.get(dbtype);
#055            if (driverInfo == null)
#056            {
#057                System.out.println("不支持数据库类型: " + dbtype);
#058                return;
#059            }
#060            try
#061            {
#062                Integer.parseInt(port);
#063            }
#064            catch (Exception ex)
#065            {
#066                System.out.println("端口号必须是整数类型.");
#067                return;
#068            }
#069            Connection conn = null;
#070            try
#071            {
#072                Class.forName(driverInfo.driver);
#073                String url = driverInfo.url;
#074                url = url.replace("{host}", host);
#075                url = url.replace("{port}", port);
#076                url = url.replace("{database}", database);
#077                conn = DriverManager.getConnection(url, username, password);
#078                conn.close();
#079                System.out.println("连接到: " + url + " 成功...");
#080            }
#081            catch (Exception ex)
#082            {
#083                System.out.println("连接出错，原因： " + ex.toString());
#084            }
#085        }
#086
#087        public static void main(String[] args)
#088        {
#089            if (args.length < 6)
#090            {
#091                System.out.println("GenericConnection usage:    ");
#092                System.out.println("GenericConnection    <dbtype> <host> <port> <database> <username> <password>");
#093                System.out.println("You need to supply the above 6 arguments to run.");
#094            }
#095
#096            verify(args[0], args[1], args[2], args[3], args[4], args[5]);
#097        }
#098
#099 }
```

在本程序中，我们通过事先注册 11 种 JDBC 驱动（第 21 到 31 行），URL 串采用模板的形式，在运行时，对 {host}、{port}、{databasa} 这三个变量进行替换。从而达到动态连接的效果，只需要告诉程序连接的数据库类型（第 34 到 47 行将类型和驱动的模板进行了映射），以及连接所需要的信息：主机名、端口号、数据库名、用户名和密码，即可完成连接的动态创建。

打包运行，将你所要测试的驱动库 jar 文件放到编译好的 class 文件的输出路径 bin 的子目录 lib 下边，在 JDK1.6 下，运行一个 ase 类型的驱动连接测试：

> D:\MyDocument\MyBooks\ASE\code\java\bin>java -cp lib*;.com.iihero.GenericConnection ase 192.168.56.1 5000 iihero spring spring1
> 连接到：jdbc:sybase:Tds:192.168.56.1:5000/iihero 成功...

> **注意**
> JDK1.6 开始，支持 classpath 的模糊匹配，所以 lib* 意味着会将 lib 目录下边所有的 jar 文件添加到 CLASSPATH 里边。

如果把 jTDS 驱动 jtds-1.2.5.jar 放到 lib 目录下，我们一样可以测试使用它连接 ASE 数据库的结果：

> D:\MyDocument\MyBooks\ASE\code\java\bin>java -cp lib*;.com.iihero.GenericConnection ase_jtds 192.168.56.1 5000 iihero spring spring1
> 连接到：jdbc:jtds:sybase://192.168.56.1:5000/iihero 成功...

再看看 Postgre SQL 数据库连接的诊断，将 PostgreSQL 数据库的 JDBC 驱动库 postgresql-9.1-901.jdbc4.jar 放到 lib 子目录下边，运行下述命令，连接本机上的 PostgreSQL 数据库：

> D:\MyDocument\MyBooks\ASE\code\java\bin>java -cp lib*;.com.iihero.GenericConnection pgsql localhost 5432 iihero spring spring1
> 连接出错，原因:org.postgresql.util.PSQLException: 连线被拒,请检查主机名称和埠号,并确定 postmaster 可以接受 TCP/IP 连线

从出错原因来看，PostgreSQL 数据库没有启动，启动成功之后再运行相同的命令，得到结果：

> D:\MyDocument\MyBooks\ASE\code\java\bin>java -cp lib*;.com.iihero.GenericConnection pgsql localhost 5432 iihero spring spring1
> 连接到：jdbc:postgresql://localhost:5432/iihero 成功...

这个诊断程序非常通用，如果遇到新的数据库类型，只需要在 DriverInfo 类里添加相应的注册信息即可。在命令行里，第一个参数只要指定正确的数据库类型即可。这个类型名在 DriverInfo 类的哈希表 Drivers 里作为键值存储起来。

16.3 使用 JDBC 操作 ASE 表数据

无论是哪种访问接口，针对表数据的操作都可以分为四种（C[reate]：insert；R[ead]：select；U[pdate]：update；D[elete]：delete），外加对表和其他数据库对象的 DDL 操作（create, drop, alter）。

下面，我们针对示例数据库 iihero 中的表，分别介绍上述几种操作的实现。

16.3.1 Select 查询操作

数据准备，这里假设要对 iihero 数据库中的表 emp 进行查询并获取结果。其基本过程如下：

Step 1 建立到 ASE 数据库 iihero 的连接（Connection）。

Step 2 创建相应的 Statement 对象。

Step 3 由 Statement 对象创建要执行的查询，生成结果集 ResultSet。

Step 4 遍历 ResultSet 对象，获取并输出最终结果。

Step 5 关闭 ResultSet、Statement 及 Connection 以释放资源。

下面是执行查询 SELECT * FROM emp WHERE empno=7499 并输出结果的完整代码清单：

```
#001  package com.iihero;
#002
#003  import java.sql.*;
#004  import javax.sql.rowset.CachedRowSet;
#005  import com.sun.rowset.CachedRowSetImpl;
#006
#007  /**
#008   * @author xionghe
#009   */
#010  public class ASEDatabase
#011  {
#012      private Connection _conn;         // 连接句柄
#013      private String     _username;     // 用户名
#014      private String     _password;     // 密码
#015      private String     _host;         // 数据库主机名/IP
#016      private int        _port;         // 端口号
#017      private String     _database;     // 目标数据库名
#018
#019      final static String DriverClassASE = "com.sybase.jdbc3.jdbc.SybDriver";
#020      final static String DriverURLASE = "jdbc:sybase:Tds:{host}:{port}/{database}";
#021
#022      public ASEDatabase(String host, int port, String database, String username, String password) throws Exception
#023      {
#024          _host = host;
#025          _port = port;
#026          _username = username;
#027          _password = password;
#028          _database = database;
#029          Class.forName(DriverClassASE);
#030          String url = DriverURLASE.replace("{host}", _host).replace("{port}", ""+_port).replace("{database}", _database);
#031          _conn = DriverManager.getConnection(url, _username, _password);
```

```
#032        }
#033
#034        // 执行一个 SQL 查询，返回一个 CachedRowSet
#035        public ResultSet executeQuery(String sql) throws Exception
#036        {
#037            Statement stmt = _conn.createStatement();
#038            ResultSet rset = stmt.executeQuery(sql);
#039            CachedRowSet   crset = new CachedRowSetImpl();
#040            crset.populate(rset);
#041            rset.close();
#042            stmt.close();
#043            return crset;
#044        }
#045
#046        public void dumpResultSet(ResultSet rset) throws Exception
#047        {
#048            if (rset == null)
#049            {
#050                System.out.println("结果集为空.");
#051                return;
#052            }
#053            int nCols = rset.getMetaData().getColumnCount();
#054            while (rset.next())
#055            {
#056                for (int i=1; i<=nCols; i++)
#057                {
#058                    Object o = rset.getObject(i);
#059                    System.out.print(o + "\t");
#060                }
#061                System.out.print("\n");
#062            }
#063        }
#064
#065        public void close()
#066        {
#067            if (_conn != null)
#068            {
#069                try
#070                {
#071                    _conn.close();
#072                }
#073                catch (Exception ex)
#074                {
#075                }
#076            }
#077        }
```

```
#078
#079        public static void main(String[] args)
#080        {
#081            try
#082            {
#083                ASEDatabase ase = new ASEDatabase("192.168.56.1", 5000, "iihero", "spring", "spring1");
#084                ResultSet rset = ase.executeQuery("SELECT * FROM emp WHERE empno=7499");
#085                ase.dumpResultSet(rset);
#086                rset.close();
#087            }
#088            catch (Exception ex)
#089            {
#090                ex.printStackTrace();
#091            }
#092        }
#093
#094 }
```

最终输出结果为：

| 7499 | ALLEN | SALESMAN | 7698 | 1981-02-20 00:00:00.0 | 1600.00 | 300.00 | 30 | comment | of |
| ALLEN | null | F | | | | | | | |

需要说明的是，函数 executeQuery()返回的不是原来的那个 ResultSet，因为那个 ResultSet 是由 Statement 生成的，它的实例可能包含相应的数据库资源，必须释放，最终返回的是一个 CachedResultSet，它派生于 ResultSet，但是完全基于内存的一个结果集。

另外，本例中的 SQL 语句虽然带有条件，但不是动态绑定的。如果参数值是完全动态的，则需要使用 PreparedStatement 接口类来准备 SQL 语句。

同样是上例，如果想动态绑定 empno 的值并获取查询的结果，可以写这样的函数：

```
public void dumpEmpByNo(int empno) throws Exception
{
    String sql = "SELECT * FROM emp WHERE empno=?";
    PreparedStatement pstmt = _conn.prepareStatement(sql);
    pstmt.setInt(1, empno);
    ResultSet rset = pstmt.executeQuery();
    CachedRowSet   crset = new CachedRowSetImpl();
    crset.populate(rset);
    rset.close();
    pstmt.close();
    dumpResultSet(crset);
}
```

PreparedStatement 接口类主要用于参数值的动态绑定，这里将输入的参数值 empno 与第一个占位符 "?" 绑定在一起，然后执行查询。动态绑定主要用于两种情形：

1. 参数值是动态确定的，不是事先预知的。
2. 同一条 SQL 语句多次执行，使用动态绑定，数据库服务器只需要解析一次，从而有可能提高效率。

值得一提的是，动态绑定还有一个最大的好处是，它可以有效地防止 SQL 注入。所谓 SQL 注入，就是指通过输入特殊的查询条件，导致能够轻易地通过后台数据库的查询验证。在如今的 Web 应用甚至普通的信息化系统当中，用户输入用户名和密码的提交过程往往意味着要执行一个 SQL 查询。

例如，我们在 iihero 数据库里有一个用户表 dru_user(username varchar(32) primary key, password varchar(32) not null)，执行下述建表语句并插入一条用户记录：

```
1> create table dru_user(username varchar(32) primary key, password varchar(32) not null)
2> go
1> insert into dru_user values('iihero', 'pwd_iihero')
2> go
(1 row affected)
```

如果使用下面的代码来验证一个用户是否合法，那将是一场灾难：

```java
public boolean isValidUser(String username, String password) throws Exception
{
    boolean res = false;
    String sql = "SELECT count(1) FROM dru_user WHERE username='" + username
        + "' AND password='" + password + "'";
    System.out.println("SQL: " + sql);
    Statement stmt = _conn.createStatement();
    ResultSet rset = stmt.executeQuery(sql);
    rset.next();
    res = rset.getInt(1) > 0;
    rset.close();
    stmt.close();
    return res;
}
```

这里只是利用 username 和 password 的值拼接成一个 SQL 查询，然后执行查询。这样很容易构成漏洞。

看看下边这两行调用，会发现，用户可以提供无效的用户名密码，同样返回 true。

```
System.out.println("用户有效? " + ase.isValidUser("iihero", "pwd_iihero"));
System.out.println("用户有效? " + ase.isValidUser("abcde", "wrong' OR '1'='1"));
```

我们看看中间的 SQL 语句输出：

```
SQL: SELECT count(1) FROM dru_user WHERE username='iihero' AND password='pwd_iihero'
用户有效? true
```

SQL: SELECT count(1) FROM dru_user WHERE username='abcde' AND password='wrong' OR '1'='1'
用户有效? true

发现用户通过第二个参数 wrong OR 1=1 完成拼接以后，形成查询条件中有一个 1=1 的恒等式，永远为 true，所以按照这种思路用户完全可以随意构造一个登录的用户名和密码，就可以绕过登录验证。

解决这种漏洞的最直接有效的方法就是使用 PreparedStatement，将上述代码稍稍改动，实现成如下形式：

```
// 避免 SQL 注入
public boolean isValidUserBeta(String username, String password) throws Exception
{
    boolean res = false;
    String sql = "SELECT count(1) FROM dru_user WHERE username=? AND password=?";
    System.out.println("SQL: " + sql);
    PreparedStatement stmt = _conn.prepareStatement(sql);
    stmt.setString(1, username);
    stmt.setString(2, password);
    ResultSet rset = stmt.executeQuery();
    rset.next();
    res = rset.getInt(1) > 0;
    rset.close();
    stmt.close();
    return res;
}
```

这样，在执行 System.out.println("用户有效? " + ase.isValidUserBeta("abcde", "wrong' OR '1'='1"))时，就会返回 false。因为它是由数据库端来将 wrong OR 1=1 这个字符串绑定到列 password 进行匹配，肯定找不到对应的记录行。

不光是 JDBC 接口会遇到 SQL 注入的漏洞，其他所有各种类型的访问接口都会有类似的情况，只要处理得当，是可以避免此类差错的。总结起来，SQL 注入漏洞大多是由直接的 SQL 语句拼接造成的，使用输入参数的动态绑定可以有效防止 SQL 注入漏洞。

16.3.2　Insert/Update/Delete 操作

上一节主要提到 SQL 查询操作，如何通过 Statement 或者 PreparedStatement 获取最终的结果集，执行 SQL 查询时，大多调用 Statement 的 executeQuery()方法。

但是 Insert/Update/Delete 操作执行的则是 executeUpdate()方法。

在上一节我们也看到了，执行动态绑定时，在传入参数和需要显式地传入参数值和参数类型时，这样的代码显得不够能用和灵活。

针对 PreparedStatement 类，它有众多 set<Type>()方法，可以根据传入的参数值动态地确定类型，从而达到动态传入参数值的目的。

相应的代码（同样位于类 ASEDatabase 当中）如下：

```
#001    public static enum NULL
#002    {
#003        // 常见类型的 NULL 值
#004        VARCHAR(Types.VARCHAR), INTEGER(Types.INTEGER), DOUBLE(Types.DOUBLE), BOOLEAN(Types.BIT), BYTE(Types.TINYINT), VARBINARY(
#005                    Types.VARBINARY), LONGVARBINARY(Types.LONGVARBINARY), TIME(Types.TIME), TIMESTAMP(Types.TIMESTAMP), DATE(
#006                    Types.DATE), BIGINT(Types.BIGINT), BINARY(Types.BINARY), FLOAT(Types.FLOAT), DECIMAL(Types.DECIMAL);
#007
#008        private int sqlType;
#009
#010        private NULL(int sqlType)
#011        {
#012            this.sqlType = sqlType;
#013        }
#014
#015        public int getSqlType()
#016        {
#017            return sqlType;
#018        }
#019    }
#020    private static void setParams(PreparedStatement stmt, Object[] params) throws SQLException
#021    {
#022        for (int i=0; i<params.length; i++)
#023        {
#024            setParam(stmt, i+1, params[i]);
#025        }
#026    }
#027
#028    // 根据 obj 的实际类型动态的设定参数值
#029    private static void setParam(PreparedStatement stmt, int i, Object obj) throws SQLException
#030    {
#031        if ( obj instanceof NULL )
#032        {
#033            stmt.setNull(i, ((NULL) obj).getSqlType());
#034        }
#035        else if ( obj instanceof String )
#036        {
#037            stmt.setString(i, (String) obj);
#038        }
#039        else if ( obj instanceof Integer )
#040        {
#041            stmt.setInt(i, ((Integer) obj).intValue());
#042        }
#043        else if ( obj instanceof Double )
```

```
#044            {
#045                    stmt.setDouble(i, ((Double) obj).doubleValue());
#046            }
#047            else if ( obj instanceof Boolean )
#048            {
#049                    stmt.setBoolean(i, ((Boolean) obj).booleanValue());
#050            }
#051            else if ( obj instanceof Byte )
#052            {
#053                    stmt.setByte(i, ((Byte) obj).byteValue());
#054            }
#055            else if ( obj instanceof ByteArrayOutputStream )
#056            {
#057                    stmt.setBytes(i, ((ByteArrayOutputStream) obj).toByteArray());
#058            }
#059            else if ( obj instanceof Time )
#060            {
#061                    stmt.setTime(i, (Time) obj);
#062            }
#063            else if ( obj instanceof Timestamp )
#064            {
#065                    stmt.setTimestamp(i, (Timestamp) obj);
#066            }
#067            else if ( obj instanceof InputStream )
#068            {
#069                    try
#070                    {
#071                            stmt.setBinaryStream(i, (InputStream) obj, ((InputStream) obj).available());
#072                    }
#073                    catch (IOException e)
#074                    {
#075                            throw new RuntimeException(e);
#076                    }
#077            }
#078            else
#079            {
#080                    stmt.setObject(i, obj);
#081            }
#082    }
#083
#084    // 完全动态参数化查询
#085    public ResultSet executeQuery(String sql, Object[] params) throws Exception
#086    {
#087            PreparedStatement stmt = _conn.prepareStatement(sql);
#088            setParams(stmt, params);
#089            ResultSet rset = stmt.executeQuery();
#090            CachedRowSet    crset = new CachedRowSetImpl();
#091            crset.populate(rset);
```

```
#092            rset.close();
#093            stmt.close();
#094            return crset;
#095        }
#096
#097        // 完全动态参数化 Insert/Update/Delete
#098        public int executeUpdate(String sql, Object[] params) throws Exception
#099        {
#100            PreparedStatement stmt = _conn.prepareStatement(sql);
#101            setParams(stmt, params);
#102            int affected = stmt.executeUpdate();
#103            stmt.close();
#104            return affected;
#105        }
#106
#107        // 相同的 SQL 语句，多次动态绑定并执行 Insert/Update/Delete 操作
#108        public int[] executeUpdate(String sql, List<Object[]> paramsList) throws Exception
#109        {
#110            int[] res = new int[paramsList.size()];
#111            PreparedStatement stmt = _conn.prepareStatement(sql);
#112            int i=0;
#113            for (Object[] params : paramsList)
#114            {
#115                setParams(stmt, params);
#116                int affected = stmt.executeUpdate();
#117                res[i++] = affected;
#118            }
#119            stmt.close();
#120            return res;
#121        }
```

首先，我们定义了各种 SQL 类型的 NULL 值，用于 PreparedStatement 的 setNull 操作。请看第 29～82 行的 setParam 操作，在调用 setNull 时，必须知道原始的传入列的列类型。只要在 Object[] 里传递 NULL.<列类型>，即可传入对应的 NULL 值。

下面看看实际的调用效果：

（1）动态 Insert 操作。

```
System.out.println("insert count = " + ase.executeUpdate("INSERT INTO multitype_t values(?,?,?,?,?,?,?,?,?,?,?,?,?)",
            new Object[]{100, "iihero", "iihero_var", 55.54f, 12345.124f, 1234567.7654f, 1503.50f, "2007-11-26 15:56:03",
            0x123EF, 1, NULL.VARBINARY, "comment for 1"
            }));
```

列 col11 被传入 NULL 值。

（2）动态 Update 操作。

```
System.out.println("update count = " + ase.executeUpdate("UPDATE multitype_t SET col11=? WHERE id=?", new Object[]
{new byte[]{30, 31, 32, 33},   100}));
```

更新了 col11 的值，传入字节数组 30,31,32,33，实际上就是 0x1e1f2021。

（3）动态 Delete 操作。

```
System.out.println("delete count = " + ase.executeUpdate("DELETE FROM multitype_t WHERE id=?", new Object[]{100 }));
```

总结上述调用，非常通用，只需要构造出对应的 Object 数组即可实现与 SQL 语句无关的动态调用。

关于动态绑定、动态 Prepare，如果想提高效率，让服务器端只对相同的 SQL 语句预编译一次，那么应该设置连接属性 DYNAMIC_PREPARE 为 true，它的默认值为 false。

至于普通的 DDL 操作，利用动态的 executeUpdate()调用可以直接完成，不再赘述。

16.3.3 事务的提交

在 ASE 的 JDBC 库中获取 Connection 之后，它的 AutoCommit 属性默认值是 true，意味着每执行一次 CUD 操作（insert、update、delete），都会隐含自动提交一起事务，没办法回退。这在生产环境当中是不能接受的。

解决办法是：

```
if (_conn.getAutoCommit())
{
    System.out.println("连接默认为： autoCommit=" + _conn.getAutoCommit());
    _conn.setAutoCommit(false);
}
```

获取连接之后，立即将 autoCommit 设置为 false。这样，就可以调用 _conn.commit()或 _conn.rollback()来提交或者回滚事务了。

16.4 BLOB/CLOB 读写

ASE 数据库有个重要的连接属性，那就是 TEXTSIZE，在其他应用程序中，尤其是 Open Client 客户端程序，其默认值是 32KB，如果要获取的 TEXT/IMAGE（TEXT 即 CLOB，IMAGE 即 BLOB）字段内容，超过了这个长度限制则会被截短，要想获取完整的内容，必须将 TEXTSIZE 设置得足够大。

在 JDBC 驱动里，该属性默认值不再是 32KB，而是 2GB，因此，我们不用为此属性设置默认值。

16.4.1 TEXT 字段的读写

首先看看 TEXT 字段的读取操作，当字段长度不长时，比如不超过 32KB 时，可以直接通过 ResultSet 接口的 getString()方法获取。如果长度很长，调用 getString()获取则效率很低。比如，数据库的 TEXT 字段存储的是某些大的文本文件的内容，这时取回可能要分块写入本地的文本文件。

虽然我们也知道 TEXT 就是 CLOB 字段类型（SQL 标准），可是 ASE 的 JDBC 驱动并未实现 ResultSet 的 getCLOB(int col)方法，其提示是：Not Support getClob(int)。

为了说明有效的用法，我们在数据库 iihero 中创建一个单独的表，脚本如下：

```
1> create table testblobclob(id int primary key, col1 image, col2 text)
2> go
```

利用上一节的程序 ASE Database 类，调用 executeUpdate()方法，插入一条记录，为 TEXT 字段插入值，这里是将文件 D:/SybaseASE/ASE-15_0/install/XIONGHE.copy.log 的内容写入字段 col2。

写入 TEXT 字段的方法代码如下所示：

```
ase.insertClob("D:/SybaseASE/ASE-15_0/install/XIONGHE.copy.log");

// 表 testblobclob(id int primary key, col1 image, col2 text)
public void insertClob(String localFile)
{
    try
    {
        String sql = "if not exists(select * from sysobjects where name='testblobclob') " +
            "execute('create table testblobclob(id int primary key, col1 image, col2 text)')";
        this.executeUpdate(sql, new Object[]{});
        sql = "insert into testblobclob values(1, 'abc', ?)";
        FileReader f = new FileReader(localFile);
        Object[] params = new Object[] { f };
        int affected = this.executeUpdate(sql, params);
        sql = "select datalength(col2) from testblobclob where id = 1";
        ResultSet rset = this.executeQuery(sql);
        this.dumpResultSet(rset);
        rset.close();
        _conn.commit();
        f.close();
    }
    catch (Exception ex)
    {
        ex.printStackTrace();
    }
}
```

这里传入的是一个 FileReader 对象，给 executeUpdate，最终会调用到 PreparedStatement 的 setCharacterStream 方法，不得不说的是，JDBC 中这个方法设计得并不十分理想，它需要用户提供这个 Reader 里要读取的字符数量，可是在 JDK 中，所有的 Reader 实现类都没有提供这个方法，于是有了下边比较迂回的实现（在 ASE Database 类的 setParam 方法里）：

```
        else if ( obj instanceof FileReader )
        {
            try
            {
                BufferedReader fr = new BufferedReader((FileReader) obj);
                char[] buf = new char[8192];
                int tmp = 0;
                int len = 0;
                fr.mark(MARK_MAX);
                while ((tmp = fr.read(buf)) > 0)
                {
                    len += tmp;
                }
                fr.reset();
                stmt.setCharacterStream(i, fr, len);
            }
            catch (Exception ex)
            {
                throw new RuntimeException(ex);
            }
        }
```

其原理是先将 FileReader 读取一遍，计算出其长度，再设法 reset，将读取文件的指针复原到最开始的位置。FileReader 类本身并不支持 reset()，BufferedReader 类才支持。并且，mark()方法的参数 readAheadLimit 指的是读取长度的限制，这里我们取 20MB 为最大限度，只要是 20MB 以内，这个记号还是有效的，还能再 reset 回来，即是此意。

使用 setCharacterStream 调用的原因是，它能充分利用 JDBC 调用，直接分块读取这个流，达到最快的速度。在本例中，输入文件的长度有 229KB。

```
1> select datalength(col2) from testblobclob
2> go

 -----------
     229622

(1 row affected)
```

读取 TEXT 字段有两种方式，如果调用 ResultSet 的 getObject 方法，返回的将是一个 String 实例，对于长度不太长的 TEXT 是可以接受的。可是如果太长，则会增加服务器端的开销。这时我们可以采用 ResultSet 的 getCharacterStream(int col)方法得到一个 Reader 对象，然后将其写入本地。

在类 ASEBLOBCLOB 的方法 readClob 方法里有一个示例实现，其代码如下：

```
#001    // 将表 testblobclob 中 id 值为<id>的列 col2 的 text 值写入本地文件 localFile 当中
#002    public void readClob(int id, String localFile)
#003    {
#004        try
#005        {
```

```
#006                String sql="SELECT col2 FROM testblobclob WHERE id=?";
#007                ResultSet rset = executeQuery(sql, new Object[] {id });
#008                if (rset.next())
#009                {
#010                    BufferedReader r = new BufferedReader(rset.getCharacterStream(1));
#011                    BufferedWriter fw = new BufferedWriter(new FileWriter(localFile));
#012                    char[] buf = new char[8192];
#013                    int len = 0;
#014                    int sum = 0;
#015                    while ( (len = r.read(buf)) > 0 )
#016                    {
#017                        fw.write(buf, 0, len);
#018                        sum += len;
#019                    }
#020                    fw.close();
#021                    r.close();
#022                    rset.close();
#023                    System.out.println("总计写入字符数为: " + sum);
#024                }
#025            }
#026            catch (Exception ex)
#027            {
#028                ex.printStackTrace();
#029            }
#030        }
```

在调用"ase.readClob(1, "C:/tmp.txt");"之后，会得到如下结果：

连接默认为: autoCommit=true
229622
总计写入字符数为: 229622

我们可以看到，整个读取 Reader，然后利用一个 BufferedWriter 不断地将读取的 Buffer 中的内容写入一个本地文件，这样的执行效率是非常快的。

16.4.2 IMAGE 字段的读写

JDBC 对 ASE 的 IMAGE 字段的读写操作与 TEXT 字段的读写操作有些类似，主要区别是，TEXT 字段调用 get/set CharacterStream，针对 Reader 对象进行操作，即操作的是字符流；而 IMAGE 字段则是调用 get/set BinaryStream，针对 InputStream 对象进行操作，即操作的是字节流，这两个是本质上的区别，不能互换。

以表 testblobclob 为例，我们来看看 IMAGE 字段的读写操作过程，其代码如下所示：

```
#001        // 将本地二进制文件插入 testblobclob 的 col1 列当中，并输出插入后的长度
#002        public void insertBlob(String localFile)
#003        {
#004            try
```

```
#005            {
#006                String sql = "if not exists(select * from sysobjects where name='testblobclob') " +
#007                        "execute('create table testblobclob(id int primary key, col1 image, col2 text)')";
#008                this.executeUpdate(sql, new Object[]{});
#009                sql = "insert into testblobclob values(2, ?, 'abc')";
#010                InputStream f = new BufferedInputStream(new FileInputStream(localFile));
#011                Object[] params = new Object[] { f };
#012                int affected = this.executeUpdate(sql, params);
#013                sql = "select datalength(col1) from testblobclob where id = 2";
#014                ResultSet rset = this.executeQuery(sql);
#015                this.dumpResultSet(rset);
#016                rset.close();
#017                _conn.commit();
#018                f.close();
#019            }
#020            catch (Exception ex)
#021            {
#022                ex.printStackTrace();
#023            }
#024        }
#025
#026    // 将表 testblobclob 中 id 值为<id>的列 col1 的 IMAGE 类型值写入本地文件 localFile 当中
#027        public void readBlob(int id, String localFile)
#028        {
#029            try
#030            {
#031                String sql="SELECT col1 FROM testblobclob WHERE id=?";
#032                ResultSet rset = executeQuery(sql, new Object[] {id });
#033                if (rset.next())
#034                {
#035                    BufferedInputStream r = new BufferedInputStream(rset.getBinaryStream(1));
#036                    BufferedOutputStream fw = new BufferedOutputStream(new FileOutputStream(localFile));
#037                    byte[] buf = new byte[8192];
#038                    int len = 0;
#039                    int sum = 0;
#040                    while ( (len = r.read(buf)) > 0 )
#041                    {
#042                        fw.write(buf, 0, len);
#043                        sum += len;
#044                    }
#045                    fw.close();
#046                    r.close();
#047                    rset.close();
#048                    System.out.println("总计写入字节数为: " + sum);
#049                }
#050            }
#051            catch (Exception ex)
#052            {
```

```
#053                    ex.printStackTrace();
#054               }
#055          }
#056
#057          ase.insertBlob("D:/SybaseASE/jConnect-6_0/classes/jconn3.jar");
#058          ase.readBlob(2, "c:/jconn3.read.jar");
```

insertBlob(String file)为写入操作，readBlob(int id, String file)为读取 IMAGE 类型值操作，并将其写入本地一个二进制文件。

最终执行结果为：

连接默认为：autoCommit=true
553569
总计写入字节数为：553569

写入操作的绑定变量的过程当中，最终会调用 PreparedStatement 的 setBinaryStream 方法。该方法是一个比较高效的方法，它最终会执行分块读取并绑定。读取操作时，会调用 ResultSet 的 getBinaryStream 方法。

在这里我们可以看到，BinaryStream 的关联类 InputStream 有 available()方法可以直接得到字节流的长度，不需要进行任何的预读取，而 CharacterStream 的关联类 Read 的任何子类都没有 length()或者 available()方法得到字符流的长度，JDBC 接口规范的原始定义当中，可能考虑到不一定要整个流的全部内容都写入数据库，可能只需要一部分。但是这样给写全部流（字符流）到数据库带来困挠，当你传入一个超大值（比如 2GB）作为长度，可能有的数据库不支持这样的无效长度操作。

16.5　调用存储过程

调用存储过程与一般的 CRUD 操作有些不同，它要使用的是 PreparedStatement 的子接口 CallableStatement，它所调用的 SQL 存储过程语句采用标准的调用格式：

{call <procedure-name>[(<arg1>,<arg2>, ...)]}

或者

{?= call <procedure-name>[(<arg1>,<arg2>, ...)]}

前一种格式是不带返回值的，后一种格式是存储过程带有返回值的。

JDBC 对存储过程的处理，要求在执行存储过程之前，必须对所有的输出型参数占位符进行提前注册，告诉服务器端要准备的返回类型，这些占位符与输入型参数占位符统一进行编号，从 1 计起。因此，两者是混合的，可能是设置输出类型，可能是直接设定值，即如果要实现完全动态的注册和动态绑定，还是比较复杂的。

存储过程调用完成之后,可能有多个结果集返回,应用程序必须要考虑到这一点。

本章将给出一个示例,介绍如何调用存储过程并获取最终结果。

示例所用的存储过程如下:

```
create proc spring.test_proc(@s_name char(30), @s_count int output) with recompile
as
select @s_count = count(a.sno) from spring.student a where a.sname = @s_name
select 'demo123'
select 'hello, 给一个中文测试结果'
go
sp_procxmode test_proc, "anymode"
go
declare @result int
exec spring.test_proc '李勇', @result output
select @result
go
```

我们可以直观地从存储过程的内容里判断出,它至少有三个返回值,一个是 output 型变量 @s_count,另一个是中间的两个 SQL 查询 select 'demo123' 和 select 'hello, 给一个中文测试结果'。

下面,我们来看看完整的存储过程调用代码:

```
#001    package com.iihero;
#002
#003    import java.sql.CallableStatement;
#004    import java.sql.ResultSet;
#005    import java.sql.Types;
#006
#007    public class ASEProcedure extends ASEDatabase
#008    {
#009        public ASEProcedure(String host, int port, String database, String username, String password) throws Exception
#010        {
#011            super(host, port, database, username, password);
#012        }
#013
#014        // 执行存储过程:test_proc, 并处理结果
#015        public void executeProc()
#016        {
#017            try
#018            {
#019                CallableStatement stmt = _conn.prepareCall("{call test_proc(?, ?)}");
#020                stmt.setString(1, "李勇");
#021                stmt.registerOutParameter(2, Types.INTEGER);
#022                boolean hasResult = stmt.execute();
#023                int rowsAffected = 0;
#024                int i=0;
#025                int j=0;
#026                do
```

```
#027                {
#028                    if (hasResult)
#029                    {
#030                        j++;
#031                        ResultSet rset = stmt.getResultSet();
#032                        if (rset.next())
#033                        {
#034                            System.out.println("第" + j + "轮结果集输出：");
#035                            System.out.println("            " + rset.getObject(1));
#036                        }
#037                        rset.close();
#038                    }
#039                    rowsAffected = stmt.getUpdateCount();
#040
#041                    i++;
#042                    System.out.println("" + i + ": affected rows: " + rowsAffected);
#043                    hasResult = stmt.getMoreResults();
#044                }
#045            while (hasResult || rowsAffected != -1);
#046            // 最后处理 out 参数的值
#047            System.out.println("@s_count=" + stmt.getInt(2));
#048            stmt.close();
#049        }
#050        catch (Exception ex)
#051        {
#052            ex.printStackTrace();
#053        }
#054    }
#055 }
#056
#057    public static void main(String[] args)
#058    {
#059        ASEProcedure ase;
#060        try
#061        {
#062            ase = new ASEProcedure("192.168.56.1", 5000, "iihero", "spring", "spring1");
#063            ase.executeProc();
#064            ase.close();
#065        }
#066        catch (Exception e)
#067        {
#068            e.printStackTrace();
#069        }
#070    }
#071 }
```

最终执行结果：

```
1: affected rows: 1
第 1 轮结果集输出：
          demo123
2: affected rows: 1
第 2 轮结果集输出：
          hello, 给一个中文测试结果
3: affected rows: 1
4: affected rows: -1
@s_count=1
```

调用存储过程的方法 executeProc() 的基本过程如下：

Step 1 初始化 CallableStatement 对象，使用带占位符的 SQL 语句进行初始化。

Step 2 注册 out 变量，绑定输入变量，所有占位符无论是 out 变量还是输入绑定变量，都统一进行编号，从 1 计起。因而你可以看到 setString(1, …) 及 registerOutParameter(2, …)。

Step 3 调用 stmt.execute() 方法，根据其返回值判断是否有结果集。

Step 4 第 26~46 行，从 stmt 对象中不断取出结果集并输出，直至没有结果集。然后调用 stmt.hasMoreResults() 来判断是否还有结果集。

存在结果集的条件是：hasResult 为 true 或者 affectedRows != -1，也就是说只有当 hasMoreResults() 为 false，并且 stmt.getUpdateCount()==-1 的时候，才真正没有结果集，这是 ASE 数据库的表现方式。其他数据库实现可能不太一样，所以要谨记。

Step 5 仅当所有的结果集处理完毕并释放结果集以后，才能去取 out 变量值，通过调用 CallableStatement 的 get<Type>(out_param) 来获取，见第 47 行。如果在提取结果集之前就取了 out 变量值，后边再取结果集时可能什么也取不到。

16.6 使用 JDBC 访问 ASE 元信息

JDBC 中有多种针对数据库的元信息，包括 DatabaseMetadata、ResultSetMetadata 等。通过对元信息的读取，能动态地获取一些重要信息，比如结果展示、确定是否支持某种功能等。

DatabaseMetadata 接口类主要用于获取数据库一级的元信息，有以下重要的元信息：

1. dataDefinitionIgnoredInTransactions()：是否忽略事务中的 DDL 语句，在 Sybase 中有一个数据库选项 ddl in transaction，是允许事务里支持 DDL 语句并可以回退。
2. doesMaxRowSizeIncludeBlobs()：定义 getMaxRowSize() 是否也包含了 BLOB 字段的实际大小。
3. getCatalogs()：返回数据库中所有的 catalog 名字，catalog 是一个数据库的逻辑概念，各 DBMS 对它的划分都不太一样，有的是按照用户名划分的，有的是按照数据库名划分的，

ASE 里返回的就是数据库名，围绕 catalog 的还有 api: getCatalogSeparator() 和 getCatalogTerm()。

4. getColumns(String catalog, String schemaPattern, String tableNamePattern, String columnNamePattern)：获取列名，按照上述条件来获取。

5. getCrossReference(String parentCatalog, String parentSchema, String parentTable, String foreignCatalog, String foreignSchema, String foreignTable)：获取主外键依赖关系。

6. getMaxProcedureNameLength()：获取最长过程名长度。

7. getMaxIndexLength()：最大索引长度。

8. getMaxSchemaNameLength()：最长 schema 名长度。

9. getMaxStatementLength()：SQL 语句最大长度。

10. getMaxStatements()：数据库同时能打开的 SQL 语句的最大个数。

11. getMaxTableNameLength()：表名最大长度。

12. getMaxTablesInSelect()：Select 语句中允许的表的最多个数。

13. getMaxUserNameLength()：用户名最大长度。

14. getPrimaryKeys(String catalog, String schema, String table)：返回主键信息。

15. getSchemas(String catalog, String schemaPattern)：返回 schema 信息，实际上就是用户名信息。

16. getTables(String catalog, String schemaPattern, String tableNamePattern, String[] types)：返回表名列表。

17. getURL()：返回该 DBMS 的 URL。

18. getUserName()：返回用户名。

DatabaseMetadata 由 Connection 调用 getDatabaseMetadata()方法直接得到，比较方便。

有关 ResultSetMetadata 接口类，由接口类 ResultSet 调用 getMetadata()方法得到，主要用于获取结果集中各列的元信息，主要包括：

1. getCatalogName(int col)：获取该列的对应表的 catalog 值。

2. getColumnClassName(int col)：获取对应列的对应类名（在调用 getObject()时）。

3. getColumnCount()：返回列数。

4. getColumnDisplaySize(int col)：返回某列的显示尺寸，通常返回该列的最大宽度。

5. getColumnName(int col)：返回某列的名字。

6. getColumnType(int col)：返回某列的类型。

7. getColumnTypeName(int col)：返回某列的类型名。

8. getPrecision(int col)：返回某列的精度。

9. getScale(int col)：返回某列的小数位数。

10. isAutoIncrement(int col)：是否自是增列，在 ASE 里意味着是否是 identity 列。

11. isCaseSensitive(int col)：是否大小写不区分。
12. isNullable(int col)：是否允许为空。
13. isSearchable(int col)：某列是否可以用于 where 中的查询条件。

获取元信息的代码比较简单，下面是相应的调用过程：

```java
#001 package com.iihero;
#002
#003 import java.sql.DatabaseMetaData;
#004 import java.sql.ResultSet;
#005 import java.sql.ResultSetMetaData;
#006
#007 public class ASEMetadata extends ASEDatabase
#008 {
#009     private DatabaseMetaData _dbMeta;
#010
#011     public ASEMetadata(String host, int port, String database, String username, String password) throws Exception
#012     {
#013         super(host, port, database, username, password);
#014         _dbMeta = _conn.getMetaData();
#015     }
#016
#017     // 输出数据库名列表
#018     public void listCatalogs() throws Exception
#019     {
#020         System.out.println("数据库名列表：");
#021         ResultSet rset = _dbMeta.getCatalogs();
#022         this.dumpResultSet(rset);
#023         rset.close();
#024     }
#025
#026     // 输出用户名列表
#027     public void listSchemas() throws Exception
#028     {
#029         System.out.println("用户名列表：");
#030         ResultSet rset = _dbMeta.getSchemas();
#031         this.dumpResultSet(rset);
#032         rset.close();
#033     }
#034
#035     //输出当前数据库 iihero 的所有表的信息
#036     public void listTables() throws Exception
#037     {
#038         System.out.println("表名列表：");
#039         ResultSet rset = _dbMeta.getTables(null, null, null, null);
#040         this.dumpResultSet(rset);
#041         rset.close();
#042     }
```

```
#043
#044        //输出表 table 里的主键信息
#045        public void getPrimrayKeys(String table) throws Exception
#046        {
#047            System.out.println("表: " + table +" 主键信息如下： ");
#048            ResultSet rset = _dbMeta.getPrimaryKeys("iihero", _dbMeta.getUserName(), table);
#049            dumpResultSet(rset);
#050            rset.close();
#051        }
#052
#053        // 列出 URL、用户名及驱动等元信息
#054        public void listOthers() throws Exception
#055        {
#056            System.out.println("url: " + _dbMeta.getURL());
#057            System.out.println("username: " + _dbMeta.getUserName());
#058            System.out.println("driver info: " + _dbMeta.getDriverName() + ", " + _dbMeta.getDriverVersion());
#059        }
#060
#061        // 从 dbMeta 里取某一个表的各列元信息
#062        public void getColumns(String table) throws Exception
#063        {
#064            System.out.println("表: " + table + " 元信息如下： ");
#065            ResultSet rset = _dbMeta.getColumns("iihero", _dbMeta.getUserName(), table, null);
#066            ResultSetMetaData meta = rset.getMetaData();
#067            int nCols = meta.getColumnCount();
#068            for (int i=0; i<nCols; i++)
#069            {
#070                System.out.print(meta.getColumnName(i+1) + "\t");
#071            }
#072            System.out.print("\n");
#073            dumpResultSet(rset);
#074            rset.close();
#075        }
#076
#077
#078        public static void main(String[] args)
#079        {
#080            try
#081            {
#082                ASEMetadata ase = new ASEMetadata("192.168.56.1", 5000, "iihero", "spring", "spring1");
#083                ase.listCatalogs();
#084                ase.listSchemas();
#085                ase.listOthers();
#086                ase.listTables();
#087                ase.getPrimrayKeys("student");
#088                ase.getColumns("student");
#089                ase.close();
#090            }
```

```
#091            catch (Exception ex)
#092            {
#093                ex.printStackTrace();
#094            }
#095        }
#096
#097 }
```

这里为了显示的目的，并没有以数组的形式返回相应结果，只是把结果集输出来。listTables、getColumns 等方法在实际应用中还是比较实用的，它使我们不用直接查询系统表，就可以得到准备的元信息。如 listTables() 的输出如下（表名列表）：

iihero	dbo	sysalternates	SYSTEM TABLE	null
iihero	dbo	sysattributes	SYSTEM TABLE	null
iihero	dbo	syscolumns	SYSTEM TABLE	null
iihero	dbo	syscomments	SYSTEM TABLE	null
iihero	dbo	sysconstraints	SYSTEM TABLE	null
iihero	dbo	sysdepends	SYSTEM TABLE	null
iihero	dbo	sysencryptkeys	SYSTEM TABLE	null
iihero	dbo	sysgams	SYSTEM TABLE	null
iihero	dbo	sysindexes	SYSTEM TABLE	null
iihero	dbo	sysjars	SYSTEM TABLE	null
iihero	dbo	syskeys	SYSTEM TABLE	null
iihero	dbo	syslogs	SYSTEM TABLE	null
iihero	dbo	sysobjects	SYSTEM TABLE	null
iihero	dbo	syspartitionkeys	SYSTEM TABLE	null
iihero	dbo	syspartitions	SYSTEM TABLE	null
iihero	dbo	sysprocedures	SYSTEM TABLE	null
iihero	dbo	sysprotects	SYSTEM TABLE	null
iihero	dbo	sysqueryplans	SYSTEM TABLE	null
iihero	dbo	sysreferences	SYSTEM TABLE	null
iihero	dbo	sysroles	SYSTEM TABLE	null
iihero	dbo	syssegments	SYSTEM TABLE	null
iihero	dbo	sysslices	SYSTEM TABLE	null
iihero	dbo	sysstatistics	SYSTEM TABLE	null
iihero	dbo	systabstats	SYSTEM TABLE	null
iihero	dbo	systhresholds	SYSTEM TABLE	null
iihero	dbo	systypes	SYSTEM TABLE	null
iihero	dbo	sysusermessages	SYSTEM TABLE	null
iihero	dbo	sysusers	SYSTEM TABLE	null
iihero	dbo	sysxtypes	SYSTEM TABLE	null
iihero	dbo	t	TABLE	null
iihero	dbo	t1222	TABLE	null
iihero	dbo	tact	TABLE	null
iihero	dbo	ttt	TABLE	null
iihero	spring	ULCustomer	TABLE	null
iihero	spring	ULEmployee	TABLE	null

iihero	spring	ULOrder	TABLE	null
iihero	spring	ULProduct	TABLE	null
iihero	spring	bonus	TABLE	null
iihero	spring	course	TABLE	null
iihero	spring	dept	TABLE	null
iihero	spring	dru_user	TABLE	null
iihero	spring	emp	TABLE	null
iihero	spring	movie	TABLE	null
iihero	spring	movieexec	TABLE	null
iihero	spring	moviestar	TABLE	null
iihero	spring	multitype_t	TABLE	null
iihero	spring	mydummy	TABLE	null
iihero	spring	salgrade	TABLE	null
iihero	spring	st_course	TABLE	null
iihero	spring	starsin	TABLE	null
iihero	spring	student	TABLE	null
iihero	spring	studio	TABLE	null
iihero	spring	t0	TABLE	null
iihero	spring	t1	TABLE	null
iihero	spring	tavg	TABLE	null
iihero	spring	tdatetime	TABLE	null
iihero	spring	testblob	TABLE	null
iihero	spring	testblobclob	TABLE	null
iihero	dbo	sysquerymetrics	VIEW	null

结果非常直观，第一列是数据库名，第二列是用户名，第三列是表名，第四列是表的类型（系统表、用户表、视图），第五列全部置空，留作扩展。

获取元信息对于 DBA 或者开发数据库的管理工具的相关人员来说，还是有参考意义的，比如，不同数据库之间的少批量数据迁移，在迁移之前就需要得到所有表的相关元信息。

16.7 JDBC 中的 ASE 数据库连接池

数据库连接池是 JDBC 访问中非常重要的一项应用，它直接影响到应用系统的性能，尤其是应用服务器的性能。

16.7.1 数据库连接池的基本原理

对于一个普通的应用服务器应用而言，当同时来自多个客户端的请求到来时，假设每个客户端请求都需要应用服务器做一到多次数据库访问，这自然就涉及到数据库连接。如果只是请求到来时才创建到后端的数据库物理连接，那么每次请求的处理时间就包括这个数据库连接的创建时间。

数据库连接的创建时间是非常耗时的网络 I/O 处理，相比于普通的 Web 请求而言，它在处理 Web 请求的整个时间段里占的比重就太高了，从而严重影响实际的响应速度。

建立连接池的基本思想就是，预先创建好固定数目的数据库连接，大多是在应用服务器启动的时候就已经创建好，这些连接始终保持活跃状态，因为是在启动阶段创建好的，并且连接资源（Connection 对象）都已经缓存到数据库源中（DataSource），当前端的请求到来时，如果需要数据库连接，直接从缓存的连接池里取出一个对应的连接就可以，在请求处理完毕时，再将连接放回连接池，可以为继续的请求继续服务。

总结起来，连接池的基本原理如下：

1. 每个连接池中的所有连接共用相同的连接信息（用户名、密码、主机、端口号及数据库名），有一个初始的最小数目的连接数目，在启动阶段就会被创建好，还有一个最大的连接数目，允许在运行期动态增长。

2. 运行期，每一个数据库请求都会从连接池中取出一个连接，处理完之后，调用相应的 close() 方法，将其重新放入连接池，并标为可用。

3. 如果池里已经没有可用的连接，新来数据库请求时，只要总的连接数目没有达到最大允许数目，就直接创建物理连接，并将其标为已用状态，待处理完毕，重新放入连接池。

4. 保持所有的数据库连接为活跃状态，这需要连接池有一个轮询处理，即在连接处于空闲状态时，每隔一定时间间隔（如 5 分钟），与后端的数据库交互一次，比如做一个最简单的查询 select 1 等，其目的就是保持这个连接是活连接。相信很多人碰到过这样一种现象：获取某一个连接以后，长时间没有执行 SQL 操作，最后再处理 SQL 操作时，连接已经"死"掉了。这是因为某些 DBMS 做了相应的处理，如果客户端连接长时间没有任何操作，会主动断掉到客户端的连接，这样一来，这个物理连接就变得不可用了。最典型的，My SQL 数据库在 1 个小时没有任何活动的情况下，就会断开客户端连接。

5. 释放资源。应用服务器关闭的阶段，连接池必须能够自动关闭并释放所有的物理连接资源。

基于上述思想，我们也可以实现一套自用的连接池。但是，目前已经有很多非常实用、功能强大的开源连接池实现。这里就不作实现方面的介绍了，后边重点介绍几种开源连接池的应用。

16.7.2 开源连接池在 ASE 数据库上的应用

这里以 Web 容器 Tomcat 为例，介绍几种最常使用开源连接池对 ASE 数据库的支持。这里采用的 Tomcat 版本是 apache-tomcat-6.0.35.zip。

1. DBCP 连接池

Tomcat 自带的连接池库 tomcat-dbcp.jar 位于 TOMCAT_HOME/lib 目录下边。

DBCP 连接池有两种配置方法，一种是全局配置，即所有 Web 应用都可以访问这个连接池；另一种是 Web 应用局部配置，这种配置里的连接池只能被这个 Web 应用本身访问，与其他 Web 应用相互隔离。一般而言，如果没有多个 Web 应用共享连接池，使用局部配置比较常见。

（1）首先，我们看看全局连接池的配置方法。为了方便验证，我们制作一个简单的 Web 应用 pool_dbcp_global，只有一个 index.jsp 文件，其内容如下：

```
#001 <%@page contentType="text/html" pageEncoding="UTF-8"%>
#002 <%@page import="java.sql.*"%>
#003 <%@page import="javax.sql.*"%>
#004 <%@page import="javax.naming.*"%>
#005 <%@page import="java.util.*"%>
#006
#007 <!DOCTYPE HTML PUBLIC "-//W3C//DTD HTML 4.01 Transitional//EN"
#008         "http://www.w3.org/TR/html4/loose.dtd">
#009
#010 <html>
#011     <head>
#012         <meta http-equiv="Content-Type" content="text/html; charset=UTF-8">
#013         <title>JSP Page</title>
#014     </head>
#015     <body>
#016         <h2>Hello World! 简单连接池测试.....</h2>
#017 <%
#018     // get the connection and create table etc.
#019     Context initContext = new InitialContext();
#020     Context envContext  = (Context)initContext.lookup("java:/comp/env");
#021     Statement stmt = null;
#022     Connection conn = null;
#023     try {
#024         DataSource ds = (DataSource)envContext.lookup("jdbc/SybaseGlobal");
#025         conn = ds.getConnection();
#026         out.println("OP: get the connection successfully......!!!!<br>");
#027         stmt = conn.createStatement();
#028         stmt.executeUpdate("create table t1234567(id int not null primary key)");
#029         out.println("OP: create table t1234567(id int not null primary key)......!!!!<br>");
#030         stmt.executeUpdate("drop table t1234567");
#031         out.println("OP: drop table t1234567......!!!!<br>");
#032         out.println("测试成功.....<br>");
#033     }
#034     catch(Exception ex) {
#035         out.println(ex.toString());
#036     }
#037     finally {
#038         if (stmt != null) {
#039             try {
#040                 stmt.close();
#041             }
#042             catch (Exception e) {
#043             }
#044         }
#045         if (conn != null) {
#046             try {
#047                 conn.close();
```

```
#048                    }
#049                    catch (Exception e) {
#050                    }
#051            }
#052       }
#053
#054
#055 %>
#056    </body>
#057 </html>
```

上述代码的功能比较简单，就是从连接池里取到连接以后，做一个建表和删表的动作。如果都顺利通过，没有任何异常，表明连接池里取到的连接是可用的连接。

最关键的代码就是第 20～25 行，第 20 行，首先取得可用的上下文 envContext，这是固定的，直接查找 java:/comp/env 获得，然后由 envContext 通过 JNDI 查找 jdbc/SybaseGlobal 得到数据源，最终通过数据源获取连接。

该 Web 应用（pool_dbcp_global）的部署文件 web.xml 需要添加以下内容：

```xml
<resource-ref>
    <description>DB Connection</description>
    <res-ref-name>jdbc/SybaseGlobal</res-ref-name>
    <res-type>javax.sql.DataSource</res-type>
    <res-auth>Container</res-auth>
</resource-ref>
```

表示需要引用到资源 jdbc/SybaseGlobal 与 JNDI 查找的资源名一致。

作为全局配置，我们需要在 Tomcat 的子目录 conf 下边的 context.xml 中添加连接池 jdbc/SybaseGlobal 的配置信息。

相应内容如下：

```xml
<Resource name="jdbc/SybaseGlobal"    auth="Container" type="javax.sql.DataSource"
    maxActive="20" maxIdle="30" maxWait="10000"
    username="spring" password="spring1"
    driverClassName="com.sybase.jdbc3.jdbc.SybDriver"
    url="jdbc:sybase:Tds:192.168.56.1:5000/iihero"/>

<!--配置连接池的部分信息,可选-->
<ResourceParams name="jdbc/SybaseGlobal">
    <!-- 回收被遗弃的（一般是忘了释放的）数据库连接到连接池中-->
    <parameter>
        <name>removeAbandoned</name>
        <value>true</value>
    </parameter>
    <!-- 数据库连接超过 60 秒,不用视为被遗弃而置回连接池中-->
    <parameter>
        <name>removeAbandonedTimeout</name>
```

```
            <value>60</value>
        </parameter>
        <!-- 将被遗弃的数据库连接的回收动作写入日志-->
        <parameter>
            <name>logAbandoned</name>
        <value>true</value>
        </parameter>
```

这样修改完以后,将 pool_dbcp_global 应用置入 tomcat/webapps 目录下边,把 ASE 数据库的 JDBC 驱动库 jconn3.jar 复制到 tomcat/lib 目录下,然后启动 Tomcat,直接访问 http://localhost:8080/pool_dbcp_global 即可得到验证的结果,如图 16-17 所示。

图 16-17　DBCP 数据库连接池验证结果

(2)下面,我们再看看 DBCP 的局部配置。

将 Web 应用 pool_dbcp_global 的 index.jsp 中的第 24 行稍作修改,改为 "DataSource ds = (DataSource)envContext.lookup("**jdbc/SybaseDB**");",表示它将引用另一个自己定义的数据源 jdbc/SybaseDB。

其 web.xml 内容也作调整,相关内容如下:

```
<resource-ref>
    <description>DB Connection</description>
    <res-ref-name>jdbc/SybaseDB</res-ref-name>
    <res-type>javax.sql.DataSource</res-type>
    <res-auth>Container</res-auth>
</resource-ref>
```

然后将 pool_dbcp_global 这个 Web 应用复制一份,改目录名为 pool_dbcp,成为新的 web 应用。

还有一个真正配置数据源资源 jdbc/SybaseDB 的地方,它不再是在 conf/context.xml 文件中,而是位于 conf/Catalina/localhost/pool_dbcp.xml 当中,即每一个 Web 应用在 conf/Catalina/localhost 目录下都有一个配置文件,添加相关内容如下:

```
<Resource name="jdbc/SybaseDB" auth="Container" type="javax.sql.DataSource"
             maxActive="20" maxIdle="30" maxWait="10000"
             username="spring" password="spring1" driverClassName="com.sybase.jdbc3.jdbc.SybDriver"
             url="jdbc:sybase:Tds:192.168.56.1:5000/iihero"/>
```

jconn3.jar 驱动文件只需要复制到 webapps\pool_dbcp\WEB-INF\lib 目录下即可。这样配置完成之后，不需要重启 Tomcat 即可进行验证，访问 http://localhost:8080/pool_dbcp/即可。

2. c3p0 连接池

c3p0 连接池来源于 sourceforge 上的开源项目 c3p0，位于 http://sourceforge.net/projects/c3p0/。这里以 c3p0-0.9.0.4.jar 为例进行说明，以局部配置为例。

设 Web 应用名为 pool_c3p0，在其 web.xml 里添加如下内容：

```
<resource-ref>
    <description>DB Connection</description>
    <res-ref-name>jdbc/SybaseC3P0</res-ref-name>
    <res-type>javax.sql.DataSource</res-type>
    <res-auth>Container</res-auth>
</resource-ref>
```

在 conf/Catalina/localhost/pool_c3p0.xml 里，添加如下注册项：

```
<Resource auth="Container"
driverClass="com.sybase.jdbc3.jdbc.SybDriver"
maxPoolSize="10"
minPoolSize="2"
acquireIncrement="2"
name="jdbc/SybaseC3P0"
user="spring"
password="spring1"
factory="org.apache.naming.factory.BeanFactory"
type="com.mchange.v2.c3p0.ComboPooledDataSource"
jdbcUrl="jdbc:sybase:Tds:192.168.56.1:5000/iihero" />
```

如果是全局配置，此部分内容则放到 conf/context.xml 文件当中。

测试验证程序只需要改动：

```
DataSource ds = (DataSource)envContext.lookup("jdbc/SybaseCP30");
```

全局配置时，要求 c3p0-0.9.0.4.jar 和 jconn3.jar 文件都放到 lib 目录下，局部配置时，这两个文件都放到 WEB-INF/lib 目录下。

3. proxool 连接池

proxool 连接池来源于开源项目 http://proxool.sourceforge.net/index.html，是性能不错的一个连接池实现。它的配置基本上是基于局部配置的。

以 proxool-0.9.1.zip 为例，我们来看看它是如何配置和使用 ASE 数据库的。首先将 proxool 相关的 jar 文件放入 Web 应用的 WEB-INF/lib 子目录当中，这些文件有：proxool-0.9.1.jar、proxool-cglib.jar 及 commons-logging-1.1.jar，前两个 proxool 的 jar 文件依赖于 commons-logging-1.1.jar，还需要将 ASE 的 JDBC 库 jconn.jar 也复制过来。

创建配置文件 WEB-INF/db.xml，其内容如下：

```xml
<?xml version="1.0" encoding="UTF-8" ?>

<proxool>
    <alias>sybase</alias>
    <driver-url>jdbc:sybase:Tds:192.168.56.1:5000/iihero</driver-url>
    <driver-class>com.sybase.jdbc3.jdbc.SybDriver</driver-class>
    <statistics>1m,15m,1d</statistics>
    <driver-properties>
    <property name="user" value="spring"/>
    <property name="password" value="spring1"/>
    </driver-properties>
    <maximum-connection-count>20</maximum-connection-count>
    <minimum-connection-count>2</minimum-connection-count>
    <maximum-active-time>60000</maximum-active-time>
    <house-keeping-test-sql>select 1</house-keeping-test-sql>
</proxool>
```

修改部署文件 web.xml,添加如下内容:

```xml
<servlet>
  <servlet-name>ServletConfigurator</servlet-name>
  <servlet-class>org.logicalcobwebs.proxool.configuration.ServletConfigurator</servlet-class>
  <init-param>
     <param-name>xmlFile</param-name>
     <param-value>WEB-INF/db.xml</param-value>
  </init-param>
  <load-on-startup>1</load-on-startup>
</servlet>

<servlet>
  <servlet-name>proxool</servlet-name>
  <servlet-class>org.logicalcobwebs.proxool.admin.servlet.AdminServlet</servlet-class>
</servlet>

<servlet-mapping>
  <servlet-name>proxool</servlet-name>
  <url-pattern>/proxool</url-pattern>
</servlet-mapping>
```

从这个 web.xml 中,我们可以看出,该连接池是通过 servlet 来加载控制的。连接名是通过这个 servlet 的名字拼接上 db.xml 文件中指定的别名 Sybase 组合而成的。

而验证程序,只需要改动获取连接的那一句:

```
conn = DriverManager.getConnection("proxool.sybase");
```

不需要从 JNDI 那里查询获取数据源。

proxool 连接池另一个比较优秀的地方是它提供了一个监控和管理的界面,如这里我们可以通过 http://localhost:8080/pool_proxool/proxool/来访问这个管理界面（pool_proxool 是 Web 名称，proxool 是用于 admin 的 servlet 名），如图 16-18 所示。

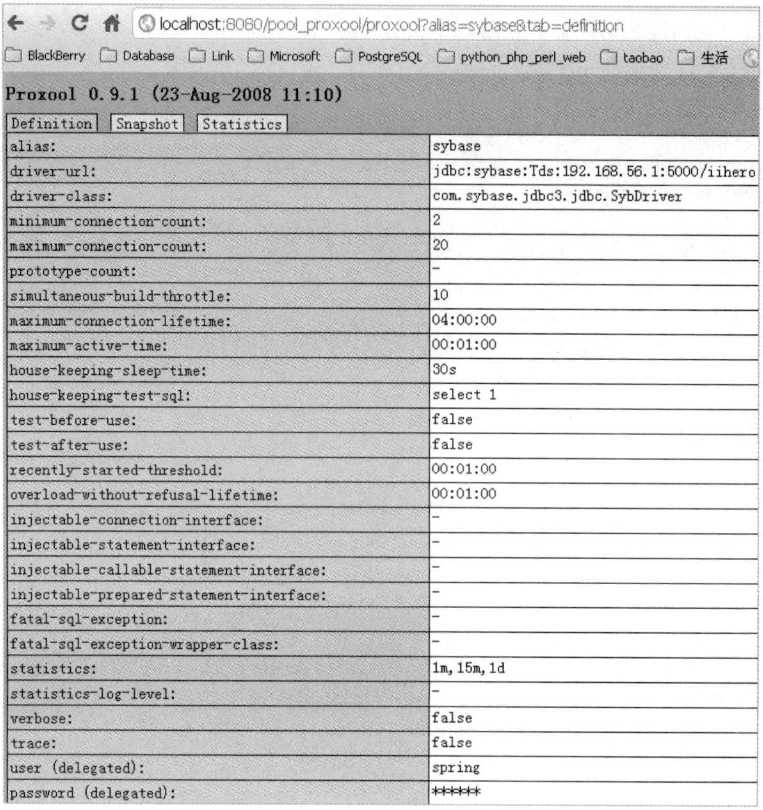

alias:	sybase
driver-url:	jdbc:sybase:Tds:192.168.56.1:5000/iihero
driver-class:	com.sybase.jdbc3.jdbc.SybDriver
minimum-connection-count:	2
maximum-connection-count:	20
prototype-count:	-
simultaneous-build-throttle:	10
maximum-connection-lifetime:	04:00:00
maximum-active-time:	00:01:00
house-keeping-sleep-time:	30s
house-keeping-test-sql:	select 1
test-before-use:	false
test-after-use:	false
recently-started-threshold:	00:01:00
overload-without-refusal-lifetime:	00:01:00
injectable-connection-interface:	-
injectable-statement-interface:	-
injectable-callable-statement-interface:	-
injectable-prepared-statement-interface:	-
fatal-sql-exception:	-
fatal-sql-exception-wrapper-class:	-
statistics:	1m, 15m, 1d
statistics-log-level:	-
verbose:	false
trace:	false
user (delegated):	spring
password (delegated):	******

图 16-18 Proxool 管理界面

这种特性对于管理和监控连接池的使用情况非常有用。

4. DBPool 连接池

还有一个轻量级的开源连接池库 DBPool 可以尝试使用，配置起来也非常方便。可以从它的官网 http://www.snaq.net/java/DBPool/DBPool-5.0.zip 上直接下载。配置起来也很方便。

首先，将下载的 DBPool-5.0.jar 以及它依赖的库 commons-logging-1.1.1.jar 放入待测应用 pool_dbpool\WEB-INF\lib 目录下，同时也将 ASE 的 JDBC 驱动 jconn3.jar 放入其中。

然后，修改部署文件 web.xml，修改它的配置文件 conf\Catalina\localhost\pool_dbpool.xml。

在 web.xml 中添加：

```
<resource-ref>
  <description>DB Connection</description>
  <res-ref-name>jdbc/DBPool_Sybase</res-ref-name>
```

```
    <res-type>javax.sql.DataSource</res-type>
    <res-auth>Container</res-auth>
</resource-ref>
```

在 pool_dbpool.xml 中修改添加：

```
<Resource factory="snaq.db.DBPoolDataSourceFactory"
        type="javax.sql.DataSource"
        name="jdbc/DBPool_Sybase"
        auth="Application"
        driverClassName="com.sybase.jdbc3.jdbc.SybDriver"
        url="jdbc:sybase:Tds:192.168.56.1:5000/iihero"
        user="spring" password="spring1"
        minPool="2" maxPool="20" maxSize="30" idleTimeout="3600"
        validationQuery="SELECT 1" />
```

修改 index.jsp 测试代码，将数据源名换为新名字 DBPool_Sybase：

DataSource ds = (DataSource)envContext.lookup("jdbc/DBPool_Sybase");

要换成全局配置也很方便，直接将 pool_dbpool.xml 中的 Resource 配置放到 conf/context.xml 中即可。只是所有的库文件（3 个 jar 文件）都要放到 tomcat/lib 目录下，以达到全局共享。

16.8　使用 Java 直接支持 ASE 中的面向对象 SQL 访问

数据库发展到今天，RDBMS（关系型数据库管理系统）和 ODBMS（对象数据库管理系统）的界限已经变得模糊了，取而代之的是 ORDBMS（对象关系型数据库管理系统），对此，Oracle 和 DB2 数据库已经给出了自己的解决方案，即采用类似以下的抽象面向对象的 SQL 语法：

```
create type Point is object
(
    x int,
    y int,
    ......
)
```

Sybase ASE 数据库发展到现在 15.X 版本，在语法上，似乎并没有对这种抽象语法的直接支持。它采取的是另一种解决方案，即可以由用户直接定义一个 Java 类，让这个 Java 类直接成为一种抽象类型。这就是 Sybase ASE 数据库的 Java 服务选件，它是一种功能非常强大的组件。

比如，我们要提供一个解决方案，直接支持二维点 Point 类型的存取，并且能直接在 SQL 中对二维点的距离坐标信息进行查询，怎么实现此功能呢？

1. 环境支持

确定你的 ASE 数据库安装并拥有 Java Options 选项，如何得知呢？从 Sybase Central 连接到 ASE 服务器，右击鼠标，选择"属性"→"版本"→"详细信息"命令，可以得到信息 ASE_JAVA，有相应的版本和有效期信息，示例如图 16-19 所示，它支持 Java，有效期到 2020 年 1 月 20 日。

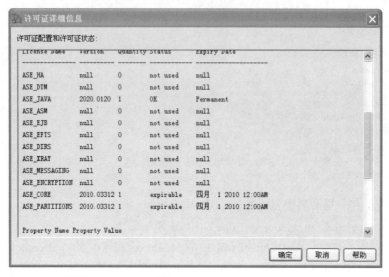

图 16-19　许可证（License）详细信息

除此以外，还得让服务器启动时也支持 Java。

```
D:\ASE150>isql -Usa -Psybase1
1> sp_configure "enable java", 1
2> go
(1 row affected)
```

对于新版本的 ASE，还要执行：

```
1> sp_configure "enable pci", 1
2> go
```

配置选项改变，因为选项是静态的，Adaptive Server 必须重启，以使改动生效。

如果将 enable java 更改为 1，ASE 使用的内存量就能增加 580KB。

(return status = 0)

运行此命令以后，重启 ASE 服务。

2．实现

首先，实现 Point 类。ASE 支持 Java 对象类型，有一个条件是，它必须实现了序列化接口。
同时，编译为 class 文件时，编译的目标版本最好是 1.1（为了保持兼容）。

Point 类的一个简单实现如下（详见 code\ JavaSQL\Point.java）：

```
#001 public class Point implements java.io.Serializable
#002 {
```

```
#003        private static final long serialVersionUID = -2550063311147977493L;
#004        public double _x;
#005        public double _y;
#006
#007        public Point()
#008        {
#009            _x = 0;
#010            _y = 0;
#011        }
#012
#013        public Point(double x, double y)
#014        {
#015            _x = x;
#016            _y = y;
#017        }
#018
#019        public Object clone()
#020        {
#021            return new Point(_x, _y);
#022        }
#023
#024        public String toString()
#025        {
#026            return "(" + _x + ", " + _y + ")";
#027        }
#028
#029        public static double distance(Point a, Point b)
#030        {
#031            return (double)Math.sqrt((a._x - b._x)*(a._x - b._x)
#032                + (a._y - b._y)*(a._y - b._y));
#033        }
#034
#035        public static void main(String[] args)
#036        {
#037            Point a = new Point(1.0, 1.0);
#038            Point b = new Point(5.0, 4.0);
#039            System.out.println("a = " + a);
#040            System.out.println("b = " + b);
#041            System.out.println(Point.distance(a, b));
#042        }
#043 }
```

在此之后，设定好 JDK 路径和 classpath 之后，编译打包：

javac -g -target 1.1 *.java

jar -cvf0 Point.jar *.class

Code\JavaSQL\compile.bat 里有完整的编译打包脚本。

3. 装载 Java 的 jar 包到数据库服务器

如图 16-20 所示，进入具体的数据库 iihero→Java 对象→Jar 文件，右击鼠标，选择"新建"命令，即可把刚才打好的 Java 包 Point.jar 装载到数据库 iihero 当中。

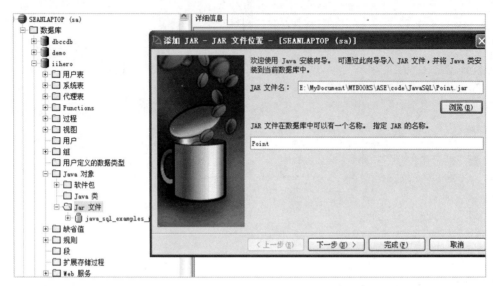

图 16-20 为 iihero 数据库添加 jar

或者直接使用命令行：

```
D:\ASE150>instjava -D iihero   -S SEANLAPTOP
-f E:\MyDocument\MYBOOKS\ASE\code\JavaSQL\Point.jar -update -j "Point" -U sa -P
```

其中，-D 指定目标数据库，-S 指定服务名，-f 指定已经编译好的 Java 包，-j 指定该包在数据库中的包名，-U 指定连接数据库的用户名，-P 指定密码。在 UNIX 平台使用命令 installjava。

4. 面向对象应用的实际效果

```
D:\ASE150>isql -Uspring -Pspring1
1> use iihero
2> go
```

创建含对象类型 Point 列的表：

```
1> create table testpoint(id int primary key, p Point)
2> go
```

使用对象特性插入数据并查询：

```
1> insert into testpoint values (1, new Point(1.0, 1.0))
2> insert into testpoint values (2, new Point(5.0, 4.0))
3> insert into testpoint values (3, new Point(6.0, 13.0))
```

```
4> go
(1 row affected)
(1 row affected)
(1 row affected)
1> select * from testpoint
2> go
 id          p
 ----------  --------------------
 1           (1.0, 1.0)
 2           (5.0, 4.0)
 3           (6.0, 13.0)
(3 rows affected)
```

直接取属性值：

```
1> select id, p>>_x, p>>_y from testpoint
2> go
 id
 ----------  ------------------  -----------------
 1           1.000000            1.000000
 2           5.000000            4.000000
 3           6.000000            13.000000

(3 rows affected)
```

查询两点间的距离：

```
1> select a.id, b.id, Point.distance(a.p, b.p) from testpoint a, testpoint b whe
re a.id < b.id
2> go
 id          id
 ----------  ----------  -------------------
 1           2           5.000000
 1           3           13.000000
 2           3           9.055385

(3 rows affected)
```

以距离作为条件查询点（查询两点间距离小于 6.0 的点对）：

```
1> select a.id, b.id from testpoint a, testpoint b where Point.distance(a.p, b.p
) <6.0 and a.id > b.id
2> go
 id          id
 ----------  ----------
 2           1
(1 row affected)
```

总结起来，使用 Java 来实现面向对象 SQL 查询存取访问，在到 ASE 数据库服务器内部，它

对 Java 类作了如下映射:

- select 出对象列的值,即 Java 对象的 toString()值。
- 静态方法可以直接在 Select 语句里当作函数进行调用,如上例中的 Point.distance()方法。
- 直接可以取类的成员值,如果那个成员是公有成员,并且有相应的 SQL 类型映射,如上例中的 p>>_x, p>>_y,使用 ">>" 操作符来提取具体的成员变量值。

按照上述原理,用户完全使用内置的 Java 包实现非常复杂的逻辑,然后在 SQL 语句中直接加以利用。尤其是那种接近于面向对象数据库的应用。

总之,使用内置 Java 支持对象类型的存取是一种很直观易用的方法,也便于用户进行扩充。从这个角度来看,Sybase ASE 实际上借助于 Java,也可以算是支持面向对象的数据存取了。

不仅如此,使用这种功能还可以写出功能强大的服务器端 Java 处理函数。有兴趣的读者可以去尝试一下。

需要说明的是,在某些版本的 15.X 中存在一个 BUG,会提示这类错误:

> The ASE PCI-Bridge is not able to process the requested dispatching. This is a fatal error, please contact your System Administrator (SA) for help.

这个 BUG 会在 591492 号 CR(Change Request)修复,必须打上新的补丁才能工作。该 CR 的描述如下:

| 591472 | Error 16022 may be reported "The ASE PCI-Bridge is not able to process the requested dispatching. This is a fatal error, etc." when the PCI is unavailable with a stack trace in the error log showing modules "jvminit", "SYB_CreateJavaVM" and "terminate_process" although the stack trace information is unnecessary. |

17

应用 PHP 访问 ASE

在 Web 应用里,除了 J2EE(现在的官方名称已经改为 JavaEE)环境在企业级应用里占据主导地位以外,就是以 PHP 为代表的开源环境,在非企业级应用里占据着绝对主导地位,它的开发效率高、灵活,总体开发费用也低于 J2EE,在大多数情况下拥有较高的性价比。

本章将对 PHP 环境里如何访问 ASE 数据库作一个系统的介绍。

PHP 语法比较简单,大部分与 C 语言相似,只是 PHP 采用的是脚本式外观,以解释的方式执行,即常以 "<?php" 打头,以 "?>" 结束,它里边包含的内容语法上与 C 语言极其相似。

17.1 PHP 运行环境搭建

PHP 运行环境中,使用比较广泛的是 Apache+PHP 及最近两年使用比较多的 Nginx+PHP,这里没有带上 My SQL 或者 PostgreSQL,是因为我们着重介绍 PHP 下边 ASE 数据库的应用,因为这两个环境在 Windows 和 Linux 下都有相应的可运行版本,配置文件也大同小异,我们就以 Windows 平台为例来作介绍。

17.1.1 Apache + PHP 运行环境

首先,当前机器要安装 Sybase ASE 数据库服务器,在生产环境中,如果要将 Web 服务器和数据库服务器进行分离,可以只安装 Sybase ASE 的 Open Client 客户端。

下载 Apache 和 PHP 的组合包 XAMPP 1.7.x,它是一个易于安装,包含 My SQL、PHP 和 Perl 的 Apache 组合包。我们可以直接从网站http://www.apachefriends.org/en/xampp-windows.html上下载一个解压缩版 xampp-win32-1.7.7-VC9.7z,最新版本是 XAMPP 1.7.7,它主要包含以下组件:

- Apache 2.2.21
- My SQL 5.5.16
- PHP 5.3.8

- phpMyAdmin 3.4.5
- FileZilla FTP Server 0.9.39
- Tomcat 7.0.21 (with mod_proxy_ajp as connector)

其中，我们只需要用到 Apache 和 PHP，Tomcat 7.x 是 XAMPP 1.7.7 新加进来的组件以前版本并不带 Tomcat。

下载完以后，将这个压缩包解压到某一个盘符的根目录下，如 D:\，这样 XAMPP 就位于 d:\xampp 目录下。进入此目录，运行命令 setup_xampp.bat，开始安装。

```
D:\xampp>setup_xampp.bat
########################################################################
# ApacheFriends XAMPP setup win32 Version                               #
#----------------------------------------------------------------------#
# Copyright (c) 2002-2012 Apachefriends 1.7.7                           #
#----------------------------------------------------------------------#
# Authors: Kay Vogelgesang <kvo@apachefriends.org>                      #
#          Carsten Wiedmann <webmaster@wiedmann-online.de>              #
########################################################################

Configure for Version 1.7.7
Configure XAMPP with awk for 'Windows_NT'
Please wait ...
Enable AcceptEx Winsocks v2 support for NT systems    DONE!

##### Have fun with ApacheFriends XAMPP! #####

Press any key to continue . . .
```

这表示组合环境初步安装完毕。如何启动呢？

单击运行程序 xampp-control.exe，显示如图 17-1 所示的控制面板。

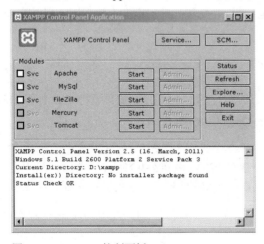

图 17-1　XAMPP 控制面板

我们只需要单击 Apache 右边的 Start 按钮即可，你也可以选中左边的 Svc 复选框，表示它将为 Apache 创建一个系统服务。Start 以后，XAMPP 处于运行状态，如图 17-2 所示。

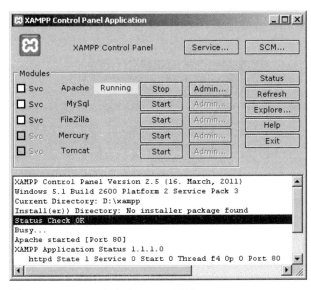

图 17-2　XAMPP 中启动 Apache 服务

这时，我们直接打开页面 http://localhost，发现它会跳转到 XAMPP 的相应目录。网站的根目录位于 D:\xampp\htdocs，下边有两个首页：index.html 和 index.php，优先加载后者。

我们把文件 index.php 重命名为 index.php_，在此目录下编写一个最简单的验证 PHP 环境的脚本 phpinfo.php，内容如下：

```
<?php
    phpinfo();
?>
```

这样，我们打开页面 http://localhost/phpinfo.php 时，如果没出现任何错误，并且有如图 17-3 所示类似的信息，即表示 PHP 运行环境是成功的。

那么 Apache 是如何自动加载到 PHP 相关模块的？并且为何 index.php 先于 index.html 被 Apache 加载呢？这些都来自于 Apache 的配置文件 D:\xampp\apache\httpd.conf 和 D:\xampp\apache\conf\extra\httpd-xampp.conf。

首先，你可以从 httpd.conf 文件中找到这样的内容：DocumentRoot "D:/xampp/htdocs"，指定这个目录是整个网站默认的根目录。

而下述内容则表示 Web 服务器处理某一个网站目录路径的加载顺序，按照 index.php、index.pl、index.cgi……这样的顺序来处理，完全可以按照自己的设计去调整这个顺序，可以将 index.html 放到最前头。

```
<IfModule dir_module>
    DirectoryIndex index.php index.pl index.cgi index.asp index.shtml index.html index.htm \
                   default.php default.pl default.cgi default.asp default.shtml default.html default.htm \
                   home.php home.pl home.cgi home.asp home.shtml home.html home.htm
</IfModule>
```

图 17-3　使用 phpinfo.php 验证 PHP 是否配置成功

该文件的第 454～455 行：

```
# XAMPP specific settings
Include "conf/extra/httpd-xampp.conf"
```

表明它包含了另外的子配置文件 extra/httpd-xampp.conf，打开这个文件，你会发现 Apache 服务器会自动加载 PHP 的相关模块：

```
<IfModule env_module>
    SetEnv MIBDIRS "D:/xampp/php/extras/mibs"
    SetEnv MYSQL_HOME "\\xampp\\mysql\\bin"
    SetEnv OPENSSL_CONF "D:/xampp/apache/bin/openssl.cnf"
    SetEnv PHP_PEAR_SYSCONF_DIR "\\xampp\\php"
    SetEnv PHPRC "\\xampp\\php"
    SetEnv TMP "\\xampp\\tmp"
</IfModule>
LoadFile "D:/xampp/php/php5ts.dll"
LoadModule php5_module "D:/xampp/php/php5apache2_2.dll"
```

```
<FilesMatch "\.php$">
     SetHandler application/x-httpd-php
</FilesMatch>
<FilesMatch "\.phps$">
     SetHandler application/x-httpd-php-source
</FilesMatch>
<IfModule php5_module>
     PHPINIDir "D:/xampp/php"
</IfModule>
```

至此 Apache + PHP 环境初步搭建完毕，随时都可以通过 xampp-control 面板，停掉 Apache 服务来停止它们的 Web 服务。

17.1.2　Nginx + PHP 运行环境

Nginx 目前已经发展到 1.3.0 版本，并且已经有可用的 Windows 版本下载，它与 PHP 结合起来非常轻便高效，相当多的超大型网站都从传统的 Apache/PHP 完全切换到 Nginx/PHP，节省了相当多的运维费用，因此，从某种意义上讲，Nginx+PHP 运行环境比 Apache+PHP 更有实际意义。

我们来看看实际的搭建过程：

首先，从 Nginx 的官网上下载 Nginx 1.3.0：http://nginx.org/en/download.html→http://nginx.org/download/nginx-1.3.0.zip，从 PHP 的官网上下载 PHP5.3.8 的线程安全的解压缩版：http://windows.php.net/downloads/releases/archives/→php-5.3.8-Win32-VC9-x86.zip（线程安全版本）。

接着，将下载到的 nginx-1.3.0.zip 解压到 D:\，php-5.3.8-Win32-VC9-x86.zip 解压到 D:\tools\php5308。

然后，进入目录 D:\tools\php5308，将 php.ini-development 复制为 php.ini，作为 PHP 的配置文件，并修改以下参数：

```
serialize_precision = 100
error_log = "D:\nginx-1.3.0\logs\php_errors.log"
extension_dir = "ext"     (这是指定扩展模块的加载路径)
upload_max_filesize = 128M
extension=php_bz2.dll
extension=php_gd2.dll
extension=php_gettext.dll
extension=php_mbstring.dll
extension=php_soap.dll
extension=php_sockets.dll
extension=php_xmlrpc.dll
extension=php_xsl.dll

date.timezone = PRC    (时区设置)
```

针对 Nginx，为了支持 PHP 的处理，也需要修改其配置文件，进入 D:\nginx-1.3.0\conf\nginx.conf。

找到 server 下边的如下内容：

```
location / {
    root    html;
    index   index.html index.htm;
}
```

将这个片段的内容改为：

```
root    html;
index   index.html index.htm index.php;
charset utf-8;
location ~ \.php$ {
    root           html;
    fastcgi_pass   127.0.0.1:9000;
    fastcgi_index  index.php;
    include        fastcgi.conf;
}
```

其中，网站的根路径由 root html 指定，表明 html 子目录是根目录。index 指定目录路径下的优先处理顺序，这里是 index.html→index.htm→index.php。为更好地支持中文，这里设定字符集为 UTF-8。Nginx 会对最后的响应处理作相应的字符集转换。

这里 PHP 以 fastcgi 的方式启动，端口为 9000。

启动 PHP，进入目录 D:\tools\php5308，运行命令行：

```
php-cgi -b 127.0.0.1:9000
```

然后再进入目录 D:\nginx-1.3.0，直接运行 nginx.exe 可执行程序，这样，在浏览器里输入 http://localhost，如果浏览器显示 "It works!"，则表明 Nginx 已经开始工作。

将前一节中的测试程序 phpinfo.php 复制到 D:\nginx-1.3.0\html 目录下，浏览器里输入 http://localhost/phpinfo.php，即可验证 PHP 的 cgi 程序是否运行正常。

注意，要得到 nginx 的默认编译选项，通过运行命令行 "nginx -V" 可以得到默认的参数信息：

```
D:\nginx-1.3.0>nginx.exe -V
nginx version: nginx/1.3.0
TLS SNI support enabled
configure arguments: --with-cc=cl --builddir=objs.msvc8 --with-debug --prefix= --conf-path=conf/nginx.conf --pid-path=logs/nginx.pid --http-log-path=logs/access.log --error-log-path=logs/error.log --sbin-path=nginx.exe --http-client-body-temp-path=temp/client_body_temp --http-proxy-temp-path=temp/proxy_temp --http-fastcgi-temp-path=temp/fastcgi_temp --http-scgi-temp-path=temp/scgi_temp --http-uwsgi-temp-path=temp/uwsgi_temp --with-cc-opt=-DFD_SETSIZE=1024 --with-pcre=objs.msvc8/lib/pcre-8.30 --with-zlib=objs.msvc8/lib/zlib-1.2.5 --with-select_module --with-http_realip_module --with-http_addition_module --with-http_sub_module --with-http_dav_module --with-http_stub_status_module --with-http_flv_module --with-http_mp4_module --with-http_gzip_static_module --with-http_random_index_module --with-http_secure_link_module --with-mail --with-openssl=objs.msvc8/lib/openssl-1.0.1c --with-openssl-opt=enable-tlsext --with-http_ssl_module --with-mail_ssl_module --with-ipv6
```

至此，Nginx + PHP 环境基本搭建完毕。

要停止 Nginx 服务程序，直接运行命令行 D:\nginx-1.3.0>nginx.exe -s stop 即可。

17.1.3 PHP 环境对 ASE 数据库的支持

前边两小节只是对 PHP 与 Apache 或 Nginx 组合的基本运行环境的配置，并未考虑到 Sybase ASE 数据库的支持，因为这个支持只与 PHP 的配置有关，因此将其独立出来。

打开 PHP 的配置文件 php.ini（对 Nginx，它位于我们解压缩后的 D:\tools\php5308 目录；对 XAMPP，它位于 D:\xampp\php\php.ini），将 Sybase ASE 的模块支持打开（取消前边的注释符号分号"；"）：
extension=php_sybase_ct.dll（去掉前边的注释符分号）

这里先以 Nginx 为例，然后重启 php-cgi.exe，可能会出现如图 17-4 所示的错误对话框。

图 17-4 错误对话框

这是因为 PHP 里提供的 php_sybase_ct.dll 编译时，依赖的库是 libcs.dll，而 ASE 数据库到 15.x 版本以后，这些 DLL 都已经改名为 libsybct.dll，有一个 syb 前缀符。

解决的办法是，进到 Sybase ASE 的目录%SYBASE%\%SYBASE_OCS%\scripts，运行命令 copylibs.bat create，如下所示：

```
C:\>cd /d %SYBASE%\%SYBASE_OCS%\scripts
D:\SybaseASE\OCS-15_0\scripts>copylibs.bat create
        1 file(s) copied.
        1 file(s) copied.
        1 file(s) copied.
        1 file(s) copied.
        1 file(s) copied.
        1 file(s) copied.
        1 file(s) copied.
```

该脚本实际上就是把几个新的动态库复制成兼容过去的 ASE 15.0 以前的文件名，便于第三方库动态加载。

之后，进入目录 D:\tools\php5308，重新执行 php-cgi -b 127.0.0.1:9000，即可顺利启动 php-cgi 程序，并成功加载 Sybase ASE 访问模块。要成功地支持中文及可能的多种字符集，ASE 数据库最好是 UTF-8 字符集，它是整个服务器级别的设置，在启动的时候，从启动的日志中就可以看出：

```
00:00000:00001:2012/06/04 11:10:00.63 server  ASE's default unicode sort order is 'binary'.
00:00000:00001:2012/06/04 11:10:00.63 server  ASE's default sort order is:
00:00000:00001:2012/06/04 11:10:00.63 server          'bin_utf8' (ID = 50)
```

```
00:00000:00001:2012/06/04 11:10:00.63 server    on top of default character set:
00:00000:00001:2012/06/04 11:10:00.63 server        'utf8' (ID = 190).
```

我们可以在网站的根路径下，放一个简单的测试程序 ase.php（使用 UTF-8 文件编码）进行验证：

```
<?php
    define('DBUSER__', 'spring');
    define('DBPASSWD__', 'spring1');
    define('DATABASE__', 'iihero');
    define('DBSERVER__', 'XIONGHE');

    sybase_min_server_severity(11);
    $link = sybase_connect(DBSERVER__, DBUSER__, DBPASSWD__, 'utf8') or die("连接失败 !");
    echo "成功连接 ASE 数据库. ";
    sybase_close($link);
?>
```

只要屏幕输出"成功连接 ASE 数据库"，则表明上述连接信息是成功的。

sybase_connect 函数的几个参数当中，第一个参数来自于 ASE 的 ini 文件（或称为 interfaces 文件）中的注册项，是一个服务名。在第 4 章"连接 ASE"里有详细介绍。

至此，支持 ASE 数据库访问的 PHP 运行环境已经成功搭建起来了。

17.2 php_sybase_ct 模块介绍

PHP 的 sybase_ct 模块主要提供了如下一些 API：

1．sybase_affected_rows

int sybase_affected_rows ([resource $link_identifier])：返回与相应连接对应的最后一次 Insert/Update/Delete 操作所影响的记录行数，参数$link_identifier 可选，如果不指定，则假定为上一次使用的连接。

2．sybase_close

bool sybase_close ([resource $link_identifier])：关闭由$link_identifier 指定的数据库连接，成功则返回 True，失败返回 False。参数$link_identifier 可选，表示一个数据库连接，如果不指定此参数，则假定为上一次使用的连接。

3．sybase_connect

resource sybase_connect ([string $servername [, string $username [, string $password [, string $charset [, string $appname [, bool $new = false]]]]]])：该函数用于创建到 ASE 数据库的连接。如果同一个脚本中有第二次使用相同参数值的调用，则直接返回第一次调用的连接。在脚本结束处理时，无论你是否调用了 sybase_close()来断开连接，该连接都会被断开，从这点来看，sybase_close 可以

不用显式地调用。

参数：
- servername：在 interfaces 或 sql.ini 文件中定义的有效的服务名。
- username：ASE 数据库用户名。
- password：用户名对应的密码。
- charset：数据库连接使用的字符集。
- appname：为数据库连接指定一个应用名，它只是偶尔有用，主要用于区分不同的连接。
- new：用于标明是否创建一个新的连接，默认值为 False。

4. sybase_data_seek

bool sybase_data_seek (resource $result_identifier , int $row_number)：ASE 的结果集移动指定的行数，后续的 sybase_fetch_row()调用将从新行的位置返回记录行。

参数：
- result_identifier：结果集变量。
- row_number：移动的行数。

返回值：成功则返回 True，失败返回 False

5. sybase_deadlock_retry_count

void sybase_deadlock_retry_count (int $retry_count)：遇到死锁时进行若干次尝试，次数由参数 retry_count 指定。默认情况下，每个死锁都会执行无限次尝试，直至进程由 ASE 数据库服务器杀死。

6. sybase_fetch_array

array sybase_fetch_array (resource $result)：是函数 sybase_fetch_row()的扩展版本，它提供了按列序号存储结果，也可以按列名作键值存储结果。

7. sybase_fetch_assoc

array sybase_fetch_assoc (resource $result)：是 sybase_fetch_row 的另一个版本，按照列名而不是列的序号作为结果行的键值。

8. sybase_fetch_field

object sybase_fetch_field (resource $result [, int $field_offset = -1])：获取字段的元信息。

参数：
- result：结果集。
- field_offset：列序号。

9. sybase_fetch_object

object sybase_fetch_object (resource $result [, mixed $object])：返回一个 object，功能上与 sybase_fetch_array 类似。

参数：
- result：结果集。

- object:指定返回的列值类型。

返回值:返回带属性的 object,如果没有对应的结果行,则返回 False。

10. sybase_fetch_row

array sybase_fetch_row (resource $result):返回结果集中的一行,并将行指针移向下一行,如果没有记录行,则返回 False。

返回值:一个数组,每列都有一个序号,从 0 计起。PHP 类型和 ASE 列数据类型的对应关系如表 17-1 所示。

表 17-1　PHP 中的变量和 ASE 数据库字段类型间的映射关系

PHP	Sybase ASE
string	VARCHAR,TEXT,CHAR,IMAGE,BINARY,VARBINARY,DATETIME
int	NUMERIC,DECIMAL,INT,BIT,TINYINT,SMALLINT
float	NUMERIC,DECIMAL,REAL,FLOAT,MONEY
NULL	NULL

11. sybase_field_seek

bool sybase_field_seek (resource $result , int $field_offset):获取列信息过程时,定位到指定的列上。成功则返回 True,否则返回 False。

12. sybase_free_result

bool sybase_free_result (resource $result):释放结果集资源,成功返回 True。只有在担心当前页面使用了太多的内存需要立即释放的时候才调用它。当页面脚本处理终止时,所有结果集会自动释放。

13. sybase_get_last_message

string sybase_get_last_message (void):返回 ASE 服务器端的最后一条消息。

14. sybase_min_client_severity

void sybase_min_client_severity (int $severity):设置客户端报错的最小严重级别,在高于此级别值时才报错。

15. sybase_min_error_severity

void sybase_min_error_severity (int $severity):设定出错的最小严重级别值。

16. sybase_min_message_severity

void sybase_min_message_severity (int $severity):设定消息的最小严重级别。

17. sybase_min_server_severity

void sybase_min_server_severity (int $severity):设定服务器端的最小严重级别。

18. sybase_num_fields

int sybase_num_fields (resource $result):返回结果集中的列的数目。

19. sybase_num_rows

int sybase_num_rows (resource $result)：返回结果集里的行数。

20. sybase_pconnect

resource sybase_pconnect ([string $servername [, string $username [, string $password [, string $charset [, string $appname]]]]])：返回持久的 ASE 数据库连接，参数意义与 sybase_connect 相同。由此函数创建的连接不会自动关闭，即便调用了 sybase_close()也不起作用。

21. sybase_query — Sends a Sybase query

mixed sybase_query (string $query [, resource $link_identifier])：发送一个查询到当前数据库连接。

返回值：如果成功执行，并且有结果集，则返回结果集；如果没有结果集，则返回 True，否则返回 False。

22. sybase_result

string sybase_result (resource $result , int $row , mixed $field)：返回结果集中指定行指定字段的值，在处理大数据量的结果集时，不推荐使用此函数。

23. sybase_select_db

bool sybase_select_db (string $database_name [, resource $link_identifier])：设定当前的数据库，后续的所有 query 都将基于这个设定的数据库。

24. sybase_set_message_handler

bool sybase_set_message_handler (callable $handler [, resource $link_identifier])：用于自定义的回调函数设置，以处理 ASE 服务器端返回的消息。

参数：

- handler：是一个函数，它需要带 5 个参数，前 4 个为 int，最后 1 个为 string，如果此函数返回 False，PHP 会返回常规的错误消息。
- link_identifier：连接对象，可选，如果不指定，则默认为上一次的连接。

25. sybase_unbuffered_query

resource sybase_unbuffered_query (string $query, resource $link_identifier [, bool $store_result])：向当前数据库发送一个查询，它与 sybase_query()不同，它只读取结果集中的第一行，而 sybase_fetch_array 等调用会读取后续的记录行，sybase_data_seek()会读取指定的记录行。该函数对于大数据量的结果集会取得更好的性能。

参数：

- query：查询语句。
- link_identifier：连接对象。
- store_result：是否存储结果集，大数据量的结果集推荐使用 False。

上述 25 个函数就是 PHP 脚本访问 ASE 数据库的核心 API 接口,它直接基于 ASE 的 Open Client

的 ct-lib 开发，因而具有很高的访问效率。

17.3　一个访问 ASE 数据库的 PHP 简单实例

这是一个简单的用户登录及显示所有产品清单的小程序。

17.3.1　数据库数据准备

这个小系统有几张表，建表及数据初始化的 ASE SQL 脚本内容 data.sql 如下：

```
create table customers
(
id int identity not null primary key,
name varchar(32) not null,
password varchar(32) not null,
address varchar(256) not null,
tel    varchar(16) not null
)
go

create table orders
(
    id int identity not null primary key,
    custid int not null,
    amount money not null
)
go

create table ordersitem
(
    orderid int not null,
    prod_id int not null,
    quantity int not null
)
go

create table products
(
    id   int identity not null primary key,
    name    varchar(64) not null,
    price money not null,
    category   varchar(16) not null,
    description varchar(128) null
)
go
```

```
insert into products(name, price, category) values('Java 编程', 55.00, '图书')
insert into products(name, price, category) values('C 编程', 32.00, '图书')
insert into products(name, price, category) values('离散数学教程', 22.00, '图书')
insert into products(name, price, category) values('ASE 数据库教程', 48.00, '图书')
insert into products(name, price, category) values('数据结构', 25.00, '图书')
insert into products(name, price, category) values('算法导论', 75.00, '图书')
insert into products(name, price, category) values('T 牌公路自行车', 774.00, '交通工具')
insert into products(name, price, category) values('永久自行车', 180.00, '交通工具')
insert into products(name, price, category) values('LG 液晶显示器', 995.00, '电子产品')
insert into products(name, price, category) values('Sandisk U 盘(16G)', 85.00, '电子产品')
go

insert into customers(name,password,address,tel) values('Wang', 'wang123', '北京丰台区 1133 号,100075', '010-88888888')
insert into customers(name,password,address,tel) values('Hero', 'he123', '北京海淀 1133 号,100075', '010-88888887')
insert into customers(name,password,address,tel) values('Joe', 'joe123', '北京丰台区 1132 号,100075', '010-88888886')
insert into customers(name,password,address,tel) values('Rose', 'rose123', '北京丰台区 1131 号,100075', '010-88888885')
go
```

其中 customers 表用于存储和验证用户登录信息，products 表用于存储所有产品的清单。

将上述脚本运行在数据库 iihero 上：

```
isql -Uspring -i D:\MyDocument\MyBooks\ASE\code\php\data.sql
```

保密字：

```
(1 row affected)
(1 row affected)
(1 row affected)
……
```

17.3.2 系统实现

这里以 Nginx 1.3.0+ PHP5.3.8 作为运行环境，首先在网站的根目录下去掉 index.html（重命名即可），增加一个登录页面，名为 index.html，其文件编码格式为 UTF-8。内容如下所示：

```
#001 <html>
#002 <head>
#003 <title>Login</title>
#004 <meta http-equiv="Content-Type" content="text/html; charset=utf8">
#005 </head>
#006
#007 <body>
#008 <form name="form" method="post" action="login.php">
#009 <table width="300" border="0" align="center" cellpadding="2" cellspacing="2">
#010 <tr>
```

```
#011 <td width="150"><div align="right">用户名：</div></td>
#012 <td width="150"><input type="text" name="username"></td>
#013 </tr>
#014 <tr>
#015 <td><div align="right">密码：</div></td>
#016 <td><input type="password" name="password"></td>
#017 </tr>
#018 <tr>
#019 </tr>
#020 </table>
#021 <p align="center">
#022 <input type="submit" name="Submit" value="Submit">
#023 <input type="reset" name="Reset" value="Reset">
#024 </p>
#025 </form>
#026 </body>
#027 </html>
```

该页面将传递 username 和 password 两个参数给 login.php 脚本。

login.php 脚本的内容如下，用于完成对 customer 的用户验证，如果验证成功，则列出可用的 products 列表，否则提示登录失败。

```
#001 <?php
#002 require ('config/server.conf');
#003 session_start();
#004
#005 $username = $_REQUEST['username'];
#006 $password = $_REQUEST['password'];
#007
#008 sybase_min_server_severity(11);
#009
#010 $link = sybase_connect(SERVER__, DBUSER__, DBPASSWD__, 'utf8') or die("Could not connect !");
#011
#012 sybase_select_db(DATABASE__, $link) or die("select db error!");
#013 $query=sybase_query("SELECT count(*) FROM customers WHERE name='$username' and password='$password'");
#014 $count=sybase_result($query, 0, 0);
#015
#016 if ($count == 0)
#017 {
#018     echo "<script language=javascript>alert('用户名密码错误!!!');history.back();</script>";
#019 }
#020 else
#021 {
#022
#023     $_SESSION['username'] = $username;
```

```
#024        $_SESSION['password'] = $password;
#025
#026        // "登录成功....";
#027        $q = sybase_query("SELECT id, name, price, category FROM products");
#028        echo "<div align=center>";
#029        echo "商品清单<br>";
#030        echo "<table width=600 align=center>";
#031        while($row = sybase_fetch_row($q))
#032        {
#033            echo "<tr>";
#034            echo "<td>".$row[0]."</td>";
#035            echo "<td>".$row[1]."</td>";
#036            echo "<td>".$row[2]."</td>";
#037            echo "<td>".$row[3]."</td>";
#038            echo "</tr>";
#039        }
#040        echo "</table>";
#041        echo "<table><a href=logout.php>退出</a></table>";
#042        echo "</div>";
#043 }
#044
#045 sybase_close($link);
#046
#047 ?>
```

第 2 行提示，要包含文件 config/server.conf，它里边存储的是数据库的连接信息。config/server.conf 内容如下：

```
<?php
    define('DBUSER__', 'spring');
    define('DBPASSWD__', 'spring1');
    define('DATABASE__', 'iihero');
    define('SERVER__', '192.168.56.1');
?>
```

login.php 中的第 10 行 sybase_connect() 调用要用到这些常量。但是这些值不能暴露给浏览器用户，即不能让用户通过浏览器http://localhost/config/[server.conf]访问得到，如果不加任何保护，用户通过这个 URL 访问就可以得到所有的连接信息，保护的方法是禁止用户通过浏览器访问 config 目录下的所有文件。

我们在 config 目录下添加一个访问规则文件 .htaccess，内容如下：

```
deny from all
ErrorDocument 403 /index.html
```

禁止所有的浏览器访问 config 目录下的内容，从而起到保护的作用。有人主张在此目录下放

置一个 index.html|index.php 等文件，那只能屏蔽对http://localhost/config的访问，不能阻止对http://localhost/config/server.conf的直接访问，更为安全的做法是将 server.conf 文件放到网站目录的外围，比如放到 d:\nginix-1.3.0\conf 目录下，login.php 中直接使用全路径来引用它。

login.php 的第 10～12 行为连接数据库，选择一个目标数据库，为后续查询作准备。第 13～24 行为用户名及密码验证，只有当 username 和 password 同时存在，才算验证通过。

第 27～42 行是展示可用产品列表。

在显示列表的同时，提供 logout 的链接 logout.php，它的内容如下：

```
<?php
session_start();
unset($_SESSION['username']);
unset($_SESSION['password']);

// 最后彻底销毁 session.
session_destroy();

echo "注销成功";
?>
```

即清除会话里的 usrename 及 password 值，并清除整个会话中的内容，强制用户重新登录。

我们在首页输入正确的用户名（Wang）和密码（wang123）（如图 17-5 所示），即可进入登录以后的页面，如图 17-6 所示。

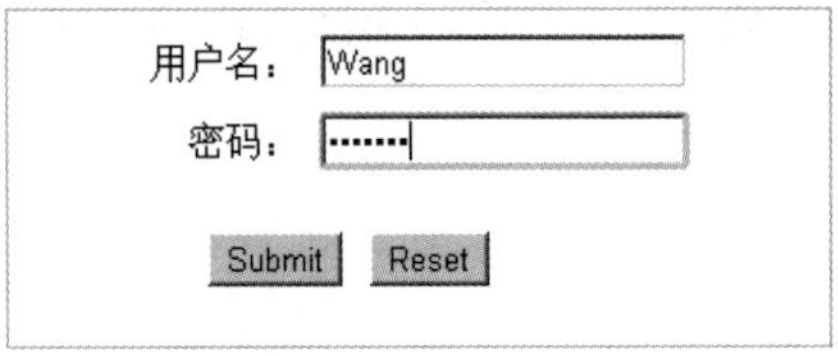

图 17-5　登录页面

商品清单

编号	产品名称	单价	产品类型
1	Java编程	55	图书
2	C编程	32	图书
3	离散数学教程	22	图书
4	ASE数据库教程	48	图书
5	数据结构	25	图书
6	算法导论	75	图书
7	T牌公路自行车	774	交通工具
8	永久自行车	180	交通工具
9	LG液晶显示器	995	电子产品
10	Sandisk U盘(16G)	85	电子产品

退出

图 17-6　登录成功以后的列表展示

上面仅仅是一个简单的示例，用以示意如何利用 php_sybase_ct 接口库来访问 ASE 数据库。

总结起来，php_sybase_ct 的 API 还是有些缺陷的，比如上边的示例，它无法阻止外边用户的 SQL 注入，我通过输入用户名 abc，密码 abc OR 1 = 1 即可绕过验证。这是因为 php_sybase_ct 库没有提供参数化绑定的方法。有没有解决方法呢？

目前比较有效的方法是使用存储过程，因为传递给存储过程的参数是动态绑定的，不过不要在存储过程当中执行动态的 SQL 语句，即生成动态的 SQL 语句去查询，这样就没有任何效果了。

比如针对 customers 表，创建如下存储过程：

```
1> create proc validate_customer @username varchar(32), @password varchar(32)
2> as
3> begin
4>    select count(*) from customers where name=@username and password=@password
5> end
6> go
```

将 login.php 中验证用户的逻辑修改为存储过程调用：

```
// $query=sybase_query("SELECT count(*) FROM customers WHERE name='$username' and password='$password'");
$query=sybase_query("exec validate_customer '$username', '$password'", $link);
```

这样就可以有效地避免 SQL 注入，因为$password 的值和$username 的值最终会通过存储过程中的绑定变更的 SQL 语句得以执行，而不是自然拼接。

补充一点，如果存储过程的实现里，使用的是 execute '<sql>' 来招待 SQL 语句，并且<sql>是通过变量拼接而成的 SQL 语句，那将不能有效地避免 SQL 注入。

18

应用 Python 访问 ASE

Python 2.5 提供了标准的 PEP 249（Python Database API Specification v2.0），可以使用基本一致的接口来访问各种类型的数据库，对应于 ASE 数据库，则有开源的 python-sybase 实现该接口，从 Web 站点 http://python-sybase.sourceforge.net 上可以下载到该模块。

18.1 安装 python-sybase 模块

首先，确保系统中已经安装好了 Python，并且也安装了 Sybase ASE 客户端，它带有 Open Client，从环境变量%SYBASE%及%SYBASE_OCS%可以判断得到（Linux/UNIX 平台下是$SYBASE 和 $SYBASE_OCS）。

另外，由于安装过程需要对其中的 C 程序进行编译和链接，Windows 平台需要拥有 Microsoft Visual Studio .NET 2003\Vc7 来进行编译。

从 http://downloads.sourceforge.net/python-sybase/python-sybase-0.38.tar.gz 下载到 python-sybase 的模块压缩包，将其解压至目录 python-sybase-0.38，发现其目录结构如下：

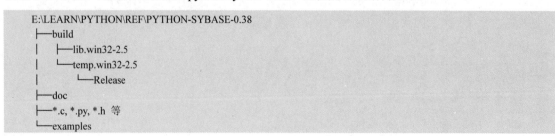

一般情况下，直接运行 python setup.py install，直接就完成了 python-sybase 模块的安装，实际情况可能有些不太一样，从它的网站上的平台支持列表，我们可以看到，目前在 Windows 平台下似乎不支持 Open Client 15.0.x，如表 18-1 所示。

表 18-1　python-sybase 支持平台及 ASE 版本

Client		Server	
OS	Libraries	OS	Libraries
Linux	Sybase ASE 15.0.1（32bits）	Linux	Sybase ASE 15.0.1
Linux 64-bits	Open Client 12.5	Linux	Sybase ASE 12.5
Linux	Sybase ASE 12.5（32bits）	Linux	Sybase ASE 12.5
Linux	Sybase ASE 11.9.2	Linux	Sybase ASE 11.9.2
Linux	Sybase ASE 11.0.3	Linux	Sybase ASE 11.0.3
Linux	FreeTDS	Linux	Sybase ASA 9.0.2
Mac OS X 10.4.x	FreeTDS 0.62	Linux	Sybase ASE 12.5
Windows 2000 Prof SP1	Sybase ASE 11.9.2	Windows NT 4.0	Sybase ASE 11.9.2
Windows NT 4.0 SP6	Sybase ASE 11.9.2	HPUX 10.20	Sybase ASE 11.9.2
Windows 98	Sybase ASE 11.9.2	NT 4.0 SP6	Open Client 11.5
Windows 95 OSR2	Sybase ASE 11.9.2	Solaris 2.6	Open Client 11.5
Solaris 10	Sybase ASE 15.0.1（32bits）	Solaris 10	Sybase ASE 15.0.1
Solaris 10	Sybase ASE 12.5（32bits）	Solaris 10	Sybase ASE 12.5
Solaris 8	Sybase ASE 15.0.1（32bits）	Solaris 8	Sybase ASE 15.0.1
Solaris 8	Sybase ASE 12.5（32bits）	Solaris 8	Sybase ASE 12.5
SunOS 5.9	Open Client 12.0	SunOS 5.9	Sybase ASE 12.5
Solaris 2.6	Sybase ASE 11.5.1	Solaris 8	Open Client 11.5
Solaris 5.6	Sybase ASE 11.0.3		
AIX 5.3	Sybase ASE 15.0.1（32bits）	AIX 5.3	Sybase ASE 15.0.1
AIX 5.2	Sybase ASE 12.5（32bits）	AIX 5.2	Sybase ASE 12.5.1
IRIX 6.5	11.5.1		
HP-UX 11	11.5.1		

要支持 Sybase ASE 15.0 及以上版本，需要手动修改 setup.py 中的有关代码。
我们从中找到下面的代码块：

```
elif os.name == 'nt':                    # win32
    # Not sure how the installation location is specified under NT
    if sybase is None:
        sybase = r'i:\sybase\sql11.5'
        if not os.access(sybase, os.F_OK):
            sys.stderr.write(
                'Please define the Sybase installation directory in'
```

```
                    'the SYBASE environment variable.\n')
            sys.exit(1)
    syb_libs = ['libblk', 'libct', 'libcs']
```

将其中的 syb_libs = ['libblk', 'libct', 'libcs']替换为 syb_libs = ['libsybblk', 'libsybct', 'libsybcs']即可。原因是在 ASE 15.0 及以上版本，这三个动态库的名字都发生了变化。我在 Sybase ASE 15.0 下安装这个包时发现了此问题。ASE 15.0 以下版本，无须修改该文件就可以安装。安装的命令是 python setup.py install。

安装完以后，在 python 目录下，我们会发现有下述安装好的文件：

Lib\site-packages\Sybase.py

Lib\site-packages\python_sybase-0.38-py2.5.egg-info

Lib\site-packages\sybasect.pyd

需要注意的是，这个安装依赖于 Visual Studio 2003 中的 vc7 编译器。如果你使用的是 Visual Studio 2005（包含 vc8）、Visual Studio 2008（包含 vc9）或者以上版本的编译器，以上安装过程将出错误。首先进入 Visual Studio 的命令行窗口（导入相关 VC 的环境变量 vcvars32.bat），由于该安装依赖于 Open Client，如果你用的不是安装版的 Open Client，要导入 Open Client 环境变量。之后进入 python-sybase-0.38 目录，运行如下命令：

```
D:\MyDocument\MyBooks\ASE\python-sybase-0.38>python setup.py install
running install
running build
running build_py
creating build
creating build\lib.win32-2.5
copying Sybase.py -> build\lib.win32-2.5
running build_ext
error: Python was built with Visual Studio 2003;
extensions must be built with a compiler than can generate compatible binaries.
Visual Studio 2003 was not found on this system. If you have Cygwin installed,
you can try compiling with MingW32, by passing "-c mingw32" to setup.py.
```

这里就提示必须使用 Visual Studio 2003 来进行编译安装。绕过此错误的临时办法是修改相关的 python 文件，修改之前注意备份。

方法是直接编辑 python\Lib\distutils\msvccomiler.py，找到：

```
# x86
if self.__version >= 7:
    self.__root = r"Software\Microsoft\VisualStudio"
    self.__macros = MacroExpander(self.__version)
else:
    self.__root = r"Software\Microsoft\Devstudio"
```

注释掉 self.__macros = MacroExpander(self.__version)这一行，同时删掉该目录下的已编译文件

msvcompiler.pyc，让它强制重新编译。之后设置两个环境变量：

  set DISTUTILS_USE_SDK=1

  set MSSdk=1

  再运行上述命令，得到下述正确的安装过程：

```
D:\MyDocument\MyBooks\ASE\python-sybase-0.38>python setup.py install
running install
running build
running build_py
running build_ext
building 'sybasect' extension
creating build\temp.win32-2.5
creating build\temp.win32-2.5\Release
c:\tools\vs9\VC\BIN\cl.exe /c /nologo /Ox /MD /W3 /GX /DNDEBUG -DWANT_BULKCOPY -DHAVE_DATETIME
-DHAVE_BLK_ALLOC -DHAVE_BLK_DESCRIBE -DHAVE_BLK_DROP -DHAVE_BLK_ROWXFER_MULT -DHAVE_
BLK_TEXTXFER -DHAVE_CT_CURSOR -DHAVE_CT_DATA_INFO -DHAVE_CT_DYNAMIC -DHAVE_CT_SEND_DATA
-DHAVE_CT_SETPARAM -DHAVE_CS_CALC -DHAVE_CS_CMP -ID:\SybaseASE\OCS-15_0\include -Id:\tools\python25\
include -Id:\tools\python25\PC /Tcblk.c /Fobuild\temp.win32-2.5\Release\blk.obj
    cl : Command line warning D9035 : option 'GX' has been deprecated and will be removed in a future release
    cl : Command line warning D9036 : use 'EHsc' instead of 'GX'
blk.c
c:\tools\vs9\VC\BIN\cl.exe /c /nologo /Ox /MD /W3 /GX /DNDEBUG -DWANT_BULKCOPY -DHAVE_DATETIME
-DHAVE_BLK_ALLOC -DHAVE_BLK_DESCRIBE -DHAVE_BLK_DROP -DHAVE_BLK_ROWXFER_MULT -DHAVE_
BLK_TEXTXFER -DHAVE_CT_CURSOR -DHAVE_CT_DATA_INFO -DHAVE_CT_DYNAMIC -DHAVE_CT_SEND_DATA
-DHAVE_CT_SETPARAM -DHAVE_CS_CALC -DHAVE_CS_CMP -ID:\SybaseASE\OCS-15_0\include -Id:\tools\python25\
include -Id:\tools\python25\PC /Tcdatabuf.c /Fobuild\temp.win32-2.5\Release\databuf.obj
    cl : Command line warning D9035 : option 'GX' has been deprecated and will be removed in a future release
    cl : Command line warning D9036 : use 'EHsc' instead of 'GX'
databuf.c
c:\tools\vs9\VC\BIN\cl.exe /c /nologo /Ox /MD /W3 /GX /DNDEBUG -DWANT_BULKCOPY -DHAVE_DATETIME
-DHAVE_BLK_ALLOC -DHAVE_BLK_DESCRIBE -DHAVE_BLK_DROP -DHAVE_BLK_ROWXFER_MULT -DHAVE_
BLK_TEXTXFER -DHAVE_CT_CURSOR -DHAVE_CT_DATA_INFO -DHAVE_CT_DYNAMIC -DHAVE_CT_SEND_DATA
-DHAVE_CT_SETPARAM -DHAVE_CS_CALC -DHAVE_CS_CMP -ID:\SybaseASE\OCS-15_0\include -Id:\tools\python25\
include -Id:\tools\python25\PC /Tccmd.c /Fobuild\temp.win32-2.5\Release\cmd.obj
    cl : Command line warning D9035 : option 'GX' has been deprecated and will be removed in a future release
    cl : Command line warning D9036 : use 'EHsc' instead of 'GX'
cmd.c
c:\tools\vs9\VC\BIN\cl.exe /c /nologo /Ox /MD /W3 /GX /DNDEBUG -DWANT_BULKCOPY -DHAVE_DATETIME
-DHAVE_BLK_ALLOC -DHAVE_BLK_DESCRIBE -DHAVE_BLK_DROP -DHAVE_BLK_ROWXFER_MULT -DHAVE_
BLK_TEXTXFER -DHAVE_CT_CURSOR -DHAVE_CT_DATA_INFO -DHAVE_CT_DYNAMIC -DHAVE_CT_SEND_DATA
-DHAVE_CT_SETPARAM -DHAVE_CS_CALC -DHAVE_CS_CMP -ID:\SybaseASE\OCS-15_0\include -Id:\tools\python25\
include -Id:\tools\python25\PC /Tcconn.c /Fobuild\temp.win32-2.5\Release\conn.obj
    cl : Command line warning D9035 : option 'GX' has been deprecated and will be removed in a future release
    cl : Command line warning D9036 : use 'EHsc' instead of 'GX'
conn.c
c:\tools\vs9\VC\BIN\cl.exe /c /nologo /Ox /MD /W3 /GX /DNDEBUG -DWANT_BULKCOPY -DHAVE_DATETIME
-DHAVE_BLK_ALLOC -DHAVE_BLK_DESCRIBE -DHAVE_BLK_DROP -DHAVE_BLK_ROWXFER_MULT -DHAVE_
```

BLK_TEXTXFER -DHAVE_CT_CURSOR -DHAVE_CT_DATA_INFO -DHAVE_CT_DYNAMIC -DHAVE_CT_SEND_DATA -DHAVE_CT_SETPARAM -DHAVE_CS_CALC -DHAVE_CS_CMP -ID:\SybaseASE\OCS-15_0\include -Id:\tools\python25\include -Id:\tools\python25\PC /Tcctx.c /Fobuild\temp.win32-2.5\Release\ctx.obj

    cl : Command line warning D9035 : option 'GX' has been deprecated and will be removed in a future release

    cl : Command line warning D9036 : use 'EHsc' instead of 'GX'

    ctx.c

    c:\tools\vs9\VC\BIN\cl.exe /c /nologo /Ox /MD /W3 /GX /DNDEBUG -DWANT_BULKCOPY -DHAVE_DATETIME -DHAVE_BLK_ALLOC -DHAVE_BLK_DESCRIBE -DHAVE_BLK_DROP -DHAVE_BLK_ROWXFER_MULT -DHAVE_BLK_TEXTXFER -DHAVE_CT_CURSOR -DHAVE_CT_DATA_INFO -DHAVE_CT_DYNAMIC -DHAVE_CT_SEND_DATA -DHAVE_CT_SETPARAM -DHAVE_CS_CALC -DHAVE_CS_CMP -ID:\SybaseASE\OCS-15_0\include -Id:\tools\python25\include -Id:\tools\python25\PC /Tcdatafmt.c /Fobuild\temp.win32-2.5\Release\datafmt.obj

    cl : Command line warning D9035 : option 'GX' has been deprecated and will be removed in a future release

    cl : Command line warning D9036 : use 'EHsc' instead of 'GX'

    datafmt.c

    c:\tools\vs9\VC\BIN\cl.exe /c /nologo /Ox /MD /W3 /GX /DNDEBUG -DWANT_BULKCOPY -DHAVE_DATETIME -DHAVE_BLK_ALLOC -DHAVE_BLK_DESCRIBE -DHAVE_BLK_DROP -DHAVE_BLK_ROWXFER_MULT -DHAVE_BLK_TEXTXFER -DHAVE_CT_CURSOR -DHAVE_CT_DATA_INFO -DHAVE_CT_DYNAMIC -DHAVE_CT_SEND_DATA -DHAVE_CT_SETPARAM -DHAVE_CS_CALC -DHAVE_CS_CMP -ID:\SybaseASE\OCS-15_0\include -Id:\tools\python25\include -Id:\tools\python25\PC /Tciodesc.c /Fobuild\temp.win32-2.5\Release\iodesc.obj

    cl : Command line warning D9035 : option 'GX' has been deprecated and will be removed in a future release

    cl : Command line warning D9036 : use 'EHsc' instead of 'GX'

    iodesc.c

    c:\tools\vs9\VC\BIN\cl.exe /c /nologo /Ox /MD /W3 /GX /DNDEBUG -DWANT_BULKCOPY -DHAVE_DATETIME -DHAVE_BLK_ALLOC -DHAVE_BLK_DESCRIBE -DHAVE_BLK_DROP -DHAVE_BLK_ROWXFER_MULT -DHAVE_BLK_TEXTXFER -DHAVE_CT_CURSOR -DHAVE_CT_DATA_INFO -DHAVE_CT_DYNAMIC -DHAVE_CT_SEND_DATA -DHAVE_CT_SETPARAM -DHAVE_CS_CALC -DHAVE_CS_CMP -ID:\SybaseASE\OCS-15_0\include -Id:\tools\python25\include -Id:\tools\python25\PC /Tclocale.c /Fobuild\temp.win32-2.5\Release\locale.obj

    cl : Command line warning D9035 : option 'GX' has been deprecated and will be removed in a future release

    cl : Command line warning D9036 : use 'EHsc' instead of 'GX'

    locale.c

    c:\tools\vs9\VC\BIN\cl.exe /c /nologo /Ox /MD /W3 /GX /DNDEBUG -DWANT_BULKCOPY -DHAVE_DATETIME -DHAVE_BLK_ALLOC -DHAVE_BLK_DESCRIBE -DHAVE_BLK_DROP -DHAVE_BLK_ROWXFER_MULT -DHAVE_BLK_TEXTXFER -DHAVE_CT_CURSOR -DHAVE_CT_DATA_INFO -DHAVE_CT_DYNAMIC -DHAVE_CT_SEND_DATA -DHAVE_CT_SETPARAM -DHAVE_CS_CALC -DHAVE_CS_CMP -ID:\SybaseASE\OCS-15_0\include -Id:\tools\python25\include -Id:\tools\python25\PC /Tcmsgs.c /Fobuild\temp.win32-2.5\Release\msgs.obj

    cl : Command line warning D9035 : option 'GX' has been deprecated and will be removed in a future release

    cl : Command line warning D9036 : use 'EHsc' instead of 'GX'

    msgs.c

    c:\tools\vs9\VC\BIN\cl.exe /c /nologo /Ox /MD /W3 /GX /DNDEBUG -DWANT_BULKCOPY -DHAVE_DATETIME -DHAVE_BLK_ALLOC -DHAVE_BLK_DESCRIBE -DHAVE_BLK_DROP -DHAVE_BLK_ROWXFER_MULT -DHAVE_BLK_TEXTXFER -DHAVE_CT_CURSOR -DHAVE_CT_DATA_INFO -DHAVE_CT_DYNAMIC -DHAVE_CT_SEND_DATA -DHAVE_CT_SETPARAM -DHAVE_CS_CALC -DHAVE_CS_CMP -ID:\SybaseASE\OCS-15_0\include -Id:\tools\python25\include -Id:\tools\python25\PC /Tcnumeric.c /Fobuild\temp.win32-2.5\Release\numeric.obj

    cl : Command line warning D9035 : option 'GX' has been deprecated and will be removed in a future release

    cl : Command line warning D9036 : use 'EHsc' instead of 'GX'

    numeric.c

    c:\tools\vs9\VC\BIN\cl.exe /c /nologo /Ox /MD /W3 /GX /DNDEBUG -DWANT_BULKCOPY -DHAVE_DATETIME -DHAVE_BLK_ALLOC -DHAVE_BLK_DESCRIBE -DHAVE_BLK_DROP -DHAVE_BLK_ROWXFER_MULT -DHAVE_

BLK_TEXTXFER -DHAVE_CT_CURSOR -DHAVE_CT_DATA_INFO -DHAVE_CT_DYNAMIC -DHAVE_CT_SEND_DATA -DHAVE_CT_SETPARAM -DHAVE_CS_CALC -DHAVE_CS_CMP -ID:\SybaseASE\OCS-15_0\include -Id:\tools\python25\include -Id:\tools\python25\PC /Tcmoney.c /Fobuild\temp.win32-2.5\Release\money.obj

  cl : Command line warning D9035 : option 'GX' has been deprecated and will be removed in a future release
  cl : Command line warning D9036 : use 'EHsc' instead of 'GX'
  money.c
  c:\tools\vs9\VC\BIN\cl.exe /c /nologo /Ox /MD /W3 /GX /DNDEBUG -DWANT_BULKCOPY -DHAVE_DATETIME -DHAVE_BLK_ALLOC -DHAVE_BLK_DESCRIBE -DHAVE_BLK_DROP -DHAVE_BLK_ROWXFER_MULT -DHAVE_BLK_TEXTXFER -DHAVE_CT_CURSOR -DHAVE_CT_DATA_INFO -DHAVE_CT_DYNAMIC -DHAVE_CT_SEND_DATA -DHAVE_CT_SETPARAM -DHAVE_CS_CALC -DHAVE_CS_CMP -ID:\SybaseASE\OCS-15_0\include -Id:\tools\python25\include -Id:\tools\python25\PC /Tcdatetime.c /Fobuild\temp.win32-2.5\Release\datetime.obj

  cl : Command line warning D9035 : option 'GX' has been deprecated and will be removed in a future release
  cl : Command line warning D9036 : use 'EHsc' instead of 'GX'
  datetime.c
  c:\tools\vs9\VC\BIN\cl.exe /c /nologo /Ox /MD /W3 /GX /DNDEBUG -DWANT_BULKCOPY -DHAVE_DATETIME -DHAVE_BLK_ALLOC -DHAVE_BLK_DESCRIBE -DHAVE_BLK_DROP -DHAVE_BLK_ROWXFER_MULT -DHAVE_BLK_TEXTXFER -DHAVE_CT_CURSOR -DHAVE_CT_DATA_INFO -DHAVE_CT_DYNAMIC -DHAVE_CT_SEND_DATA -DHAVE_CT_SETPARAM -DHAVE_CS_CALC -DHAVE_CS_CMP -ID:\SybaseASE\OCS-15_0\include -Id:\tools\python25\include -Id:\tools\python25\PC /Tcsybasect.c /Fobuild\temp.win32-2.5\Release\sybasect.obj

  cl : Command line warning D9035 : option 'GX' has been deprecated and will be removed in a future release
  cl : Command line warning D9036 : use 'EHsc' instead of 'GX'
  sybasect.c
  c:\tools\vs9\VC\BIN\link.exe /DLL /nologo /INCREMENTAL:NO /LIBPATH:D:\SybaseASE\OCS-15_0\lib /LIBPATH:d:\tools\python25\libs /LIBPATH:d:\tools\python25\PCBuild libsybblk.lib libsybct.lib libsybcs.lib /EXPORT:initsybasect build\temp.win32-2.5\Release\blk.obj build\temp.win32-2.5\Release\databuf.obj build\temp.win32-2.5\Release\cmd.obj build\temp.win32-2.5\Release\conn.obj build\temp.win32-2.5\Release\ctx.obj build\temp.win32-2.5\Release\datafmt.obj build\temp.win32-2.5\Release\iodesc.obj build\temp.win32-2.5\Release\locale.obj build\temp.win32-2.5\Release\msgs.obj build\temp.win32-2.5\Release\numeric.obj build\temp.win32-2.5\Release\money.obj build\temp.win32-2.5\Release\datetime.obj build\temp.win32-2.5\Release\sybasect.obj /OUT:build\lib.win32-2.5\sybasect.pyd /IMPLIB:build\temp.win32-2.5\Release\sybasect.lib
   Creating library build\temp.win32-2.5\Release\sybasect.lib and object build\temp.win32-2.5\Release\sybasect.exp
  running install_lib
  copying build\lib.win32-2.5\sybasect.pyd -> d:\tools\python25\Lib\site-packages
  copying build\lib.win32-2.5\sybasect.pyd.manifest -> d:\tools\python25\Lib\site-packages
  running install_egg_info
  Removing d:\tools\python25\Lib\site-packages\python_sybase-0.38-py2.5.egg-info
  Writing d:\tools\python25\Lib\site-packages\python_sybase-0.38-py2.5.egg-info

18.2 使用 Python 连接 ASE

这里我们将通过简单的示例介绍如何使用 Python 来连接 ASE 数据库。

```
#001 #!/usr/bin/python
#002 # -*- coding: gbk -*-
#003
#004 import Sybase
#005
```

```
#006 class TestConn:
#007     def __init__(self):
#008         pass
#009     def test(self):
#010         db = Sybase.connect('sean-laptop', 'sa', 'sybase1', 'sybsystemprocs', datetime='python')
#011         print "连接成功!"
#012         db.close()
#013         print "连接关闭"
#014
#015 if __name__ == '__main__':
#016     t = TestConn()
#017     try:
#018         t.test()
#019     except Exception, ex:
#020         print Exception,":",ex
#021
#022     print "finished !"
```

通过一行简单的调用 Sybase.connect()即可连接 Sybase ASE。函数 connect 的 4 个参数分别为数据源（数据库服务名）、用户名、密码、数据库名：

Connect(dsn, user, passwd, database, …)

dsn 变量值（服务名）应该来源于我们前边介绍的 sql.ini 或者 interfaces 文件中注册的数据库服务名。

它返回的是一个 Connection 对象。通过 Connection 的 close()方法即可关闭数据库连接。Python 有自己的异常机制，如果遇到异常，可以输出异常信息并退出。

18.3 使用 Python 访问 ASE 数据库表

我们通过一个综合性的例子来看看 Python 如何访问 ASE 的数据库表，执行常规的 CRUD（Create、Select、Update、Delete）操作。

```
#001 #!/usr/bin/python
#002 # -*- coding: gbk -*-
#003
#004 import Sybase
#005
#006 class HexdbError(Exception): pass
#007
#008 class TestDML:
#009     def __init__(self, host, user, password, dbname=''):
#010         self._host = host
#011         self._user = user
#012         self._password = password
```

```
#013            self._dbname = dbname
#014            pass
#015
#016        def dropTable(self, cursor, tname):
#017            try:
#018                cursor.execute('drop table ' + tname)
#019            except Exception:
#020                print "exception:", Exception
#021                pass
#022
#023        def test(self):
#024            db = Sybase.connect(self._host, self._user, self._password, self._dbname, datetime='python')
#025            print "begin create table t1"
#026
#027            cursor = db.cursor()
#028            self.dropTable(cursor, 't1')
#029            cursor.execute('create table t1(id int primary key, col2 varchar(32))')
#030            cursor.close()
#031            db.close()
#032            print "connect closed, table t1 created."
#033
#034        def testInsert(self):
#035            db = Sybase.connect(self._host, self._user, self._password, self._dbname, datetime='python')
#036            cursor = db.cursor()
#037            cursor.execute("insert into t1 values(1, '中文')")
#038            cursor.close()
#039            cursor = db.cursor()
#040            cursor.execute("insert into t1 values(@id, @col2)", {"@id":100, "@col2":"test 英文"})
#041            cursor.close()
#042            db.commit()
#043            print "insert successfully!"
#044            db.close()
#045
#046        def testUpdate(self):
#047            db = Sybase.connect(self._host, self._user, self._password, self._dbname, datetime='python')
#048            cursor = db.cursor()
#049            cursor.execute("update t1 set col2=@col2 where id=@id", {"@id":100, "@col2":"test 英文_update"})
#050            cursor.close()
#051            db.commit()
#052            print "update rows with id=100 successfully!"
#053            db.close()
#054
#055
#056        def testFetch(self):
#057            db = Sybase.connect(self._host, self._user, self._password, self._dbname, datetime='python')
#058            cursor = db.cursor()
#059            cursor.execute('select * from t1')
#060            for row in cursor.fetchall():
```

```
#061                print row[0], row[1]
#062                cursor.close()
#063                db.commit()
#064                print "fetch data finished!"
#065                db.close()
#066
#067
#068
#069 if __name__ == '__main__':
#070     t = TestDML('XIONGHE', 'spring', 'spring1', 'iihero')
#071     try:
#072         t.test()
#073         t.testInsert()
#074         t.testUpdate()
#075         t.testFetch()
#076     except Exception, ex:
#077         print Exception,":",ex
#078
#079     print "finished DML."
```

这个示例完整地演示了对表 t1 的创建、插入、更新、删除及查询操作。

执行一个 SQL 语句并获取相应结果的基本过程如下：

Step 1 获取数据库连接。

> db = Sybase.connect(self._host, self._user, self._password, self._dbname, datetime='python') 得到连接

Step 2 由数据库连接，生成 cursor 对象。

> cursor = db.cursor

Step 3 使用 cursor 执行相应的 SQL 语句。

当需要传递参数给 SQL 语句时，推荐使用动态绑定，例如：

> cursor.execute("update t1 set col2=@col2 where id=@id", {"@id":100, "@col2":"test 英文_update"})

这里分别向@col2 和@id 传递两个参数值。

Step 4 如果是查询操作，调用 cursor.fetchall()操作获取结果。

Step 5 关闭 cursor。

Step 6 关闭数据库连接。

19 使用 ADO.NET 访问 ASE

ASE ADO.NET 接口库主要用作 Sybase ASE 数据库 ADO.NET 的提供者，主要方便开发人员通过.NET 来访问 ASE 数据库，支持 C#、VB.NET、C++的.NET 托管扩展，以及 J#四种编程语言访问 ASE。

从本质上来说，ASE ADO.NET 数据提供者是一个公共语言运行库（CLR）的类库，该库实现了所有的 ADO.NET 访问接口。由于库的实现基于托管代码和.NET 框架，因而支持.NET 下的多种编程语言。

ASE 的 ADO.NET 接口相比以前的其他接口，有一个很重要的优势：ADO.NET 接口实现在速度上比以前的 OLE DB 接口（主要用于 ADO）要快，在.NET 环境中，它不同于以前的库接口实现，它是直接与后端的 ASE 服务器通信，没有使用任何桥接技术，即不依赖于其他的库接口[①]。

ASE ADO.NET 实现库主要包括下述几个动态库文件：

- sybase.Data.AseClient.dll，主要用作客户端访问时的接口层，即客户端只需要直接引用该库即可。
- sybdrvado115.dll，是 ASE ADO.NET 实现时真正要用到的基础文件，客户端不需要直接引用该库。
- sbgse2.dll 和 sybdrvssl.dll，主要用于 SSL（安全套接层）的支持，即支持对来往通信的加密处理。
- sybdrvkrb.dll，该库主要用于 Kerberos 用户口令验证。

---

① 其实，在 ASE 的其他库中，大多要依赖于 OpenClient 的 ctlib 接口库，它的 dll 文件位于%SYBASE%\OCS-15_0\dll 下，它是目前 ASE 中效率最高的底层访问接口。

如果我们使用 Visual Studio[.net]自带的工具 depends（它一般位于路径%VCHome%\Common7\Tools\Bin 目录下，该目录通常也定义为%VS71COMNTOOLS%或%VS8COMNTOOLS%），打开动态库 sybdrvado115.dll，发现其他依赖的动态库只有几个，如图 19-1 所示。

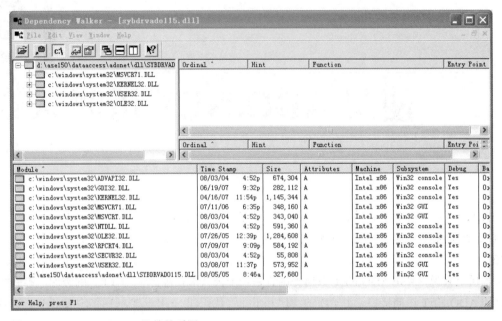

图 19-1　sybdrvado115.dll 依赖关系图

由图 19-1 可以看到，sybdrvado115.dll 并没有直接依赖于 Open Client 的任何库。

对一般应用而言，我们只需要了解第一个库 Sybase.Data.AseClient.dll 和第二个库 sybdrvado115.dll。

实际上，ASE 的 ADO.NET 的实现是直接通过底层的 TDS 协议（Sybase 数据库专用的通信协议）与 Server 端实现交互的，并没有调用 Open Client 的底层 API。它不再需要任何配置文件和 Open Client 库，使用起来比较便利，这是一个很大的进步。

19.1　ASE ADO.NET 运行时环境

要想使用 ADO.NET 开发 ASE 数据库应用，首先系统必须拥有或者已经安装好.NET 框架，至少是 1.1 版本。另外，必须安装好 ASE 的 ADO.NET 提供者（Provider），它在 ASE 15.0、ASE 12.5.x 的 SDK 安装包或者 ASE 数据库的安装包里都有这个选项，安装完以后，在 ASE 的安装目录下会发现有如下子目录结构：

```
D:\ASE150>dir dataaccess
 驱动器 D 中的卷没有标签
 卷的序列号是 107B-71AD

 D:\ASE150\dataaccess 的目录

2007-12-02  18:06    <DIR>          .
2007-12-02  18:06    <DIR>          ..
2007-12-02  18:07    <DIR>          ADONET
2007-12-02  18:06    <DIR>          ODBC
2007-12-02  18:06    <DIR>          OLEDB
```

即 dataaccess 目录下有一个子目录 ADONET，前边介绍的 5 个 DLL 文件就位于 ADONET\dll 下边。

ASE 的 ADO.NET 实现对应的 ADO.NET 接口及其具体实现类参照表 19-1。

表 19-1　ASE ADO.NET 具体实现类及命名空间

命名空间 System.Data	System.Data.AseClient
IDbConnection	AseConnection
IDbCommand	AseCommand
IDbTransaction	AseTransaction
IDataReader	AseDataReader
IDataParameter, IDbDataParameter	AseParameter
IDataAdapter, IDbDataAdapter	AseDataAdapter
DataSet	

我们都知道，MS SQL Server 和 OLEDB 的 ADO.NET 实现分别使用了命名空间 System.Data.SqlClient 和 System.Data.OleDb，Sybase ASE 则使用了独立的命名空间 System.Data.AseClient，以示区分。

ADO.NET 主要提供以下两种数据访问方式：

- 基于流的数据访问。
  使用的是具体的 DataReader 类（如 AseDataReader）来获取来自数据库服务器端只读、单向的数据，应用程序在遍历数据时，必须事先已经与数据库建立好连接。

- 离线（断连状态下）数据访问。
  使用的是 DataSet 类，是一种内存数据缓存，借助于它可以在断开数据库连接的情况下工作。使用 DataSet 类时，通常会创建一个具体的 DataSet 对象——AseDataSet 对象，并使用 AseDataAdapter 为其填充数据，最后可以将填充的数据提交给数据库或存为本地文件。

运行时的基本结构图如图 19-2 所示。

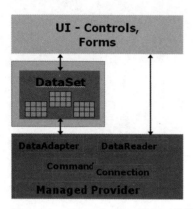

图 19-2　ASE ADO.NET 运行时结构示意图

下面将以 C#语言为例，分节介绍 ASE ADO.NET 接口库的详细使用方法，同时也会介绍如何将开发出的应用进行打包部署。

19.2　连接 ASE 数据库

本节将通过一个简单的例子，介绍如何通过 ADO.NET 连接 ASE 数据库并执行查询，获取查询结果。最后，会介绍如何将这个开发好的应用进行打包部署到新的机器上。

ASE ADO.NET 最基本最常用的几个类是：

- AseConnection，主要用于建立 ASE 的数据库连接。
- AseCommand，主要用于执行 SQL 语句，可以是 DQL、DDL 或者 DML 操作。
- AseDataReader，主要用于 SQL 查询以后生成的结果集进行遍历访问。

基本调用流程如下：

Step 1　根据具体的 ASE 数据库连接串信息，创建一个 AseConnection 实例。

Step 2　通过已创建的 AseConnection 实例及具体的 SQL 查询语句，创建一个 AseCommand 对象。

Step 3　由 AseCommand 对象调用 ExecuteReader()函数，生成 AseDataReader 结果集，或者执行 ExecuteNonQuery()函数完成 Insert/Update/ Delete 操作。

Step 4　通过 AseDataReader 来遍历最后的结果集。

完整的实例如下（见代码清单中的文件 dotnet\Demo\DemoMain 目录下的 DemoSelect.cs）：

```
#001 using System;
#002 using System.Collections;
#003 using System.Data;
#004 using Sybase.Data.AseClient;
```

```
#005
#006 public class DotnetDemo1
#007 {
#008     private string host;
#009     private int port;
#010     private string user;
#011     private string password;
#012     public DotnetDemo1(string host, int port, string user, string password)
#013     {
#014         this.host = host;
#015         this.port = port;
#016         this.user = user;
#017         this.password = password;
#018     }
#019
#020     public void SimpleTest()
#021     {
#022         AseConnection conn = new AseConnection( "Data Source='" + host + "';Port='" + port + "';UID='" + user + "';PWD='" + password + "';Database='iihero';" );
#023         AseCommand cmd = null;
#024         AseDataReader reader = null;
#025
#026         try
#027         {
#028             conn.Open();
#029             cmd = new AseCommand( "select sname from spring.student", conn );
#030             reader = cmd.ExecuteReader();
#031             while( reader.Read() )
#032             {
#033                 Console.WriteLine(reader.GetString( 0 ));
#034             }
#035         }
#036         catch( AseException ex )
#037         {
#038             Console.WriteLine(ex.Message);
#039         }
#040         finally
#041         {
#042             if (reader != null && !reader.IsClosed)
#043                 reader.Close();
#044             if (cmd != null)
#045                 cmd.Dispose();
#046             if (conn != null && conn.State != ConnectionState.Closed)
#047                 conn.Close();
#048         }
#049     }
#050
#051     public static void Main(string[] argv)
#052     {
```

```
#053            DotnetDemo1 t = new DotnetDemo1("sean-laptop", 5000, "spring", "spring1");
#054            t.SimpleTest();
#055        }
#056 }
```

在本例中，第 1～4 行声明了程序要用到的命名空间，或者称为包，第 4 行的 Sybase.Data.AseClient 直接来源于 DLL 文件：Sybase.Data.AseClient.dll。

你需要在你的.NET 工程里显式地引用这个 DLL 文件。即打开 solution→project→references 文件，添加 reference，找到 Sybase.Data.AseClient.dll 所在目录，添加即可。

第 12 行，DotnetDemo1(string host, int port, string user, string password)构造函数，主要用于构造数据库连接时要用到的主机名、端口号、用户名及密码信息。

第 20～49 行，是测试函数 SimpleTest()的实现。它主要用于执行对表 spring.student 的查询，获取字段 sname 的所有值。第 42～47 行，用于释放结果集、command 对象及数据库连接。

第 22 行，用于创建一个 AseConnection 对象，关键在于它的连接串信息，这里完整的连接串内容为：

Data Source='sean-laptop';Port='5000';UID='spring';PWD='spring1';Database='iihero';

每个属性信息都使用分号（;）结尾，甚至最后一个属性也如此。

Data Source 属性用于指定 ADO.NET 数据源的主机名，即 ASE 数据库的机器名或者 IP 地址。这里 Data Source 属性在 ASE DOTNET 中有几个同义词，它们分别是：Server、DataSource、Address、Addr、Network Address、Server Name。

- Port：指定数据源的主机端口号，对应的同义词有 Server Port。
- UID：指定要访问的数据库的用户名，对应的同义词有 UserID、User。
- PWD：指定要访问的数据库的用户密码，对应的同义词有 Password。
- Database：指定上述用户要访问的数据库名，如果不指定，则访问该用户的默认数据库。对应的同义词有 Initial Catalog。

这些连接的属性信息只要设置正确就可以连接 ASE 数据库了。

使用 csc 编译命令，生成可执行文件 demo1.exe：

```
E:\MyDocument\MYBOOKS\ASE\code>csc /debug:full /out:demo1.exe dotnet_demo1.cs
/reference:D:\ASE150\DataAccess\ADONET\dll\Sybase.Data.AseClient.dll
Microsoft (R) Visual C# 2005 Compiler version 8.00.50727.1433
for Microsoft (R) Windows (R) 2005 Framework version 2.0.50727
Copyright (C) Microsoft Corporation 2001-2005. All rights reserved.

E:\MyDocument\MYBOOKS\ASE\code>demo1.exe
李勇
刘晨
王敏
张铁林
```

- /debug: full 用于生成带有调试信息的 exe 程序。
- /out: demo1.exe 表示链接生成 demo1.exe 二进制文件。
- /reference: D:\ASE150\DataAccess\ADONET\dll\Sybase.Data.AseClient.dll 表示编译链接过程中要引用到 ASE 的 ADO.NET 库文件 Sybase.Data.AseClient.dll。

需要补充说明的是，还有几个重要的连接属性参数在实际的应用开发中可能要用到：

1. AnsiNull：默认值为 0，值为 1 时，则严格按照 AnsiNull 的定义来规定 NULL 的行为，这个属性值很重要。在 ASE 中，其默认值为 0，这意味着，=NULL 和 ISNULL 语义上是一样的。这将与其他数据库中的 NULL 语义不一样。要想保持一样的行为，需要将此值设为 1。

2. Pooling：默认值为 True，表示要使用连接池。如果不想使用连接池特性，可以将其设为 False。

3. QuotedIdentifier：默认值为 0。它表示标识符可以通过加上双引号来作为对象使用，而文本则可以用单引号分割。在 ASE 中，通过 set quoted_identifier on 就可以使用加引号的方式把某些标识符当作对象来使用。

例如：

```
1> set quoted_identifier on
2> go
1> create table tt("select" int)
2> go
1> insert into tt values(1)
2> go
(1 row affected)
1> select "select" from tt
2> go
  select
  -----------
  1

(1 row affected)
1> set quoted_identifier off
2> go
1> select "select" from tt
2> go

  ------
  select
(1 row affected)
1> create table tt("select" int)
2> go
Msg 102, Level 15, State 1:
Server 'XIONGHE', Line 1:
Incorrect syntax near '('.
```

当 quoted_identifier 为 off 时，select 被当作是普通字符串（这就是区别），也不能当作字段名建表了。

4. TextSize：这个值表示往 ASE 发送或从 ASE 接收二进制或文本数据的最大长度，默认值是 32KB。如果你想设为 64KB，可以将 TextSize 设为 65536。

19.3 创建删除表

创建删除表操作也称为 DDL（数据定义语言）操作，是比较常用的一种 SQL 操作。使用 ADO.NET，可以非常方便地执行表的创建与删除操作。

一个典型的建表 SQL 语句如下：

```
/* 1. 雇员数据库 */
create table emp
(
    empno int not null,
    ename varchar(32) null,
    job varchar(32) null,
    mgr int null,
    hiredate datetime null,
    sal decimal(8, 2) null,
    comm decimal(8, 2) null,
    deptno decimal(2) null,
    comments text null,
    photo image null,
    gender char(1) not null
)
```

我们只要基于建表或删表 SQL 语句及 Connection 对象，来创建一个 AseCommand 对象，并执行其 ExecuteNonQuery()方法，即可完成 DDL 操作，一个建表的示例代码如下所示：

```
#001 public void CreateTable()
#002 {
#003     AseConnection conn = new AseConnection( "Data Source='" + host + "';Port='" + port + "';UID='" + user + "';PWD='" + password + "';Database='iihero';" );
#004     AseCommand cmd = null;
#005     String sql = "create table emp ("
#006         +      "empno int not null,"
#007         +      "ename varchar(32) null,"
#008         +      "job varchar(32) null,"
#009         +      "mgr int null,"
#010         +      "hiredate datetime null,"
#011         +      "sal decimal(8, 2) null,"
#012         +      "comm decimal(8, 2) null,"
#013         +      "deptno decimal(2) null,"
```

```
#014                    +     "comments text null,"
#015                    +     "photo image null,"
#016                    +     "gender char(1) not null"
#017                    +")";
#018        try
#019        {
#020            conn.Open();
#021            AseTransaction tx = conn.BeginTransaction();
#022            cmd = new AseCommand(sql, conn);
#023            int rowsAffected = cmd.ExecuteNonQuery();
#024            Console.WriteLine(rowsAffected.ToString() + " row(s) affected");
#025            tx.Commit();
#026        }
#027        catch( AseException ex )
#028        {
#029            Console.WriteLine(ex.Message);
#030        }
#031        finally
#032        {
#033            if (cmd != null)
#034                cmd.Dispose();
#035            if (conn != null && conn.State != ConnectionState.Closed)
#036                conn.Close();
#037        }
#038 }
```

如果想要执行删除表 emp 的操作，只需要将字符串变量 sql 赋值为 drop table emp 即可。

实际上，只要是 DDL 操作，包括 ALTER TABLE、CREATE USER 等 DDL 语句，都可以采取类似的方法处理。

19.4 插入数据

使用 ADO.NET 操纵 ASE 数据库的一个常用操作是往数据库表中插入数据。我们首先看看如何插入普通的结构化数据。以 iihero 数据库中的表 emp 为例，表 emp 定义如下：

```
/* 1. 雇员数据库 */
create table emp
(
    empno int not null,
    ename varchar(32) null,
    job varchar(32) null,
    mgr int null,
    hiredate datetime null,
    sal decimal(8, 2) null,
```

```
    comm decimal(8, 2) null,
    deptno decimal(2) null,
    comments text null,
    photo image null,
    gender char(1) not null
)
```

19.4.1 使用 DataSet 类来插入数据

我们来看看下边的代码示例,它将向表 emp 里插入一条记录:

```
#001  public void SimpleTest()
#002  {
#003      AseConnection conn = new AseConnection( "Data Source='" + host + "';Port='" + port + "';UID='" + user + "';PWD='" + password + "';Database='iihero';" );
#004      AseCommand cmd = null;
#005      AseDataReader reader = null;
#006      AseDataAdapter da = null;
#007      DataSet ds = null;
#008
#009      try
#010      {
#011          conn.Open();
#012          AseTransaction tx = conn.BeginTransaction();
#013          da = new AseDataAdapter("select * from emp", conn);
#014          AseCommandBuilder cb = new AseCommandBuilder(da);
#015          ds = new DataSet();
#016          da.Fill(ds);
#017
#018          DataTable dt = ds.Tables[0];
#019          //dt.Rows.Add(new object[] {1});
#020          DataRow r = dt.NewRow();
#021          r["empno"] = 1;
#022
#023          r["ename"] = "张无忌";
#024          r["job"] = "SALESMAN";
#025          r["mgr"] = 7839;
#026          r["hiredate"] = "1981-02-20";
#027          r["sal"] = 8000.56;
#028          r["comm"] = 800;
#029          r["deptno"] = 30;
#030          r["gender"] = "M";
#031
#032          dt.Rows.Add(r);
#033          ds.WriteXml("abc.xml");     // 将 ds 中的数据写到文件 abc.xml 当中
#034          int updatedRows = da.Update(dt);
```

```
#035
#036                Console.WriteLine("inserted rows = " + updatedRows);
#037                tx.Commit();
#038            }
#039            catch( AseException ex )
#040            {
#041                Console.WriteLine(ex.Message);
#042            }
#043            finally
#044            {
#045                if (ds != null)
#046                    ds.Dispose();
#047                if (da != null)
#048                    da.Dispose();
#049                if (conn != null && conn.State != ConnectionState.Closed)
#050                    conn.Close();
#051            }
#052 }
```

这里我们没有使用任何 Insert 语句。第 13 行，通过一个 Select 语句创建了一个 DataAdapter 对象 da，然后通过 da 创建一个 DataSet 对象 ds，用于存储表数据的同时，会建立到 ASE 数据库的关联。

第 14 行，是为 DataAdapter 创建一个具体的 SQL 命令对象，这行代码必不可少，因为后边的更新操作会隐式地调用这个 AseCommandBuilder。

一个 DataSet 对象包含零到多个 DataTable 对象，每个 DataTable 相当于一张保存到内存中的表。因此，第 18～32 行实际上是对内存中的表数据进行操作。直到第 34 行，执行了 DataAdapter 的更新操作以后，数据插入才算完成。第 34 行，da.Update(dt)会返回实际上更新的记录的行数。

完整的代码清单见 dotnet 目录下的 DemoInsert.cs，执行该程序，会发现 abc.xml 里有 emp 表的所有数据：

```
#001 <?xml version="1.0" standalone="yes"?>
#002 <NewDataSet>
#003   <Table>
#004     <empno>7369</empno>
#005     <ename>SMITH</ename>
#006     <job>CLERK</job>
#007     <mgr>7902</mgr>
#008     <hiredate>1980-12-17T00:00:00.0000000+08:00</hiredate>
#009     <sal>800</sal>
#010     <deptno>20</deptno>
#011     <comments>comment of smith</comments>
#012     <gender>M</gender>
#013   </Table>
#014   <Table>
```

```
#015        <empno>7499</empno>
#016        <ename>ALLEN</ename>
#017        <job>SALESMAN</job>
#018        <mgr>7698</mgr>
#019        <hiredate>1981-02-20T00:00:00.0000000+08:00</hiredate>
#020        <sal>1600</sal>
#021        <comm>300</comm>
#022        <deptno>30</deptno>
#023        <comments>comment of ALLEN</comments>
#024        <gender>F</gender>
#025      </Table>
#026      <Table>
#027        <empno>7521</empno>
#028        <ename>WARD</ename>
#029        <job>SALESMAN</job>
#030        <mgr>7698</mgr>
#031        <hiredate>1981-02-22T00:00:00.0000000+08:00</hiredate>
#032        <sal>1250</sal>
#033        <comm>500</comm>
#034        <deptno>30</deptno>
#035        <comments>comment of WARD</comments>
#036        <gender>M</gender>
#037      </Table>
#038      <Table>
#039        <empno>7566</empno>
#040        <ename>JONES</ename>
#041        <job>MANAGER</job>
#042        <mgr>7839</mgr>
#043        <hiredate>1981-04-02T00:00:00.0000000+08:00</hiredate>
#044        <sal>2975</sal>
#045        <deptno>20</deptno>
#046        <comments>comment of JONES</comments>
#047        <gender>F</gender>
#048      </Table>
#049      <Table>
#050        <empno>7654</empno>
#051        <ename>MARTIN</ename>
#052        <job>SALESMAN</job>
#053        <mgr>7698</mgr>
#054        <hiredate>1981-09-28T00:00:00.0000000+08:00</hiredate>
#055        <sal>1250</sal>
#056        <comm>1400</comm>
#057        <deptno>30</deptno>
#058        <comments>comment of MARTIN</comments>
#059        <gender>M</gender>
#060      </Table>
#061      <Table>
#062        <empno>7698</empno>
```

```
#063        <ename>BLAKE</ename>
#064        <job>MANAGER</job>
#065        <mgr>7839</mgr>
#066        <hiredate>1981-05-01T00:00:00.0000000+08:00</hiredate>
#067        <sal>2850</sal>
#068        <deptno>30</deptno>
#069        <comments>comment of BLAKE</comments>
#070        <gender>M</gender>
#071      </Table>
#072      <Table>
#073        <empno>7782</empno>
#074        <ename>CLARK</ename>
#075        <job>MANAGER</job>
#076        <mgr>7839</mgr>
#077        <hiredate>1981-06-09T00:00:00.0000000+08:00</hiredate>
#078        <sal>2450</sal>
#079        <deptno>10</deptno>
#080        <comments>comment of CLARK</comments>
#081        <gender>M</gender>
#082      </Table>
#083      <Table>
#084        <empno>7788</empno>
#085        <ename>SCOTT</ename>
#086        <job>ANALYST</job>
#087        <mgr>7566</mgr>
#088        <hiredate>1982-12-09T00:00:00.0000000+08:00</hiredate>
#089        <sal>3000</sal>
#090        <deptno>20</deptno>
#091        <comments>comment of SCOTT</comments>
#092        <gender>M</gender>
#093      </Table>
#094      <Table>
#095        <empno>7839</empno>
#096        <ename>KING</ename>
#097        <job>PRESIDENT</job>
#098        <hiredate>1981-11-17T00:00:00.0000000+08:00</hiredate>
#099        <sal>5000</sal>
#100        <deptno>10</deptno>
#101        <comments>comment of KING</comments>
#102        <gender>M</gender>
#103      </Table>
#104      <Table>
#105        <empno>7844</empno>
#106        <ename>TURNER</ename>
#107        <job>SALESMAN</job>
#108        <mgr>7698</mgr>
#109        <hiredate>1981-09-08T00:00:00.0000000+08:00</hiredate>
#110        <sal>1500</sal>
```

```
#111        <comm>0</comm>
#112        <deptno>30</deptno>
#113        <comments>comment of TURNER</comments>
#114        <gender>M</gender>
#115    </Table>
#116    <Table>
#117        <empno>7876</empno>
#118        <ename>ADAMS</ename>
#119        <job>CLERK</job>
#120        <mgr>7788</mgr>
#121        <hiredate>1983-12-01T00:00:00.0000000+08:00</hiredate>
#122        <sal>1100</sal>
#123        <deptno>20</deptno>
#124        <comments>comment of ADAMS</comments>
#125        <gender>M</gender>
#126    </Table>
#127    <Table>
#128        <empno>7900</empno>
#129        <ename>JAMES</ename>
#130        <job>CLERK</job>
#131        <mgr>7698</mgr>
#132        <hiredate>1981-12-03T00:00:00.0000000+08:00</hiredate>
#133        <sal>950</sal>
#134        <deptno>30</deptno>
#135        <comments>comment of JAMES</comments>
#136        <gender>M</gender>
#137    </Table>
#138    <Table>
#139        <empno>7902</empno>
#140        <ename>FORD</ename>
#141        <job>ANALYST</job>
#142        <mgr>7566</mgr>
#143        <hiredate>1981-12-03T00:00:00.0000000+08:00</hiredate>
#144        <sal>3000</sal>
#145        <deptno>20</deptno>
#146        <comments>comment of FORD</comments>
#147        <gender>M</gender>
#148    </Table>
#149    <Table>
#150        <empno>7934</empno>
#151        <ename>MILLER</ename>
#152        <job>CLERK</job>
#153        <mgr>7782</mgr>
#154        <hiredate>1982-01-23T00:00:00.0000000+08:00</hiredate>
#155        <sal>1300</sal>
#156        <deptno>10</deptno>
#157        <comments>comment of MILLER</comments>
#158        <gender>M</gender>
```

```
#159        </Table>
#160        <Table>
#161          <empno>1</empno>
#162          <ename>张无忌</ename>
#163          <job>SALESMAN</job>
#164          <mgr>7839</mgr>
#165          <hiredate>1981-02-20T00:00:00.0000000+08:00</hiredate>
#166          <sal>8000.56</sal>
#167          <comm>800</comm>
#168          <deptno>30</deptno>
#169          <gender>M</gender>
#170        </Table>
#171 </NewDataSet>
```

同时会输出更新记录的行数：inserted rows = 1。

在这里，由于使用了"select * from emp"来创建 DataAdapter，它会使用整个 emp 表的数据来填充后来的 DataSet 对象。因此，当 emp 表数据比较大时，显然，这个填充过程及后来的更新操作都会耗时。那么，有没有方法进行改进呢？

事实上，我们可以修改用于创建 DataAdapter 的 SQL 语句，将前边的代码片段修改如下：

```
da = new AseDataAdapter("select * from emp where empno<0", conn);
AseCommandBuilder cb = new AseCommandBuilder(da);
ds = new DataSet();
da.Fill(ds);

DataTable dt = ds.Tables[0];
//dt.Rows.Add(new object[] {1});
DataRow r = dt.NewRow();
r["empno"] = 2;

r["ename"] = "张无忌";
r["job"] = "SALESMAN";
r["mgr"] = 7839;
r["hiredate"] = "1981-02-20";
… …
```

做出上述修改以后，我们发现，生成的 DataSet 对象里的表数据就非常小了，从 abc.xml 文件里的内容也可以判断出来（abc.xml 里只包含待插入的一条记录）：

```
<?xml version="1.0" standalone="yes"?>
<NewDataSet>
  <Table>
    <empno>2</empno>
    <ename>张无忌</ename>
    <job>SALESMAN</job>
    <mgr>7839</mgr>
    <hiredate>1981-02-20T00:00:00.0000000+08:00</hiredate>
```

```xml
        <sal>8000.56</sal>
        <comm>800</comm>
        <deptno>30</deptno>
        <gender>M</gender>
    </Table>
</NewDataSet>
```

通过 where 里的条件，几乎可以确保每次拿到的 DataSet 中的表都是空表，从而确保数据插入相对高效。上边的示例只是往 DataSet 里添加一条记录，其实，我们可以连续添加多条记录，然后一次 Update 调用即可完成所有插入操作。

19.4.2 使用 Insert 语句来插入数据

使用 Insert 语句来插入表数据，是比较传统的方法，通常，要指定一个 Insert 语句。如向表 T(id int primary key, name varchar(32)) 插入数据，Insert 语句为：

```
Insert into T values(1, "Harry")
```

这是一个完全静态的 SQL 插入语句，在程序中创建对应的 Command 对象，然后执行这个 SQL 语句即可完成插入操作，其示例代码（dotnet\Demo\DemoMain 目录下的 DemoInsertBySQL.cs 方法 SimpleInsert）如下：

```csharp
#001    public void SimpleInsert()
#002    {
#003        AseConnection conn = new AseConnection( "Data Source='" + host + "';Port='" + port + "';UID='" + user + "';PWD='" + password + "';Database='iihero';" );
#004        AseCommand cmd = null;
#005        String sql = "insert into spring.emp values(" +
#006            "4, '张无忌', 'CLERK', 7698, '2008-03-11', 8000, 800, 20, null, null, 'M')";
#007        try
#008        {
#009            conn.Open();
#010            AseTransaction tx = conn.BeginTransaction();
#011            cmd = new AseCommand(sql, conn);
#012            int rowsAffected = cmd.ExecuteNonQuery();
#013            Console.WriteLine(rowsAffected.ToString() + " row(s) affected");
#014            tx.Commit();
#015        }
#016        catch( AseException ex )
#017        {
#018            Console.WriteLine(ex.Message);
#019        }
#020        finally
#021        {
#022            if (cmd != null)
#023                cmd.Dispose();
#024            if (conn != null && conn.State != ConnectionState.Closed)
```

```
#025                conn.Close();
#026            }
#027        }
```

这里，我们只是往表 emp 当中插入一条记录，并且使用的是静态的 SQL 插入语句，当插入大量数据时，就需要不断地通过新的静态 SQL 语句创建一个新的 AseCommand 对象，执行 AseCommand 类的 ExecuteNonQuery 方法，总体执行效率不高。

AseCommand 类实现 ADO.NET 中的 IDbCommand 接口，有三个重要的 Execute 方法：

- ExecuteReader()：用于执行一个 SQL 查询语句（大多是 Select 语句，也有可能是返回记录集的存储过程，统称为 DQL 操作），返回一个只读的 DataReader 对象，对应于 ASE 数据库中的一个单向游标。
- ExecuteNonQuery()：执行一个 SQL 命令，返回受影响的记录行数。这类 SQL 命令通常是 Update、Delete、Insert、Create、Drop 命令，上例中，我们就用到了 Insert 命令。
- ExecuteScalar()，执行一个命令，返回单个值。这个函数通常用于获取统计查询的结果，或者想获取某结果集的第一行第一列的值。

如果想通过 AseCommand 类完成相对高效的批量插入功能，就需要使用 AseCommand 的参数动态绑定功能，通过动态地设置绑定的参数变量的值，就不需要每次创建新的 AseCommand 对象了。

相对于静态的 Insert 语句：Insert into T values(1, "Harry")，它对应的带参数动态绑定形式的 SQL 命令为：Insert into T values(@id, @name)，"@"后接一个字符串表示一个占位符。

下边的示例代码片断通过参数化的 SQL 语句，实现向表 emp 里添加多条记录的功能：

```
#001    public void Insert()
#002    {
#003        AseConnection conn = new AseConnection( "Data Source='" + host + "';Port='" + port + "';UID='" + user + "';PWD='" + password + "';Database='iihero';" );
#004        AseCommand cmd = null;
#005        String sqlInsert = "insert into spring.emp values(@empno, @ename, @job, @mgr, @hiredate, @sal, @comm, @deptno, null, null, @gender)";
#006        try
#007        {
#008            conn.Open();
#009            AseTransaction tx = conn.BeginTransaction();
#010            cmd = new AseCommand(sqlInsert, conn);
#011            AseParameter[] pas = new AseParameter[9];
#012            pas[0] = cmd.Parameters.Add("@empno", 5);
#013            pas[1] = cmd.Parameters.Add("@ename", "张无忌");
#014            pas[2] = cmd.Parameters.Add("@job", "CLERK");
#015            pas[3] = cmd.Parameters.Add("@mgr", 7698);
#016            pas[4] = cmd.Parameters.Add("@hiredate", "2008-03-11");
#017            pas[5] = cmd.Parameters.Add("@sal", 8000);
```

```
#018            pas[6] = cmd.Parameters.Add("@comm", 800);
#019            pas[7] = cmd.Parameters.Add("@deptno", 20);
#020            pas[8] = cmd.Parameters.Add("@gender", "M");
#021            int rowsAffected = cmd.ExecuteNonQuery();
#022            Console.WriteLine(rowsAffected.ToString() + " row(s) affected");
#023            for (int i=6; i< 10; ++i)
#024            {
#025                pas[0].Value =   i;
#026                pas[1].Value =   "张无忌";
#027                pas[2].Value =   "CLERK";
#028                pas[3].Value =   7698;
#029                pas[4].Value =   "2008-03-11";
#030                pas[5].Value =   8000;
#031                pas[6].Value =   800;
#032                pas[7].Value =   20;
#033                pas[8].Value =   "M";
#034                cmd.ExecuteNonQuery();
#035                Console.WriteLine(rowsAffected.ToString() + " row(s) affected");
#036            }
#037            tx.Commit();
#038        }
#039        catch( AseException ex )
#040        {
#041            Console.WriteLine(ex.Message);
#042        }
#043        finally
#044        {
#045            if (cmd != null)
#046                cmd.Dispose();
#047            if (conn != null && conn.State != ConnectionState.Closed)
#048                conn.Close();
#049        }
#050    }
```

在这里，我们只创建了一个 AseCommand 对象。首先，创建和初始化参数化 SQL 语句 insert into spring.emp values(@empno, @ename, @job, @mgr, @hiredate, @sal, @comm, @deptno, null, null, @gender)需要的 9 个参数，然后调用一次 ExecuteNonQuery，以后每次插入都只需要将这 9 个参数的值重新设定一下，再次调用 ExecuteNonQuery 即可。

添加动态参数的方法如下：

```
pas[0] = cmd.Parameters.Add("@empno", 5);
```

这是比较简单的创建方式，Add 方法的第一个参数是占位符名，第二个参数是该动态参数的值。当然，我们也可以在每插入一条记录之前，都重新设定这些参数值。

```
#001    public void Insert2()
#002    {
#003        AseConnection conn = new AseConnection( "Data Source='" + host + "';Port='" + port + "';UID='" + user + "';PWD='" + password + "';Database='iihero';" );
#004        AseCommand cmd = null;
#005        String sqlInsert = "insert into spring.emp values(@empno, @ename, @job, @mgr, @hiredate, @sal, @comm, @deptno, null, null, @gender)";
#006        try
#007        {
#008            conn.Open();
#009            AseTransaction tx = conn.BeginTransaction();
#010            cmd = new AseCommand(sqlInsert, conn);
#011            AseParameter[] pas = new AseParameter[9];
#012            int rowsAffected = 0;
#013            for (int i=6; i< 10; ++i)
#014            {
#015                pas[0] = cmd.Parameters.Add("@empno", i);
#016                pas[1] = cmd.Parameters.Add("@ename", "张无忌");
#017                pas[2] = cmd.Parameters.Add("@job", "CLERK");
#018                pas[3] = cmd.Parameters.Add("@mgr", 7698);
#019                pas[4] = cmd.Parameters.Add("@hiredate", "2008-03-11");
#020                pas[5] = cmd.Parameters.Add("@sal", 8000);
#021                pas[6] = cmd.Parameters.Add("@comm", 800);
#022                pas[7] = cmd.Parameters.Add("@deptno", 20);
#023                pas[8] = cmd.Parameters.Add("@gender", "M");
#024                rowsAffected = cmd.ExecuteNonQuery();
#025                // cmd.Parameters.Clear();
#026                Console.WriteLine(rowsAffected.ToString() + " row(s) affected");
#027            }
#028            tx.Commit();
#029        }
#030        catch( AseException ex )
#031        {
#032            Console.WriteLine(ex.Message);
#033        }
#034        finally
#035        {
#036            if (cmd != null)
#037                cmd.Dispose();
#038            if (conn != null && conn.State != ConnectionState.Closed)
#039                conn.Close();
#040        }
#041    }
```

第 25 行，将 cmd.Parameters()清空，我们将这一行的注释去掉了，针对 AseCommand 对象，它是多余的，这与它的具体实现有关。实际上，如果不清空，它的 Parameters 集合是一个键值对集合（如 HashMap、TreeMap 等数据结构），针对相同的键只有一个元素。也就是说，一个 Add("@id", value)调用会自动覆盖已有的@id 参数，如果没有这个参数，则自动创建一个。

因此，这种用法会与具体的 ADO.NET 实现有关。如果具体的 ADO.NET 实现里并不采用键值对集合来表示，而只是普通的数组，那么可能就需要调用 cmd.Parameters()来清空参数列表了。

由于重复执行初始化操作，效率应略低于本节中介绍的前一种方法。

19.4.3 BLOB/CLOB 数据的插入操作

BLOB 数据属于非结构化二进制大数据对象，通常无法直接使用 Select 语句查询得到 BLOB 字段的内容。CLOB 数据属于非结构化文本大数据对象，同样无法直接使用 Select 语句查询得到 CLOB 字段的全部内容。

因此，相对于简单的结构化数据类型，针对它们的操作要相对复杂一些。

在 ASE 数据库中，BLOB 数据使用 IMAGE 字段类型来表示，CLOB 数据使用 TEXT 字段类型来表示。

为简化问题，这里假设有表 blobtab 和 clobtab，定义如下：

```
CREATE TABLE blobtab
(id int primary key, col2 image)
go

CREATE TABLE clobtab
(id int primary key, col2 text)
go
```

下边的示例将向这两个表中各插入一行，分别插入长度为 2MB 的二进制文本内容。

```
#001  using System;
#002  using System.Collections;
#003  using System.Data;
#004  using Sybase.Data.AseClient;
#005
#006
#007  class DemoBlob
#008  {
#009      private string host;
#010      private int port;
#011      private string user;
#012      private string password;
#013      public DemoBlob(string host, int port, string user, string password)
#014      {
#015          this.host = host;
#016          this.port = port;
#017          this.user = user;
#018          this.password = password;
#019      }
```

```
#020    public void WriteBlob()
#021    {
#022        AseConnection conn = new AseConnection( "Data Source='" + host + "';Port='" + port
#023            + "';UID='" + user + "';PWD='" + password + "';Database='iihero';TextSize=64000" );
#024        AseCommand cmd = null;
#025        String sql = "insert into blobtab values(@id, @blob)";
#026        try
#027        {
#028            conn.Open();
#029            AseTransaction tx = conn.BeginTransaction();
#030            cmd = new AseCommand(sql, conn);
#031            cmd.Parameters.Add("@id", 1);
#032            byte[] b = new byte[2048000];
#033            for (int i=0; i<b.Length; ++i) b[i] = (byte)(i%128);
#034            AseParameter p = cmd.Parameters.Add("@blob", AseDbType.Image, b.Length);
#035            p.Value = b;
#036            int rowsAffected = cmd.ExecuteNonQuery();
#037            Console.WriteLine(rowsAffected.ToString() + " row(s) affected on table: blobtab");
#038            tx.Commit();
#039            cmd.Dispose();
#040
#041            sql = "insert into clobtab values(@id, @clob)";
#042            tx = conn.BeginTransaction();
#043            cmd = new AseCommand(sql, conn);
#044            cmd.Parameters.Add("@id", 1);
#045            char[] a = new char[2048000];
#046            for (int i=0; i<a.Length; ++i) a[i] = (char)('a' + i%26);
#047            p = cmd.Parameters.Add("@clob", AseDbType.Text, a.Length);
#048            p.Value = a;
#049            rowsAffected = cmd.ExecuteNonQuery();
#050            Console.WriteLine(rowsAffected.ToString() + " row(s) affected on table: clobtab");
#051            tx.Commit();
#052
#053            cmd = new AseCommand("select col2 from blobtab", conn);
#054            IDataReader dr = cmd.ExecuteReader();
#055            if (dr.Read())
#056            {
#057                byte[] res = new byte[2048000];
#058                long len = dr.GetBytes(0, 0, res, 0, res.Length);
#059                Console.WriteLine("get blob leng = " + len);
#060            }
#061            dr.Close();
#062        }
#063        catch( AseException ex )
#064        {
#065            Console.WriteLine(ex.Message);
#066        }
#067        finally
```

```
#068        {
#069            if (cmd != null)
#070                cmd.Dispose();
#071            if (conn != null && conn.State != ConnectionState.Closed)
#072                conn.Close();
#073        }
#074    }
#075
#076 }
```

我们会发现这里 TextSize=64000，对于插入操作没有影响，一次可以插入 2MB 的数据，但是读取时，却只能读取 64KB 的数据，因此，如果要想读取 2MB 的数据，必须将 TextSize 设置为 2M=2097152。

19.5 更新数据

更新数据与插入数据非常相似，可以直接通过 Update 语句，动态绑定要更新的字段。

例如，雇员 SMITH 的原薪水为 800 美元，现在要将薪水调整为 1800 美元，使用下边的 SQL 语句：

```
"update emp set sal=@sal where ename=@ename"
```

在程序中对@sal 和@ename 使用参数进行动态绑定即可。

完整的例程如下（dotnet\DemoUpdate.cs）：

```
#001 using System;
#002 using System.Collections.Generic;
#003 using System.Text;
#004
#005 using System.Data;
#006 using Sybase.Data.AseClient;
#007
#008 class DemoUpdate
#009 {
#010     private string host;
#011     private int port;
#012     private string user;
#013     private string password;
#014     public DemoUpdate(string host, int port, string user, string password)
#015     {
#016         this.host = host;
#017         this.port = port;
#018         this.user = user;
#019         this.password = password;
#020     }
```

```
#021
#022        public void Update()
#023        {
#024            AseConnection conn = new AseConnection("Data Source='" + host + "';Port='" + port + "';UID='" + user + "';PWD='" + password + "';Database='iihero';");
#025            AseCommand cmd = null;
#026            String sqlInsert = "update emp set sal=@sal where ename=@ename";
#027            try
#028            {
#029                conn.Open();
#030                AseTransaction tx = conn.BeginTransaction();
#031                cmd = new AseCommand(sqlInsert, conn);
#032                cmd.Parameters.Add("@sal", 1800);
#033                cmd.Parameters.Add("@ename", "SMITH");
#034                int rowsAffected = cmd.ExecuteNonQuery();
#035                Console.WriteLine(rowsAffected.ToString() + " row(s) affected");
#036                tx.Commit();
#037            }
#038            catch (AseException ex)
#039            {
#040                Console.WriteLine(ex.Message);
#041            }
#042            finally
#043            {
#044                if (cmd != null)
#045                    cmd.Dispose();
#046                if (conn != null && conn.State != ConnectionState.Closed)
#047                    conn.Close();
#048            }
#049        }
#050 }
```

第 34 行，rowsAffected = cmd.ExecuteNonQuery()表示执行 Update 操作以后，表中已经更新的行数。如果没有符合更新条件的行，则返回 0。本例中返回 1，表示有一条记录成功的获得更新。

19.6 调用存储过程

使用.NET 调用存储过程与调用普通的 SQL 语句有些类似，可以使用变量进行动态绑定。它可以像普通的 SQL 语句那样发出 commandText，也可以采用标准的存储过程调用方式的语法：{ call <proc> { ?, ? .... } }。

下边是一个简单的示例，存储过程 test_proc 基于本书的样例数据库 iihero。

test_proc 内容如下：

```
create proc spring.test_proc(@s_name char(30), @s_count int output) with recompile
as
```

```
select @s_count = count(a.sno) from spring.student a where a.sname = @s_name
select 'demo123'
go
declare @result int
exec spring.test_proc '李勇', @result output
select @result
go
```

执行上述存储过程及其调用，可以直接得到如下结果：

```
1> declare @result int
2> exec test_proc '李勇', @result output
3> go


 -------
  demo123

(1 row affected)
(return status = 0)

Return parameters:
 -----------
  1
```

如果使用.NET 程序实现调用，其代码如下（DemoMain\DemoProc.cs）：

```
#001 using System;
#002 using System.Collections.Generic;
#003 using System.Text;
#004 using System.Data;
#005 using Sybase.Data.AseClient;
#006 using System.Data.Common;
#007
#008 class DemoProc
#009 {
#010     private string host;
#011     private int port;
#012     private string user;
#013     private string password;
#014     public DemoProc(string host, int port, string user, string password)
#015     {
#016        this.host = host;
#017        this.port = port;
#018        this.user = user;
#019        this.password = password;
#020     }
#021
#022     public void TestProcNonStandard()
#023     {
```

```
#024        AseConnection conn = new AseConnection("Data Source='" + host + "';Port='" + port + "';UID='" + user +
"';PWD='" + password + "';Database='iihero';");
#025        DbCommand cmd = null;
#026        try
#027        {
#028            conn.Open();
#029            cmd = conn.CreateCommand();
#030            cmd.CommandText = "test_proc";        //非标准方式
#031            cmd.CommandType = System.Data.CommandType.StoredProcedure;
#032            DbParameter param = cmd.CreateParameter();
#033            param.ParameterName = "@s_name";
#034            param.DbType = System.Data.DbType.String;
#035            param.Direction = System.Data.ParameterDirection.Input;
#036            param.Value = "李勇";
#037            cmd.Parameters.Add(param);
#038            DbParameter param2 = cmd.CreateParameter();
#039            param2.ParameterName = "@s_count";
#040            param2.DbType = System.Data.DbType.Int32;
#041            param2.Direction = System.Data.ParameterDirection.Output;
#042            cmd.Parameters.Add(param2);
#043            DbDataReader rs = cmd.ExecuteReader();
#044            while (rs.Read())
#045            {
#046                Console.WriteLine(rs.GetString(0));
#047            }
#048            rs.Close();
#049            // output param2
#050            Console.WriteLine("@s_count = " + param2.Value);
#051            cmd.Dispose();
#052            conn.Close();
#053        }
#054        catch (Exception ex)
#055        {
#056            System.Console.WriteLine(ex.Message + "/n" + ex.InnerException);
#057        }
#058    }
#059
#060    public void TestProc()
#061    {
#062        AseConnection conn = new AseConnection("Data Source='" + host + "';Port='" + port + "';UID='" + user +
"';PWD='" + password + "';Database='iihero';");
#063        DbCommand cmd = null;
#064        try
#065        {
#066            conn.Open();
#067            cmd = conn.CreateCommand();
#068            cmd.CommandText = "{ call test_proc(?, ?) }";        //标准方式调用
#069            // 或者使用 DbCommand cmd = new AseCommand("{ call test_proc(?, ?) }", conn);
```

```
#070            DbParameter param = cmd.CreateParameter();
#071            param.ParameterName = "@s_name";
#072            param.DbType = System.Data.DbType.String;
#073            param.Direction = System.Data.ParameterDirection.Input;
#074            param.Value = "李勇";
#075            cmd.Parameters.Add(param);
#076            DbParameter param2 = cmd.CreateParameter();
#077            param2.ParameterName = "@s_count";
#078            param2.DbType = System.Data.DbType.Int32;
#079            param2.Direction = System.Data.ParameterDirection.Output;
#080            cmd.Parameters.Add(param2);
#081            DbDataReader rs = cmd.ExecuteReader();
#082            while (rs.Read())
#083            {
#084                Console.WriteLine(rs.GetString(0));
#085            }
#086            rs.Close();
#087            // output param2
#088            Console.WriteLine("@s_count = " + param2.Value);
#089            cmd.Dispose();
#090            conn.Close();
#091        }
#092        catch (Exception ex)
#093        {
#094            System.Console.WriteLine(ex.Message + "/n" + ex.InnerException);
#095        }
#096    }
#097
#098    //
#099    //DemoProc t = new DemoProc("192.168.0.8", 5000, "spring", "spring1");
#100    //t.TestProcNonStandard();
#101    //t.TestProc();
#102 }
```

执行结果：

demo123
@s_count = 1

我们可以比较一下，标准方式就是在生成 AseCommand 实例时，使用标准的存储过程调用语法 "cmd.CommandText = "{ call test_proc(?, ?) }";"，这时不需要指定 Command 类型，数据库后端直接能判断出它是存储过程调用。

非标准方式在上述代码第 30～31 行，cmd 的语句直接就是 test_proc，这时必须指定它的 Command 类型为 System.Data.CommandType.StoredProcedure。

在实际开发过程中，推荐使用标准方式。

19.7 获取结果集或表的元信息

除了通常的 CRUD 操作以外，有时候需要动态获取一个表或者结果集中各列的元信息，比如列的数据类型、长度、列名等。如何通过.NET 程序实现呢？

由 AseCommand 类实现的 IDbCommand 接口中，有一个重要的方法——ExecuteReader，它有一个重载形式，接受一个 CommandBehaviour 类型的参数，该参数可以指定返回结果集的不同方式，如下语句只要求返回一行数据：

rs = cmd.Executereader(CommandBehavior.SingReader);

CommandBehavior 枚举值如下：

- SingleRow, 提示查询应返回一行。默认行为是返回一个或多个结果集。
- SingleResult, 希望返回一个标量值。
- KeyInfo, 返回列和主键信息，用于结果集的 GetSchema 方法来获取列模式信息。
- SchemaOnly, 用于获取对应结果集的列名。
- SequentialAccess, 允许按列顺序地访问返回行中的数据，常用于大二进制或文本字段。
- CloseConnection, 结果集关闭时，连接也关闭。

我们已知 iihero 示例数据库中 student 表的表结构如下：

```
create table student
(
    sno int not null primary key,
    sname varchar(32) not null,
    sgender char(1) not null,
    sbirth datetime not null,
    sage numeric(2) null,
    sdept varchar(128) null
)
Go
```

如何在程序中读取 student 表中各列的元信息呢？

其实现代码如下（DemoMain\DemoMetadata.cs）：

```
#001 using System;
#002 using System.Collections.Generic;
#003 using System.Text;
#004 using System.Data;
#005 using Sybase.Data.AseClient;
#006 using System.Data.Common;
#007
#008 class DemoMetadata
#009 {
```

```
#010        private string host;
#011        private int port;
#012        private string user;
#013        private string password;
#014
#015        public DemoMetadata(string host, int port, string user, string password)
#016        {
#017            this.host = host;
#018            this.port = port;
#019            this.user = user;
#020            this.password = password;
#021        }
#022
#023        public void TestMetadata()
#024        {
#025            AseConnection conn = new AseConnection("Data Source='" + host + "';Port='" + port + "';UID='" + user + "';PWD='" + password + "';Database='iihero';");
#026            AseCommand cmd = null;
#027            String sql = "SELECT * from student";
#028            try
#029            {
#030                conn.Open();
#031                AseTransaction tx = conn.BeginTransaction();
#032                cmd = new AseCommand(sql, conn);
#033
#034                AseDataReader rs = cmd.ExecuteReader(CommandBehavior.SchemaOnly);
#035                DataTable dt = rs.GetSchemaTable();
#036                int n = 0;
#037                foreach (DataRow dtrow in dt.Rows)
#038                {
#039                    Console.WriteLine("第" + (++n) + " 列信息: ~~~~~");
#040                    foreach (DataColumn dtcol in dt.Columns)
#041                    {
#042                        Console.Write(dtcol.ColumnName + "=" + dtrow[dtcol]+ "\t");
#043                    }
#044                    Console.Write("\n");
#045                }
#046                rs.Close();
#047                tx.Commit();
#048            }
#049            catch (AseException ex)
#050            {
#051                Console.WriteLine(ex.Message);
#052            }
#053            finally
#054            {
#055                if (cmd != null)
#056                    cmd.Dispose();
```

```
#057                    if (conn != null && conn.State != ConnectionState.Closed)
#058                        conn.Close();
#059            }
#060        }
#061
#062 }
```

我们主要通过第 35 行的"**DataTable dt = rs.GetSchemaTable();**"来得到完整的元信息。执行结果如下：

第 1 列信息:~~~~
ColumnName=sno ColumnOrdinal=0 ColumnSize=4   NumericPrecision=10   NumericScale=0 IsUnique=False IsKey=False   BaseServerName= BaseCatalogName=iihero BaseColumnName=sno   BaseSchemaName=spring   BaseTableName=student   DataType=System.Int32   AllowDBNull=False   ProviderType=4 IsAliased=False IsExpression=False   IsIdentity=False   IsAutoIncrement=False   IsRowVersion=False   IsHidden=False IsLong=False   IsReadOnly=False
第 2 列信息:~~~~
ColumnName=sname   ColumnOrdinal=1 ColumnSize=32   NumericPrecision=0   NumericScale=0 IsUnique=False IsKey=False   BaseServerName= BaseCatalogName=iihero BaseColumnName=sname   BaseSchemaName=spring   BaseTableName=student   DataType=System.String   AllowDBNull=False   ProviderType=12 IsAliased=False   IsExpression=False   IsIdentity=False   IsAutoIncrement=False   IsRowVersion=False   IsHidden=False IsLong=False   IsReadOnly=False
第 3 列信息:~~~~
ColumnName=sgender   ColumnOrdinal=2 ColumnSize=1   NumericPrecision=0   NumericScale=0 IsUnique=False IsKey=False   BaseServerName= BaseCatalogName=iihero BaseColumnName=sgender BaseSchemaName=spring   BaseTableName=student   DataType=System.String   AllowDBNull=False   ProviderType=1 IsAliased=False   IsExpression=False   IsIdentity=False   IsAutoIncrement=False   IsRowVersion=False   IsHidden=False IsLong=False   IsReadOnly=False
第 4 列信息:~~~~
ColumnName=sbirth   ColumnOrdinal=3 ColumnSize=23   NumericPrecision=0   NumericScale=0 IsUnique=False IsKey=False   BaseServerName= BaseCatalogName=iihero BaseColumnName=sbirth   BaseSchemaName=spring   BaseTableName=student   DataType=System.DateTime   AllowDBNull=False   ProviderType=93 IsAliased=False IsExpression=False   IsIdentity=False   IsAutoIncrement=False   IsRowVersion=False   IsHidden=False IsLong=False   IsReadOnly=False
第 5 列信息:~~~~
ColumnName=sage ColumnOrdinal=4 ColumnSize=2   NumericPrecision=2   NumericScale=0 IsUnique=False IsKey=False   BaseServerName= BaseCatalogName=iihero BaseColumnName=sage   BaseSchemaName=spring   BaseTableName=student   DataType=System.Decimal AllowDBNull=True   ProviderType=2 IsAliased=False IsExpression=False   IsIdentity=False   IsAutoIncrement=False   IsRowVersion=False   IsHidden=False IsLong=False   IsReadOnly=False
第 6 列信息:~~~~
ColumnName=sdept   ColumnOrdinal=5 ColumnSize=128   NumericPrecision=0   NumericScale=0 IsUnique=False IsKey=False   BaseServerName= BaseCatalogName=

```
iihero    BaseColumnName=sdept       BaseSchemaName=spring    BaseTableName=student
DataType=System.String    AllowDBNull=True    ProviderType=12 IsAliased=False
IsExpression=False    IsIdentity=False    IsAutoIncrement=False    IsRowVer
sion=False    IsHidden=False    IsLong=False    IsReadOnly=False
```

虽然最终结果也输出了 student 表各列的所有元信息，但是我们可能并不需要这么多属性。对于我们来说，最有用的也就是列名、列序号、列类型、列长度、精度、小数位数、是否属于键这 7 方面的信息，基本上也就能描述这个表的表结构了。如果只取上述 7 项属性，那么，我们只需要 ColumnName、ColumnOrdial、ProviderType、ColumnSize、NumericPrecision、NumericScale、IsKey 就够了。

表 19-2 列出了这些元信息属性的具体含义。

表 19-2 元信息相关含义

DataReader 列	说明
ColumnName	列的名称；它可能不唯一。如果无法确定该名称，则返回空值。此名称始终反映最近对当前视图或命令文本中的列进行的重命名
ColumnOrdinal	列的序号。它对于行的书签列（如果有的话）为零。其他列从一开始编号。该列不能包含空值
ColumnSize	列中值的最大可能长度。对于采用固定长度数据类型的列，它是该数据类型的大小
NumericPrecision	如果 ProviderType 是数值数据类型，则它是列的最大精度。精度取决于列的定义。如果 ProviderType 不是数值数据类型，则它为空值
NumericScale	如果 ProviderType 是 DBTYPE_DECIMAL 或 DBTYPE_NUMERIC，则为小数点右侧的位数；否则为空值
IsUnique	**true**：基表（BaseTableName 返回的表）中的任何两行在该列中都不能有相同的值。如果此列本身为一个键，或如果有一个仅应用于该列的 UNIQUE 类型约束，那么 IsUnique 一定为 **true**。 **false**：在基表中该列可以包含重复值。该列的默认值为 **false**
IsKey	**true**：此列是行集合中可唯一地标识该行的一组列中的一列。IsKey 设置为 **true** 的一组列必须唯一地标识行集合中的一个行。不要求此列集是最小列集。这组列可以从基表主键、唯一约束或唯一索引生成。 **false**：唯一地标识该行时不需要此列
BaseServerName	SQLDataReader 使用的 Microsoft SQL Server 实例的名称
BaseCatalogName	包含列的数据存储区中的目录的名称。如果不能确定基目录名称，则为 NULL。该列的默认值为空值
BaseColumnName	数据存储区中列的名称。如果使用别名，它可能不同于在 ColumnName 列中返回的列名称。如果无法确定基列名称，或者如果行集合列从数据存储区中的列导出但不等于该列，则为空值。该列的默认值为空值
BaseSchemaName	包含列的数据存储区中的架构的名称。如果无法确定基架构名称，则为空值。该列的默认值为空值
BaseTableName	包含列的数据存储区中的表或视图的名称。如果无法确定基表名称，则为空值。该列的默认值为空值
DataType	映射到列的 .Net Framework 类型
AllowDBNull	如果使用者可以将此列设置为空值，或者如果提供者不能确定使用者是否可以将此列设置为空值，则设置该值。否则不设置该值。即使列无法设置为空值，它仍可能包含空值

续表

DataReader 列	说明
ProviderType	列的数据类型的指示符。如果不同行的列数据类型不同,则它必须为 Object。该列不能包含空值
IsAliased	如果列名称为别名,则为 **true**;否则为 **false**
IsExpression	如果此列为表达式,则为 **true**;否则为 **false**
IsIdentity	如果此列为标识列,则为 **true**;否则为 **false**
IsAutoIncrement	**true**:该列以固定的增量向新行赋值。**false**:该列不以固定的增量向新行赋值。该列的默认值为 **false**
IsRowVersion	如果列包含无法写入的不变的行标识符,并且除了标识行外没有其他有意义的值,则设置该值。
IsHidden	如果列是隐藏的,则为 **true**;否则为 **false**
IsLong	如果列包含二进制长对象(BLOB)(它包含非常长的数据),则设置该值。非常长的数据的定义针对于提供程序
IsReadOnly	如果不能修改该列,则为 **true**;否则为 **false**

我们可以将代码中的第 37~45 行修改为:

```
foreach (DataRow dtrow in dt.Rows)
{
    Console.WriteLine("第" + (++n) + " 列信息: ~~~~");
    foreach (DataColumn dtcol in dt.Columns)
    {
        String col = dtcol.ColumnName;
        if (col.Equals("ColumnName") || col.Equals("ColumnOrdial") || col.Equals("ColumnSize") ||
            col.Equals("NumericPrecision")
            || col.Equals("NumericScale") || col.Equals("ProviderType") || col.Equals("IsKey"))
        {
            Console.Write(dtcol.ColumnName + "=" + dtrow[dtcol] + "\t");
        }
    }
    Console.Write("\n");
}
```

这样输出的各列信息为:

第 1 列信息: ~~~~
ColumnName=sno   ColumnSize=4   NumericPrecision=10   NumericScale=0   IsKey=False   ProviderType=4
第 2 列信息: ~~~~
ColumnName=sname   ColumnSize=32   NumericPrecision=0   NumericScale=0   IsKey=False   ProviderType=12
第 3 列信息: ~~~~
ColumnName=sgender   ColumnSize=1   NumericPrecision=0   NumericScale=0   IsKey=False   ProviderType=1
第 4 列信息: ~~~~
ColumnName=sbirth   ColumnSize=23   NumericPrecision=0   NumericScale=0   IsKey=False   ProviderType=93
第 5 列信息: ~~~~
ColumnName=sage   ColumnSize=2   NumericPrecision=2   NumericScale=0   IsKey=False   ProviderType=2

第 6 列信息：～～～～

ColumnName=sdept　　　ColumnSize=128　NumericPrecision=0　　NumericScale=0　　IsKey=False　　ProviderType=12

这里比较遗憾的是 ProviderType 里输出的是数值，而不是字符串描述的实际类型。比如这里的 sno 列本是 int 型，输出为 4。Sname 列为 varchar 型，输出为 12。如果要得到具体是哪种类型的字符串描述，可能还得借助于具体的存储过程。

比如，如何知道 ProviderType=93 的实际类型名是什么呢？

可以通过 datatype 值反查 type_name 值的存储过程有：sp_odbc_datatype_info、sp_oledb_datatype_info 及 sp_jdbc_datatype_info。在这里，可以使用的就是存储过程 sp_odbc_datatype_info 和 sp_jdbc_datatype_info。

前者执行方式是 exec sp_odbc_datatype_info　93，它可以得到 93 对应的类型名，如图 19-3 所示。

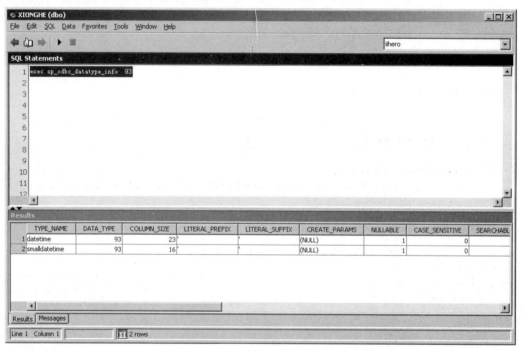

图 19-3　sp_odbc_datatype_info 的执行结果

类型名可以是 datetime 和 smalldatetime。而如果通过预调用 sp_jdbc_datatype_info，可以得到 ASE 中全部的 datatype 和 type_name 映射表。其内容如下：

Exec sp_jdbc_datatype_info

结果如下：

TYPE_NAME	DATA_TYPE
'bit'	-7

'tinyint'	-6
'bigint'	-5
'unsigned bigint'	-5
'image'	-4
'varbinary'	-3
'binary'	-2
'text'	-1
'unitext'	-1
'char'	1
'nchar'	1
'numeric'	2
'decimal'	3
'money'	3
'smallmoney'	3
'int'	4
'unsigned int'	4
'smallint'	5
'unsigned smallint'	5
'real'	7
'double precision'	8
'float'	8
'nvarchar'	12
'varchar'	12
'date'	91
'time'	92
'datetime'	93
'smalldatetime'	93
'extended type'	1111

这些内容可以只查询一次，在获取 Connection 以后，缓存到一个 Map 里，这样就能得到真正的列类型名了。这里留给读者去实现。

19.8　ASE ADO.NET 应用程序的发布

当你开发完一个 ADO.NET 的 ASE 客户端应用程序，如何将其发布出去呢？
我们再看看 %SYBASE%\DataAccess\ADONET 目录下的内容：

```
c:\>dir/b D:\SybaseASE\DataAccess\ADONET\dll
AdoNetRegistrar.exe
AseGacUtility.exe
csi
ja
locales
msvcp71.dll
msvcr71.dll
policy.1.15.Sybase.Data.AseClient
```

```
policy.1.15.Sybase.Data.AseClient.dll
sbgse2.dll
Sybase.AdoNet2.AseClient.dll
Sybase.Data.AseClient.dll
sybdrvado115a.dll
sybdrvado20.dll
sybdrvkrb.dll
sybdrvssl.dll
```

我们的应用程序主要就是依赖于这些动态库，尤其要注意的是，尽量引用 Sybase.AdoNet2.AseClient.dll，它比 Sybase.Data.AseClient.dll 功能更完备，也更易于使用。

发布时，我们将%SYBASE%\DataAccess\ADONET\dll 整个目录复制到目标机器上，并将%SYBASE%\DataAccess\ADONET\dll 添加到系统环境变量 PATH 中，这样，应用程序就能找到所依赖的库的路径了。

总结起来，应用程序依赖环境有两种：

1. DOTNET Framework 2.0，这是基础。
2. ASE ADO.NET 相关 DLL 目录，并添加到 PATH 系统环境变量当中。

相比于 Oracle 而言，Sybase ASE 的.NET 运行时环境依赖就小得多了，只是几个 DLL 而已。Oracle 数据库则需要一个大的客户端运行时库，或至少要安装一个 Oracle Instantclient。相比而言，Sybase ASE 的.NET 库无疑更加灵活。

20 Sybase ASE 功能包生成

任何一个大型商业软件都是多个组件组合而成的，ASE 数据库服务器也不例外。对于 Sybase ASE 来说，除了数据库服务器进程本身受 License 控制以外，其他自带的客户端运行时库、客户端管理工具及一些接口库是不受 License 控制的。因此，可以单独打包出来，提供给开发环境使用，这里按照组件由易到难、由小到大的顺序，依次介绍一些重要的功能包的提取。这些功能包无论是对 DBA 还是对 ASE 的开发人员来说，都是非常有意义的。

这里因为要用提取大量文件或者文件夹，所以需要预备一些工具，可以安装一个开源的免费压缩工具 7zip，找到其中的 7z.dll 和 7z.exe，将其放到目标目录当中。本章都以 Windows 平台为例进行介绍。其他平台的原理基本相似。

20.1 JUtils 工具包生成

在介绍 JDBC 那一章也有提及，只是介绍如何利用已有环境去运行 JUtils，那么如何将其提取出来，变成独立运行的工具包呢？

直接运行下边的脚本：

```
@echo off

set JCONNECT=%SYBASE%\jConnect-6_0

md jutils-2_0

xcopy /S /Y /Q %SYBASE%\jutils-2_0    jutils-2_0
copy /Y %JCONNECT%\classes\jconn3.jar .\jutils-2_0\
copy /Y %JCONNECT%\classes\jTDS3.jar .\jutils-2_0\
```

这样，JUtils 及其依赖的包都放到一个目录 JUtils-2.0 中，余下的工作是修改 jisql 及 Ribo 的启动脚本。

1. 修改 jutils-2.0\jisql\jisql.bat。

   set JAVA_HOME=D:\SybaseASE\Shared\JRE-6_0_6_32BIT（在新环境中，需要修改此值）

   set JDBC_HOME=..\（该值将固定不变）

   测试时，双击 jisql.bat 即可。

2. 修改 jutils-2.0\ribo\ribo.bat。

   set JAVA_HOME=D:\SybaseASE\Shared\JRE-6_0_6_32BIT（新环境中，需要修改此值）

   set RIBO_HOME=.\（该值固定不变）

   测试时，运行 ribo.bat -gui 即可。

3. 更改完环境变量之后，就可以用 7z.exe 打包了。

   直接运行命令 7z.exe a jutils.zip jutils-2_0，即可将 jutils-2_0 打包成 jutils.zip。这个包自动也包含了 ASE 的 JDBC 库文件 jconn3.jar。

20.2 ODBC、OLEDB 及 ADO.NET 包

ASE 数据库将这三个数据访问驱动包都放到 %SYBASE%\DataAccess 目录下，因此可以为它们一起打包。

首先，将整个 DataAccess 目录复制到一个新位置，然后压缩即可。

这个压缩包要想在新机器上得以使用，针对 ODBC 驱动和 OLEDB 驱动，必须进行 COM 组件注册：

```
ODBC 注册：
regsvr32 <DataAccess_DIR>\ODBC\dll\sybdrvodb.dll
OLEDB 注册：
regsvr32 <DataAccess_DIR>\OLEDB\dll\sybdrvoledb.dll
完整的提取脚本内容如下，详见 code\shipping\make_odbcoledbadonet.bat
@echo off

md DataAccess
xcopy /S /Y /Q %SYBASE%\DataAccess DataAccess
7z a DataAccess.zip DataAccess
@rem ODBC registration
@rem    regsvr32 <DataAccess_DIR>\ODBC\dll\sybdrvodb.dll
@rem OLEDB registration
@rem    regsvr32 <DataAccess_DIR>\OLEDB\dll\sybdrvoledb.dll
```

20.3 Open Client 库

ASE 中的 Open Client 库体积相对比较庞大，它需要提取如下文件和目录：
1. SYBASE.bat：环境变量文件。
2. charsets：字符集目录。
3. collate：排序映射目录。
4. ini interfaces：文件目录。
5. locales：语言及字符集映射设置目录。
6. OCS-15_0 openclient：动态库目录。

写成批处理脚本，内容如下（详见代码 shipping\make_openclient.bat）：

```
@echo off

md openclient
md openclient\charsets
md openclient\collate
md openclient\ini
md openclient\locales
md openclient\OCS-15_0

xcopy /S /Y /Q %SYBASE%\charsets    openclient\charsets
xcopy /S /Y /Q %SYBASE%\collate     openclient\collate
xcopy /S /Y /Q %SYBASE%\ini         openclient\ini
xcopy /S /Y /Q %SYBASE%\locales     openclient\locales
xcopy /S /Y /Q %SYBASE%\OCS-15_0    openclient\OCS-15_0
copy /Y %SYBASE%\SYBASE.bat   openclient\
7z a openclient.zip openclient
```

SYBASE.bat 中对环境变量的定义，在 openclient.zip 解压至新位置中，要进行相应调整。

比如，在新机器上，如果将 openclient.zip 解压至 d:\，那么 SYBASE 这个环境变量应该设到 d:\openclient 中。运行命令行，只能在当前的命令行会话期内有效。可以直接修改注册表，如果只想影响当前用户，添加用户环境变量即可。如果要影响所有的其他用户，则需要添加系统环境变量。

添加当前用户环境变量的注册表位置在 HKEY_CURRENT_USER\Environment 下。而添加系统环境变量的注册表位置在 HKEY_LOCAL_MACHINE\SYSTEM\CurrentControlSet\ Control\Session Manager\Environment 下。

可以使用一个命令来添加这个新的 SYBASE 环境变量：

reg add "HKCU\Environment" /v SYBASE /t REG_SZ /d d:\openclient /f（用户环境变量）

但是这样添加的环境变量并不会立即生效，它只是往注册表里添加了一项内容。要想使其生效，需要广播一条消息：WM_SETTINGCHANGE，内容为 Environment，可以写一个简单的程序段来实现，如下所示（详见 code\shipping\populateEnv\test.cpp）：

```
#001 #pragma once
#002
#003 #define WIN32_LEAN_AND_MEAN        // Exclude rarely-used stuff from Windows headers
#004 #define _WIN32_WINNT 0x0500        // add this line
#005
#006 #include <stdio.h>
#007 #include <stdlib.h>
#008 #include <time.h>
#009 #include <conio.h>
#010 #include <string.h>
#011 #include <windows.h>
#012
#013
#014
#015 int main(int argc, char* argv[]){
#016     LRESULT res = ::SendMessage(HWND_BROADCAST,WM_SETTINGCHANGE,0,(LPARAM)TEXT ("Environment"));
#017     printf("Message of settingchange sent, ret=%d, the variables set will be available now.\n", res);
#018     return 0;
#019 }
```

假设生成的可执行程序为 enable_env.exe，执行完以后，环境变量立即生效，无须重启机器。

Open Client 运行库可以独立部署在一台机器上，作为 ASE 的客户端提供访问。PHP、Python 及基于 Open Client C-API 开发的应用程序依赖于这个运行库运行。

20.4　Sybase Central 客户端工具生成

Sybase Central 是 Sybase ASE 数据库的最重要的基于 Java 界面的管理客户端，集成了它所有的日常数据库管理与访问功能。Sybase 其他数据库产品（SQL Anywhyere（ASA）、Sybase IQ）也有自己的 Sybase Central 客户端工具。它们虽然外观一样，但是实际运行内容确不相同。虽然它可以独立安装，但是需要独立的安装包，并且由于注册表及安装位置等原因，很容易造成冲突。对于开发人员和 DBA 来讲，从已经安装好的 ASE 数据库里直接提取 Sybase Central 客户端，显得非常有意义。

在安装完 Sybase ASE 之后，与 Sybase Central 客户端组件相关的目录为：
1. ASEP：该目录存放的是 Sybase Central 要用到的 ASE 插件。
2. Charsets：该目录存放的是所有字符集码表及对应排序的元文件。
3. Collate：该目录存放的是各字符集与 Unicode 之间的排序规则文件。

4. DBISQL：该目录存放的是 dbisql 工具，它会内嵌到 Sybase Central 当中。
5. ini：含义很明确，该目录主要存放 interfaces 文件，用于数据库服务名的登记。
6. locales：该目录用于存储各操作系统下 ASE 支持的国家、语言及字符集列表。
7. shared：这个目录是 Sybase Central 组件的核心共享目录，它的实现基本上都位于这个目录当中。它下边又分几个子目录：

1）java：里边主要是 Sybase Central 要用到的 Java 类库。
2）javahelp-2_0：帮助文件的 jar 包。
3）JRE-6_0_6_32BIT：Java 虚拟机运行时库，如果本机已经安装有 jdk 1.6，要想节省空间，这个目录是可以移去的。
4）Lib：这个子目录存放的是 JDBC 驱动及 ddlgen.jar 运行时库。
5）Sybase Central 6.0.0：Sybase Central 的实现部分，这部分是最核心的部分。里边有两个重要的配置文件：①配置文件：.scRepository600；②配置文件 scjview.ini，位于 win32 子目录当中，整个 Sybase Central 的可执行文件也位于这个 win32 子目录当中。

下面我们看看这两个配置文件的用法。

1. scjview.ini 文件

主要包括以下几个参数：

JRE_DIRECTORY=D:\Sybase15\Shared\JRE-6_0_6_32BIT
VM_ARGUMENTS=-Xmx500m;-Xms50m;-Djava.security.policy="D:\Sybase15\Shared\Sybase Central 6.0.0\SybaseCentral.policy"
JAR_PATHS="D:\Sybase15\Shared\Sybase Central 6.0.0";D:\Sybase15\Shared\java;D:\Sybase15\Shared\JavaHelp-2_0
ADDITIONAL_CLASSPATH=
LIBRARY_PATHS=D:\Sybase15\Shared\win32
APPLICATION_ARGUMENTS=-screpository=D:\Sybase15\Shared\Sybase Central 6.0.0;-installdir=D:\Sybase15\Shared\Sybase Central 6.0.0

参数说明：

- JRE_DIRECTORY：用于指定 Java 虚拟机 jre 目录的位置，你可以将其指定到机器上 jdk 的 jre 子目录。比如，已经有一个 jdk 1.6 位于 C:\shared\jdk1.6.0_16，你可以设定 JRE_DIRECTORY= c:\shared\jdk1.6.0_16\jre。
- VM_ARGUMENTS：用于指定 jvm 的相关参数，如上例：-Xmx500m;-Xms50m; -Djava.security.policy="D:\Sybase15\Shared\Sybase Central 6.0.0\SybaseCentral.policy"，只要将 policy 文件的位置 D:\Sybase15\Shared\Sybase Central 6.0.0\SybaseCentral.policy 修改为你实际要发布的位置就可以。
- JAR_PATHS：用于指定 jar 文件的搜索路径。
- ADDITIONAL_CLASSPATH：用于指定额外的类加载路径。
- LIBRARY_PATHS：指定的是 DLL 搜索路径。

- APPLICATION_ARGUMENTS：用于指定 Sybase Central 另一个配置文件.scRepository 的目录位置及安装目录。

这个配置文件提取完以后，当你要发布到一个不同于 D:\sybase15 的目录时，直接将 D:\Sybase15 替换为新的目录即可，也可以用脚本实现。

2. .scRepository600

这个配置文件里存储的主要是 Sybase Central 中的 ASE 插件及 SYSAM 插件的相关 CLASSPATH 信息。例如：

```
Providers/ASE1503/Classpath=D:\\Sybase15\\ASEP\\lib\\ASEPlugin.jar
Providers/ASE1503/Name=Adaptive Server Enterprise
Providers/ASE1503/AdditionalClasspath=D:\\Sybase15\\shared\\lib\\jconn3.jar; ……
```

同样，在发布到新目录当中时，将 D:\\Sybase15 替换为新目录即可。

这里还有一个配置文件同样需要调整，即 DBISQL\bin\dbisql.ini，它的参数说明形式与 scjview.ini 完全相似。

在提取完 Sybase Central 功能组件之后，我们只需要调整这三个配置文件中的发布目录的值，即可得到一个功能齐全的 Sybase Central 客户端。

我们可以用一段简短的 VBS 脚本来实现解压到新目录，并自动修改上述三个配置文件，使 Sybsae Central 能够运行。其内容如下（详见代码 code\shipping\make_sybasecentral.bat）：

```
#001 @echo off
#002
#003 set target_dir=%1
#004 set target_scj=%target_dir%\SybaseCentralASE15
#005 set target_scj2=%target_dir%\\SybaseCentralASE15
#006 echo you will install the sybase central to the new dir: %target_dir%
#007
#008 7z x -y -o%target_dir% SybaseCentralASE15.zip
#009
#010
#011 call replace_path.vbs %target_dir%\SybaseCentralASE15\DBISQL\bin\dbisql.ini %target_dir%\SybaseCentralASE15
#012 call replace_path.vbs "%target_scj%\Shared\Sybase Central 6.0.0\win32\scjview.ini" %target_dir%\SybaseCentralASE15
#013 call replace_path.vbs "%target_scj%\Shared\Sybase Central 6.0.0\.scRepository600" %target_dir%\SybaseCentralASE15
```

其中 replace_path.vbs 内容如下，只是实现将一个文本文件中的字符串进行替换的功能，其内容如下（详见代码 code\shipping\replace_path.vbs）：

```
#001 ' read the file
#002
#003 set objArgs=Wscript.arguments
```

```
#004
#005 ' objArgs(0):     source file
#006 ' objArgs(1):     the target string to replace d:\sybase15
#007
#008 set fso = createobject("scripting.filesystemobject")
#009 set stream = fso.opentextfile(objArgs(0),1)
#010 content = stream.readall()
#011 call stream.close()
#012
#013 ' replace the string
#014 content = replace(content,"D:\Sybase15", objArgs(1))
#015 newobjArgs = replace(objArgs(1), "\", "\\")
#016 content = replace(content,"D:\\Sybase15", newobjArgs)
#017
#018 ' save the file
#019 set stream = fso.opentextfile(objArgs(0), 2)
#020 call stream.write(content)
#021 call stream.close()
```

制作完 Sybase Central ASE 15.zip，将它与 install_sybasecentral.bat 及 replace_path.vbs 放到相同的目录下，执行 install_sybasecentral.bat <new_dir> 即可把 Sybase Central ASE 15 装到一个新的目录当中，并且能确保 Sybase Central 可用。

这里作为示例，将 Sybase Central 直接发布到 E:\temp：

D:\MyDocument\MyBooks\ASE\code\shipping>install_sybasecentral.bat E:\temp
you will install the sybase central to the new dir: e:\temp

进入目录 E:\temp\SybaseCentralASE15\Shared\Sybase Central 6.0.0\win32，我们直接执行文件 scjview.exe，会出现如图 20-1 所示的界面。

图 20-1 自发布 Sybase Central SySAM 插件无法加载

这个错误可以忽略，因为我们本来就没有打算把 SySAM 插件安装到 Sybase Central 中，单击 OK 按钮，即进入正常的 Sybase Central 主界面，如图 20-2 所示。

上边的 SySAM 错误也是可以避免的，我们打开配置文件 E:\temp\SybaseCentralASE15\Shared\Sybase Central 6.0.0\.scRepository600。

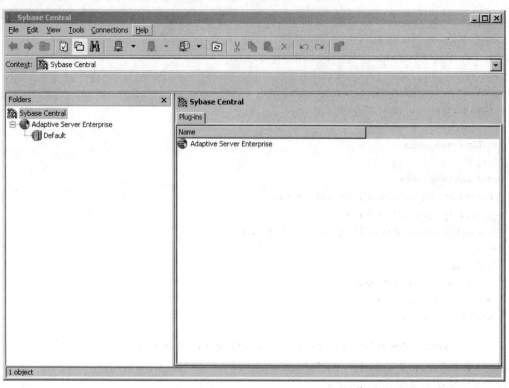

图 20-2　Sybase Central 主界面

只要把 Providers/SYSAM20/Version=15.0.3 等相关的 SYSAM20 片段注释掉，就不会加载 SYSAM 插件了。

21

Sybase ASE 发展历史及版本演进

Sybase 公司成立于 1984 年 11 月，总部设在美国加州的 Emeryville（现为美国加州的 Dublin 市）。作为全球最大的独立软件厂商之一，Sybase 公司致力于帮助企业等各种机构进行应用、内容及数据的管理和发布。Sybase 的产品和专业技术服务，为企业提供集成化的解决方案和全面的应用开发平台。

公司名称 Sybase 取自 system 和 database 相结合的含义。Sybase 公司的创始人之一 Bob Epstein 是 Ingres 大学版（与 System/R 同时期的关系数据库模型产品）的主要设计人员。

公司的第一个关系数据库产品是 1987 年 5 月推出的 Sybase SQL Server1.0。Sybase 首先提出 Client/Server 数据库体系结构的思想，并率先在 Sybase SQL Server 中实现。

提到 Ingres 数据库，不得不提一下 Michael Stonebraker，他是 Ingres 的创始人。他是加州大学伯克利分校的教授、著名的数据库学者，他在 1992 年提出对象关系数据库模型。Stonebraker 教授领导了称为 Postgres 的后 Ingres 项目。这个项目的成果是非常巨大的，在现代数据库的许多方面都做出了大量的贡献。Stonebraker 教授还做出了一件造福全人类的事情，那就是把 Postgres 放在了 BSD 版权的保护下。如今 Postgres 名字已经变成了 PostgreSQL，功能也是日渐强大，你可以自由地浏览它的代码库，可以实时了解一线工程的思路。

1987 年左右，Sybase 联合了微软，共同开发了 SQL Server。原始代码的来源与 Ingres 有些渊源。后来 1994 年两家公司合作终止。此时，两家公司都拥有一套完全相同的 SQL Server 代码。可以认为，Stonebraker 教授是目前主流数据库的奠基人。

Ingres 是比较早的数据库系统，开始于加利福尼亚大学柏克利分校的一个研究项目，该项目开始于 20 世纪 70 年代早期，在 80 年代早期结束。像柏克利大学的其他研究项目一样，它的代码使用 BSD 许可证。从 80 年代中期开始，在 Ingres 基础上产生了很多商业数据库软件，包括 Sybase、Microsoft SQL Server、NonStop SQL、Informix 和许多其他的系统。在 80 年代中期启动的后继项目 Postgres 产生了 PostgreSQL 和 Illustra，无论从何种意义上来说，Ingres 都是历史上最有影响的计算机研究项目之一。

总之，Ingres 数据库可以看作是后 Sybase ASE 数据库的先导。

1994 年，Sybase 和微软合作终止，Sybase 继续开发，将 Sybase SQL Server 往各个平台移植，版本也是跳跃式地变化，从 4.2 很快就到了 11.0。

基于这种原因，Sybase ASE 和 MS SQL Server 两个数据库的语法非常相似，系统数据库名基本相同。两者采用的都是 TDS 协议进行通信，只是版本不同而已（注意：ASE 采用的是 tds5.0，而 MS SQL Server 采用的则是 tds8.0）。

Sybase SQL Server 后来为了与微软的 MS SQL Server 相区分，改名叫 Sybase ASE（Sybase Adaptive Server Enterprise）。出于习惯，大多数人仍称 ASE 数据库为 Sybase 数据库，虽然 Sybase 旗下除了 ASE 以外，还有若干种其他数据库产品。

以下是 ASE 各个历史版本的演进过程：

Sybase SQL Server 3.0：1988 年，第一个公开发行版。

Sybase SQL Server 4.0：1990 年发布。

Sybase SQL Server 4.2：1991 年发布。

Sybase SQL Server 4.8：1992 年发布。

Sybase SQL Server 4.9：1992 年发布，同年发行了 4.9.1；版本 4.9.2 在 1993 年发布。

Sybase System X, aka Sybase SQL Server 10.0：发布于 1993 年。

Sybase SQL Server 11.0：1995 年发布。

Adaptive Server Enterprise 11.5：1997 年发布。

Adaptive Server Enterprise 11.9：1998 年发布。实际上，11.9.0 并未公开发布，11.9.2 才作为正式版本进行公开发布。版本 11.9.3 是 Sybase 第一个 64 位版本的 ASE 发行版本。

Adaptive Server Enterprise 12.0：1999 年发布。

Adaptive Server Enterprise 12.5：2001 年发布。

Adaptive Server Enterprise 12.5.0.1：2002 年发布，同年发布了 12.5.0.2。

Adaptive Server Enterprise 12.5.0.3：2003 年发布，同年发布了 12.5.1。

Adaptive Server Enterprise 12.5.2：2004 年发布，同年发布了 12.5.3。

Adaptive Server Enterprise 12.5.3a：发布于 2004/2005 年。

Adaptive Server Enterprise 12.5.4：发布于 2006 年。

Adaptive Server Enterprise 15.0：发布于 2005 年，同年发布了 12.5.3。

Adaptive Server Enterprise 15.0.1：发布于 2006 年。

Adaptive Server Enterprise 15.0.2：发布于 2007 年 6 月。

Adaptive Server Enterprise 15.0.3：发布于 2008 年 12 月。

Adaptive Server Enterprise 15.5：发布于 2010 年 6 月。

Adaptive Server Enterprise 15.7：发布于 2012 年 2 月。

2010 年 05 月 13 日，SAP 公司以 58 亿美元收购软件制造商 Sybase。正处于 ASE 15.5 发行之前。

在 Sybase ASE 发行过程中，采用了几个有关发行版本（Release）的术语：

- EBF：官方软件发行版本。EBF 号唯一性地标志了在某个 OS 平台上的发行版本。
- ESD：官方软件发行版本，它解决了当前发行版本中的一些 Bug。ESD（ESD #）通常是跨平台的，并且不包含新的特性或功能。ESD 是针对某一主版本（Major Version）的累积软件包，它必须在 GA 发行版本上安装。
- GA：GA（General Availability）是一个 Sybase 软件版本全新的、初始听发行版本。除 Bug 修复之外，它通常包含一些新特性。
- IR：IR（Interim Release，中间版本）包含了在此之前的所有 ESD 发行版本的 Bug 修复，以及其他 Bug 修复。IR 发行版本必须基于与之对应的 GA 发行版本的基础上安装。比如，要安装 12.5.0.3，就必须先安装 12.5GA（ESD）。
- MR：MR（Maintenance Release，维护版本）是 GA 发行版本的完整替换包。由于它是完整的发行包，因此可以直接安装。通常此种类型的发行版本包含一些新特性及 Bug 修复。